MAINTENANCE ENGINEERING

(Principles, Practices and Management)

MAINTENANCE ENGINEERING

(Principles, Practices and Management)

For Maintenance Engineering, Maintenance Technology or Maintenance Management at B.E. / B.Tech. Level and also Useful for Postgraduate Students of different Indian Universities

Er. Sushil Kumar Srivastava

B.E. (Mech. Engg.), M.I.E., Member-F.P.S.

**Director, OTMEC, Durgapur (WB)*
**Ex. General Manager, I.G.S.L., Kolkata-91*
**Ex. Asstt. General Manager, I/C (Mech. and Fuel)*
Steel Authority of India Limited
Alloy Steels Plant
Durgapur

S. CHAND
PUBLISHING

S Chand And Company Limited

(ISO 9001 Certified Company)

S Chand And Company Limited

(ISO 9001 Certified Company)

Head Office: Block B-1, House No. D-1, Ground Floor, Mohan Co-operative Industrial Estate, New Delhi – 110 044 | Phone: 011-66672000

Registered Office: A-27, 2nd Floor, Mohan Co-operative Industrial Estate, New Delhi – 110 044
Phone: 011-49731800

www.**schandpublishing.com**; e-mail: **info@schandpublishing.com**

Branches

Ahmedabad	:	Ph: 27542369, 27541965; ahmedabad@schandpublishing.com
Bengaluru	:	Ph: 22354008, 22268048; bangalore@schandpublishing.com
Bhopal	:	Ph: 4274723, 4209587; bhopal@schandpublishing.com
Bhubaneshwar	:	Ph: 2951580; bhubaneshwar@schandpublishing.com
Chennai	:	Ph: 23632120; chennai@schandpublishing.com
Guwahati	:	Ph: 2738811, 2735640; guwahati@schandpublishing.com
Hyderabad	:	Ph: 40186018; hyderabad@schandpublishing.com
Jaipur	:	Ph: 2291317, 2291318; jaipur@schandpublishing.com
Jalandhar	:	Ph: 4645630; jalandhar@schandpublishing.com
Kochi	:	Ph: 2576207, 2576208; cochin@schandpublishing.com
Kolkata	:	Ph: 23357458, 23353914; kolkata@schandpublishing.com
Lucknow	:	Ph: 4003633; lucknow@schandpublishing.com
Mumbai	:	Ph: 25000297; mumbai@schandpublishing.com
Nagpur	:	Ph: 2250230; nagpur@schandpublishing.com
Patna	:	Ph: 2260011; patna@schandpublishing.com
Ranchi	:	Ph: 2361178; ranchi@schandpublishing.com
Sahibabad	:	Ph: 2771238; info@schandpublishing.com

First Edition 1998
Subsequent Editions and Reprints 2002, 2006, 2007
Revised Edition 2010
Reprints 2011, 2012, 2013, 2014 (Twice), 2015, 2018
Reprint 2020

ISBN : 978-81-219-2644-7 **Product Code :** H3MEG62PAQC10ENZX10O

PRINTED IN INDIA

By Vikas Publishing House Private Limited, Plot 20/4, Site-IV, Industrial Area Sahibabad, Ghaziabad – 201 010 and Published by S Chand And Company Limited, A-27, 2nd Floor, Mohan Co-operative Industrial Estate, New Delhi – 110 044

PREFACE TO THE REVISED EDITION

Since the book Maintenance Engineering and Management was prepared in the year 2005, the author felt the need of some revision and augmentation in order to make the book more useful, broad-based and interesting and also to keep pace with the time. Accordingly the book has been named as **"Maintenance Engineering (Principles, Practices and Management)"** and tries to include all the aspects of Maintenance Engineering, i.e. its principles, its practices including techniques, technologies and methodologies and its management. However, some examples and experience from the author's earlier working experiences has been retained as those are representative and are relevant as on today.

Some engineering fields are changing dramatically from year to year, with radical breakthroughs in technologies happening often. These fields may have very many papers and books published each year on the latest practices. But maintenance engineering is a field which, for the most part, has not fundamentally changed much over the years and there are not many sources for the latest information on best practices.

But recently, maintenance engineering has, more and more, put an emphasis on true reliability. Business and establishments, which are asset intensive, such as manufacturing, rely more on reliability centered maintenance engineering (including CMMS, TPM and EAM etc) for the purpose of improving capacity utilization, without capital investment.

As such, this book is thoroughly revised and augmented and has been designed to be used as **a text book and/or reference book for many disciplines/areas, such as Maintenance Engineering, Maintenance Technology or Maintenance Management etc at appropriate levels in engineering colleges, universities, institutions and polytechnics etc and** also to practicing engineers and technologists. Accordingly, few sample questions are given at the end of each chapter. The chapters and topics, covered in this book, are expected to encompass the syllabus that may be needed by various colleges/ institutions in maintenance field. It also serves as a reference book for students of all other engineering disciplines (potential maintenance-men) in universities, colleges and institutions and also vast numbers of engineers, managers, supervisors, technologists and other persons working in or associated with maintenance and upkeep of machines, equipments and systems in any shop, plant or industry.

As in previous edition, in order to maintain the simplistic character of the book, mathematical treatment has been reduced to minimum and figures, tables and block diagrams have been increased to explain the topics/ subjects. However, readers can develop mathematical approaches based on the information in this book.

The author's 34 years long experience with maintenance of plants and equipments and constant exposure to acute never ending problems of equipment failures, repairs, monitoring and upkeep has added practicality to this book. The aim of this book is to pass on invention, ingenuity and large dose of basic maintenance principles and methods to users. This book may be reader's key, guide and chief support in tempestuous battle of maintenance in the days and years ahead.

The author, again, feels highly obliged to S. Chand and Company Ltd., New Delhi, for bringing out this book in such a short period and in such a beautiful form.

Er. Sushil Kumar Srivastava

PREFACE TO THE SECOND EDITION

(MAINTENANCE ENGINEERING AND MANAGEMENT)

The initial edition of the book was based on informations available and technologies and methodologies commonly used till 1995. Since then, quite a few improvements have taken place and new technologies and methodologies etc have come up in related fields. As such, need was felt to upgrade and augment the book in the form of thoroughly revised edition and change the name to Maintenance Engineering & Management.

The book has been designed to be used as a text book for many engineering disciplines as Maintenance Engineering, Maintenance Technology or Maintenance Management at degree/diploma level and also useful for postgraduate study in most Indian universities, institutions and polytechnics. As such, few sample questions are given at the end of each chapter. The chapters and topics, covered in this book, are expected to encompass the syllabus of various colleges/ institutions in maintenance field. It also serves as a reference book for students of all other engineering disciplines in universities, colleges and institutions and also all engineers, managers, supervisors, technologists and other persons working in or associated with maintenance and upkeep of machines, equipments and systems in any shop, plant or industry.

The numbers of chapters have been retained as fifteen. Chapters 1, 2, 3, 4, 5,6, 7, 11, 13, 14 and 15 have been completely re-written with almost all maintenance connected topics and subjects, as on use till today (2004 end). Chapters 8, 9, 10 and 12 have been retained from earlier addition, with suitable addition at the beginning and/ or end of each chapter.

As in previous edition, in order to maintain the simplistic character of the book, mathematical treatment has been reduced to minimum and figures, tables and block diagrams have been increased to explain the topics/ subjects. However, readers can develop mathematical approaches based on the informations in this book.

The author, again, feels highly obliged to S. Chand & Company Ltd., New Delhi, for bringing out this book in such a short period and in such a beautiful form.

Er. Sushil Kumar Srivastava

ABBREVIATIONS USED

AGM/DGM	Assistant General Manager/ Dy. General Manager
AMR	Additions, Modifications and Replacements
AVI	Automatic Visual Inspection
BIS	Bureau of Indian Standards
BITE	Built-in Test Equipments
BOP	Balance of Payment
CAM	Computer Aided Manufacturing
CBM(S)	Condition Based Maintenance (System)
CBT	Computer Based Training
CM	Condition Monitoring
CMMS	Computerized Maintenance Management System
CPA	Critical Path Analysis
CPM	Critical Path Method
CT	Current Transformer
EAM	Enterprise Asset Management
EDM	lectro-Disintegration Machine
EHD	Electro-Hydro-Dynamic
EMP	Essential Maintenance Practice
ETA	Event Tree Analysis
FA	Failure Analysis
FEA	Finite Element Analysis
FFT	Fast Fourier Transform
FIFO	First-in First-out
FMEA	Failure Mode and Effect Analysis
FMECA	Failure Mode Effect and Criticality Analysis
FRACAS	Failure Reporting, Analysis, and Corrective Action System
FRP	Fibre Reinforced Polymers
FTA	Fault Tree Analysis
GM	General Manager
HRD	Human Resource Development
HVAC	Heating, Ventilation and Air Conditioning
KPI	Key Performance Indicators
LAN	Local Area Network
LOB	Line Of Balance (method)

MCA	Motor circuit analysis
MCC	Motor Control Centre
MCSA	Motor Current Signature Analysis
MFP	Multifactor Productivity
MI	Maintainability Index
MP	Motor Prevention
M-PET	Maintenance Productivity Enhancement Tool
MRO	Maintenance, repair, overhaul
MTBF	Mean Time Between Failure
MTBM	Mean Time Between Maintenance
MTBS	Mean Time Between Shutdown
MTD	Mean Time Down
MTS	Maintenance Time Standard
MTTF	Mean Time To Failure
MTTR	Mean Time To Repair
NDT	Non Destructive Testing
OEE	Overall Equipment Effectiveness
OEM	Original Equipment Manufacturer
OJT	On-job Training
OR	Operation Research
PC	Personal Computer
PD	Particle Discharge
PdM	Predictive Maintenance
PERT	Programme Evaluation and Review Technique
PHD	Plasto-Hydro-Dynamic
PM	Preventive Maintenance
PQRM	Productivity, Quality, Reliability and Maintainability
QAMM	Quality Assured Maintenance Management
RAM	Reliability, Availability and Maintainability
RAS	Reliability, Availability and Serviceability
RBD	Reliability Block Diagram
RBM	Reliability Based Maintenance
RCA	Root Cause Analysis
RCFA	Root Cause Failure Analysis
RCM	Reliability Centered Maintenance
REAP	Risk-based Evaluation And Prioritizing
RNDT	Radiographic Non Destructive Testing
ROI	Return –on Investment

RTD	Resistor-temperature-devices
SMP	Standard Maintenance Practice
SOP	Standard Operating Practice
SOA(P)	Spectroscopic Oil Analysis (Process)
SWOT	Strength, Weakness, Opportunity, Threat
TAN	Total Acid Number
TBN	Total Base Number
TDR	Time Domain Reflectometry
TMM (TM2)	Total Maintenance Management
TPM	Total Productive Maintenance
TQM	Total Quality Management
TQP	Total Quality Process
UTR	Up Time Ration
VED	Vital, Essential and Desirable
WBT	Web Based Training

* * *

CONTENTS

★ ★ ★

CHAPTER

1 INTRODUCTION AND MAINTENANCE OVERVIEW

DEFINITION

Though, in general, the word "maintenance" may mean many things, e.g. care or upkeep of machinery or property; or the state of being maintained (the maintenance of friendly relations); or means of upkeep, support, or subsistence or livelihood, i.e. to provide a comfortable maintenance or alimony or child support; or, an officious meddling in a law suit etc, we are here concerned here with the first meaning, i.e. care and upkeep of any machinery, equipment, facility or property.

There is an old saying among maintenance personnel: "Engineering has it for a year, but maintenance has to live with it for 20 years." This goes with the engineering saying: "Get it out of the door; we can always fix it on someone else's work order. Historically, the maintenance function was seen as a non-core service organization that did not contribute to competitiveness. Benchmarks for maintenance were isolated measurements of tasks - that is, task orientation rather than business goal orientation.

The old concept of Maintenance was "Maintenance is about preserving physical asset. However, the new concept is that "Maintenance is about preserving the functions of assets". Today the engineers are giving rise to concepts like "asset care", which, as the name implies, seeks to care for assets. They have also led maintenance strategists to believe that maintenance is all about preserving the inherent reliability or built-in capability of any asset. When we maintain an asset, the state which we wish to preserve must be one in which it continues to do whatever its users want it to do. This in turn implies that we have to focus our attention on maintaining what each asset does rather than on what it is.

Today, maintenance is a 'science', an 'art' and also a 'philosophy'. It is science because its execution relies, sooner or later, on most or all of the science. It is an art because seemingly identical problems regularly demand and receive varying approaches and actions and because some managers, supervisors or maintenance technicians display greater aptitude for it than others show or attain. It is a philosophy because it is a discipline that can be applied intensively, modestly or not at all, depending upon a wide range of variables.

1.1 GROWTH OF MAINTENANCE ENGINEERING

Actually, maintenance is an age-old function, which developed and progressed, knowingly or unknowingly, along with the operation of the equipments/ machines. In early ages, maintenance was not a separate identity but the job of maintenance was considered as part and parcel of operator's job. This was possible because of simplicity and openness of the machines. Even the ancient carvings on the Egyptian tomb "Ra-Em-Ka" (about 2600 ~ 1700 BC) shows that a sledge is being used to transport a stone monument and a man is pouring some liquid, probably to lubricate the runners of the sledge for easy movement. That shows that even lubrication (basic maintenance function) was done by the same operating group.

However, with the growth of industrialization, the complexity and sophistication of the machines increased and the machines became less open. This started creating problems for the operating personnel and the concept of maintenance as a separate identity and separate discipline started. Maintenance Prevention (MP) was probably known first known in 1960s from a factory magazine in United States, which said that Maintenance Prevention meant design and manufacture or purchase of equipments which are almost free of maintenance.

The growth and speed of inflation of 1970s gave rise to new awareness of the rising cost of downtimes or loss of use. Also the asset and equipment/ component replacement cost became so inflated and increased that well managed maintenance programmes, to enhance the life of the existing equipments/ components became the essential aspects of all maintenance strategies.

Till in recent past, though maintenance functions got their recognition, they still did not gain a composite discipline and remained as mechanical maintenance, electrical maintenance, instrumentation maintenance, civil maintenance etc. Be it a centralized or decentralized maintenance, the individual maintenance wings merged only at a very senior level and very often, the senior level was involved to sort out the overlapping problems. Some may argue that it helped preservation of skill, but it is a myth and did not serve the basic aim or improving availability with reliability. Recently, in many organization, maintenance engineering has been taken as a composite discipline, including all types of repairs, overhauls, inventory management, root cause analysis, reliability and performance monitoring etc.

Today, in most industries, Plant Engineering is probably the biggest force to increase productivity (other than the motivated operating force) and maintenance engineering/ management is the most important component of Plant Engineering. Maintenance is an investment that buys/ gives more production time. With the increased complexity, sophistication and automation of the equipments and systems, a very serious burden falls on maintenance engineers regarding the quality and quantity of maintenance, maintenance aids and their documentation etc. Again, we can not cope up today's jobs and problems with yesterday's tools and techniques. As such continuous development is needed in maintenance areas also (involving educational and research institutions also) for developing new tools and techniques to tackle today's need and also anticipated need of tomorrow.

1.2 PURPOSE OF MAINTENANCE

The main purpose of maintenance in an industrial perspective is to reduce the business risks. In general, operation and maintenance is synonymous with high level of availability, reliability and assets' operability linking directly with production capacity, productivity and business profit.

Today, the role/purpose of maintenance is no more limited to immediate reaction to emergencies and over-power problems. We should work for 'Maintenance Excellence', which is a subset of reliability excellence and redefines traditional roles and responsibilities as well as maintenance processes that are that are necessary to assure asset reliability, maximum asset useful life and best life cycle asset cost. Under the maintenance excellence umbrella, the maintenance functions become an equal partner within the corporation's functions. It runs like any other for-profit business in a fully integrated plant organization.

In information scenario, the main purpose of the Identification and Maintenance Functions (I&M functions) is to support the end user during various scenarios of a device's life cycle be it configuration, commissioning, parameterization, diagnosis, repair, firmware update, asset management, audit trailing, and alike. Well-defined uniform parameters and rules should enable the manufacturers to offer devices that behave in a uniform manner and that should enable the end user to act quick and straight.

The need and purpose of maintenance can, further, be derived from previous paragraphs.

1.3 MAINTENANCE FUNCTIONS

To get a smooth, reliable and cost effective maintenance in a company, different functions in the company has to be coordinated and not only the maintenance department alone. Primarily it affects questions of economical, technical and organizational nature at a system level. Maintenance Function is the more dominant amongst those. Earlier the objective of maintenance function was considered to optimize plant availability at minimum cost. Today it is being considered as "Maintenance affects all aspects of business effectiveness and risk—safety, environmental integrity, energy efficiency, product quality and customer service, not just plant availability and cost.

Maintenance Functions can be grouped in two categories—Basic Functions (job-wise) and Composite Functions.

1.3.1 Basic Functions

These normally include following types of Jobs/ functions (Sl Nos1 to 4 are normally main functions and others are associated functions)—

1. Replace: To remove an unserviceable item and install a serviceable counterpart in its place.

2. Repair: The application of maintenance services, including fault location / troubleshooting, removal installation, and disassembly/assembly procedures, and maintenance actions to identify troubles and restore serviceability to an item by correcting specific damage, fault, malfunction, or failure in a part, subassembly, module (component or assembly), end item, or system.

3. Overhaul: That maintenance effort (service/ action) prescribed to restore an item to a completely serviceable/operational condition as required by maintenance standards in appropriate technical publications. Overhaul is normally the highest degree of maintenance performed by the industry. Overhaul does not normally return an item to like new condition.

4. Rebuild: It consists of those services/ actions necessary for the restoration of unserviceable equipment to a like new condition in accordance with original manufacturing standards. Rebuild is the highest degree of materiel maintenance applied to any equipment. The rebuild operation includes the act of returning to zero those age measurements (hours/ Kms, etc.) considered in classifying equipment/components. Reconditioning, renovation or other such terms are also used for rebuild.

5. Service/ Lubricate: Operations required periodically to keep an item in proper operating condition, i.e., to clean (includes decontaminate, when required), to preserve, to drain, to paint, or to replenish fuel, lubricants, chemical fluids, or gases. Lubrication is introduction of any of various substances between sliding surfaces to reduce wear and friction.

6. Inspect/ check: To determine the serviceability of an item by comparing its physical, mechanical, and/or electrical characteristics with established standards through examinations.

7. Test: To verify serviceability by measuring the mechanical or electrical characteristics of an item and compare those characteristics with prescribed standards.

8. Adjust: To maintain, within prescribed limits, by bringing into proper or exact position, or by setting the operating characteristics to specified parameters.

9. Align: To adjust specified variable elements of an item to bring about optimum or desired performance.

10. Calibrate: To determine and cause corrections to be made or to be adjusted on instruments or test, measuring, and diagnostic equipment used in precision measurement. consists of comparisons of two instruments, one of which is a certified standard of known accuracy, to detect and adjust any discrepancy in the accuracy of the instrument being compared.

11. Measure: The act or process of ascertaining the extent, dimensions, or quantity of something; e.g. measuring pressure or flow or heat or electrical current etc with suitable gadget/ instrument.

12. Winding: A symmetrically laid, electrically conducting current path in any device; e.g. electrical motor/ armature winding.

13. Install: The act of emplacing, seating, or fixing into position an item, part, or module (component or assembly) in a manner to allow the proper functioning of an equipment or system, etc.

1.3.2 Composite Functions

A Maintenance Engineer shall possess significant knowledge of statistics and probability and logistics in addition to fundamental knowledge of the operation of the equipment and machinery he/she is responsible for. Some of his/her composition functions are given below—

1. Protecting the buildings, structures and plants/ factories,
2. Reducing downtimes and increasing equipment availability; also helping in increasing equipment utilization,
3. Analyze repetitive failure and arrange their elimination,
4. Forecast maintenance spares, tools and consumables and help their procurement and inventory management,
5. Controlling and directing labour forces,
6. Economy in maintenance department,
7. Maximizing utilization of available resources,
8. Ensuring safety of installation and reducing environmental pollution,
9. Cost reduction and cost control; also helping in costing of individual jobs and departments,
10. Preparing maintenance budgets,
11. Waste reduction and waste recovery,
12. Improving technical communication,
13. Training of maintenance personnel on related jobs; Etc.

As another way of classification, maintenance functions can be of two types— Primary Functions and Secondary Functions.

1.3.3 Primary Functions (Maintenance Fundamentals)

These includes functions like—

1. Select the right bearings, fasteners, chains, sprockets, belts, gears, cylinders, actuators and other devices; their lubrication, upkeep and timely replacements. Ensure good working of their monitoring systems.

2. Equipment Inspection and lubrication;
3. Monitor oils, lubricants and lubrication systems and cooling and circulating systems.
4. Electrical Control System—Interpret electrical symbols, ladder diagrams, and electrical diagrams and how to use them as road maps to isolate electrical problems. Read and monitor electrical symbols and the grammar of schematic diagrams and use these tools to pinpoint electrical problems
   - Identify the path of an electrical problem and quickly pinpoint and correct it using tools such as a voltmeter, and an electrical control diagram.
   - Prevent overloads (electrical and mechanical etc) with controls that will shut off expensive drive motors before they burn up.
   - Identify motor types and correctly specify the exact one for each load instead of burning up motors that are too small or wasting money on motors with excess capacity.
   - Check operation of transformers, electric motors, controls, motor starters, solid state relays and motor starters and other such devices.
5. Install and monitor proper ventilation and environmental control systems to maximize service.
6. Utility (steam, gas, air etc) generation and distribution,
7. Alterations, up-gradation and new installations,
8. Step-by-step troubleshooting procedures for all the components of the establishment.

1.3.4 Secondary Functions

These may include the followings–

1. Store Keeping and Inventory management,
2. Plant protection-
3. Waste disposal,
4. Salvaging and recycling; e.g. recycling of used industrial oil or mechanical/ electrical scraps etc. Occasionally waste disposal and salvaging go together. This does not include reconditioning of mechanical/ electrical spares which goes in the primary functions.
5. Plant and equipment insurance administration,
6. Other services as may come up.

N.B.—The composite function, primary function and secondary function may appear to be overlapping to some extent, but there is distinct difference between them.

1.4 MAINTENANCE LEVELS

If we try to define level of maintenance, it may be very vague or vast. Normally three classifications, on which maintenance levels can be based, are—-**time for maintenance** (duration of maintenance, opportunistic maintenance etc), **frequency of maintenance** (daily, weekly, monthly, yearly etc) **and quality of maintenance.**

Other way for classifying maintenance may be Organizational Level and Intermediate Level. The organizational level covers tasks of the daily servicing, maintenance checks, inspections for condition, exchange of components, and quick, simple repairs as specified in the equipment's maintenance manuals etc. The work takes place at the operators' site. The intermediate level covers repairs on/ off the equipment, extended periodical inspections as specified in the maintenance manual. To fulfill these tasks, maintenance facility, qualified personnel, test equipment and special tools, as well as training aids and/or training/ model equipments are required.

However, the maintenance levels can easily be confused with maintenance types and strategies, which would be discussed in details in chapter 5.

1.5 MAINTENANCE OBJECTIVES AND RESPONSIBILITIES

Maintenance purposes, functions, objectives and responsibilities are inter-related and are overlapping to some extent. From the various aspects and paragraphs mentioned earlier in this chapter, maintenance objectives for a big industry, like a steel plant, can be the followings—

1. To maintain plants and equipments at its maximum operating efficiency, reducing downtimes and ensuring operational safety,
2. To safeguard investments by minimizing rate of deterioration and achieving this at optimum cost through budgeting and controls,
3. To help management in taking decisions on replacements or new investments and actively participate in specification preparation, equipment selection, its erection and commissioning etc,
4. Help in implementation of suitable procedures for procurement, storage and consumption of spares, tools and consumables etc (inventory control etc),
5. Standardization of spares and consumables, in conformity with plant, national and international standards and help in adoption of these standards by all users in the plant,
6. Running of centralized services like steam generation and distribution, water supply, air supply and fuel supply etc,
7. Running of captive workshops for repairs and conditioning and also for making some new spares; etc.

Maintenance Responsibilities may be considered as little elaboration of maintenance objectives. As illustrations, few responsibilities are mentioned below–

1. Proper care, up-keep and maintenance of equipments to achieve desired availability,
2. Ensure maintainability of old and new installations,
3. Identify and correct chronic and costly equipment problems,
4. Design, implement and monitor an effective and economically justified preventive, predictive and or condition based maintenance programme,
5. Comprehensive lubrication programme; monitor centralized oil, lubrication, hydraulic, pneumatic and other circulatory systems.
6. Monitor various on-site and off-site condition monitoring and built-in test equipments,
7. Protection against environment; completion of environmental Health and Safety (EHS) tasks, including preliminary hazard analysis (PHA) etc,
8. Proper documentation and reporting of maintenance data and systematic analysis of the data and feedback,
9. Coordinate with operation for safe operating practices,
10. Help and coordinate project engineering for development and implementing new projects, additions, modifications, replacements and augmentations etc,
11. Review major purchases to assure correct specification and design; Also help proper inventory control,

12. Failure analysis,
13. Budgeting maintenance costs and periodically monitoring and control; etc, etc.

All the maintenance functions and responsibilities can be grouped as–

- **MRO** (Maintenance, repair, overhaul), including lubrication, or
- '**Monitoring**' (e.g. various condition monitoring gadgets or systems), or
- '**Operation**' (e.g. operating repair machine shop or steam/ air/ gas etc generation and distribution) or
- '**Engineering**' (e.g. failure analysis, reliability and maintainability issues, design and improvements etc)
- '**Procurement**'(e.g. preparing specifications and helping procurement and inventory control for spares, tools and consumables, assessing vendor's capabilities etc.
- **Loss Elimination** Develop, implement and monitor a viable process or processes to continuously eliminate losses and wastes from all functional areas, e.g. procurement, operation, maintenance and the like, within the industry or facility.

For achieving these objectives, various functions and actions are to be taken, which would be taken subsequently in this book.

1.5.1. Maintenance Benefits

Pursuant to earlier mentioned functions and objectives, the achieved benefits may briefly be summarized as below–

1. Lower production unit cost,
2. Reduced maintenance cost and reduced lifecycle cost,
3. Better process stability,
4. Extended equipment life and lesser asset replacement,
5. Reduced maintenance part inventory,
6. Reduced overtime and reduced out-sourcing,
7. Improved employee safety and improved housekeeping,
8. Reduced risk of environmental issues,
9. Improved sense of employee pride/ownership,
10. Reduced nonconformance reports and warranty claims, and reduced lapses in statutory obligations; etc.

1.6 SYSTEM APPROACH TO MAINTENANCE FUNCTIONS

In an industry, various maintenance functions, mentioned above, have to be integrated within themselves and with other activities /operations/ strategies so that the maintenance becomes an effective one and delivers desired results. Fig. 1.1 (System Approach to Maintenance Functions) shows how the various maintenance aspects are interlinked. Almost all the items mentioned in this figure will be discussed subsequently in this book.

1.7 MAINTENANCE INTERDEPENDENCY

Because of integrated nature of functions/ activities of industries operation and management, maintenance typically can't control its own destiny. Instead, it is dependent on industry's other factors/ disciplines. Maintenance has different level of controls over different factors/ disciplines.

Maintenance has major (around 80%) control over following 5 factors and effectiveness, which rest predominantly under the control of maintenance management. It is not 100% as these still have to be within the constraint of total plant culture.

1. Work control and maintenance practices,
2. Inspection and supervision
3. Planning and scheduling,
4. Work measurement,
5. Maintenance asset's history and documentation.

Maintenance management has lesser control (around 50%) over following factors/ disciplines, related to maintenance. In these factors/ disciplines, maintenance has direct input, but may be severely constrained by corporate or industry's restrictions.

1. Aims and goals,
2. Organizational structure,
3. Organizational culture and behaviour,
4. Training and motivation,
5. Budgets and budgetary control,
6. Scheduling and coordination,
7. Annual plans and Master plans,
8. Facilities and equipments,
9. Computerization,
10. Extent of out-sourcing, etc.

Maintenance has little control (around 30%) over some of the other factors/ disciplines, such as those mentioned below.

1. Governing principles,
2. Overall Material management and inventory control,
3. Cost distribution,
4. Reports to management, etc.

1.8 ALTERNATE SYSTEMS TO MAINTENANCE RESPONSIBILITIES

Many other terms are used in industries and other establishments, which cover somewhat similar functions and responsibilities as mentioned earlier in this chapter; a few of which are given herewith–

1. Asset Care: Develop, implement and monitor/ oversee viable processes which will ensure that appropriate level of maintenance and operation care will be provided to all physical assets.

2. Life Cycle Asset Management (Maintenance Engineering for Product Life Cycle): Develop, implement and monitor/ oversee viable processes which will ensure optimum lifecycle management and care. It covers maintenance engineering design methods, maintainability predictions, analysis, testing and demonstration, and operational systems engineering principles and applications

3. Configuration Management: It ensures that all physical assets of the plant, industry or facility are designed, installed, operated, maintained and monitored in such a way that will provide maximum useful life and best lifecycle cost.

4. Risk Management of Assets: It increases the probability that industrial or manufacturing assets or processes will operate or perform when needed, by defining strategies to prevent failures, detect the onset of the failures and take early action and minimize all risks associated with the plant, industry or facility.

However, all the four systems mentioned above have lot of similarity and the subsequent chapters of this book will be applicable to all.

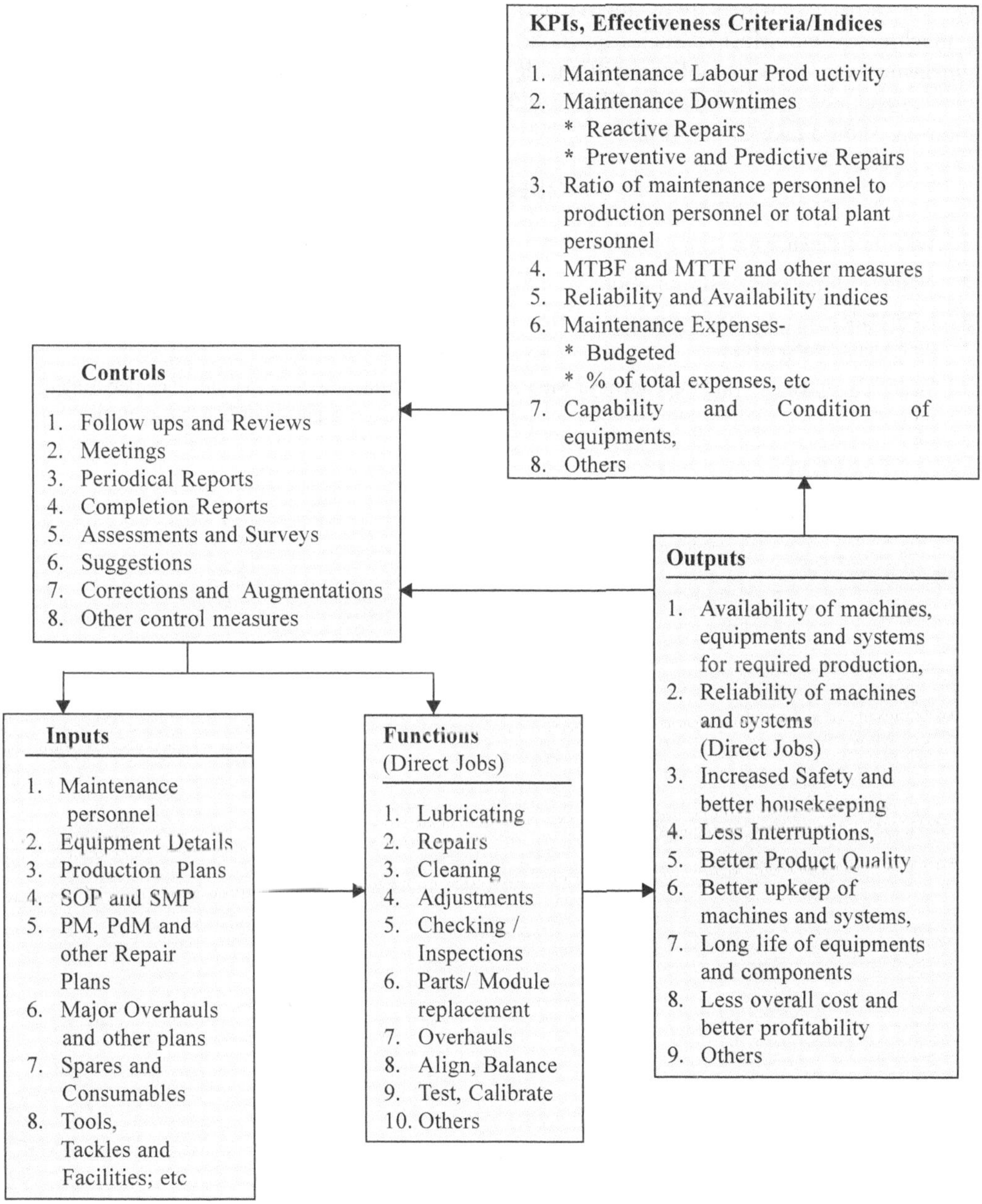

Fig. 1.1. System Approach to Maintenance and Activities

1.9 MAINTENANCE AS DISCIPLINE

Maintenance engineering has not yet got its due importance and recognition as an independent discipline, on par with other disciplines as Mechanical, Electrical, and Instrumentation engineering etc at graduate level study in engineering. Though a maintenance engineer is supposed to have some basic knowledge of mechanical, electrical, civil and instrumentation engineering, he/she definitely has to know many features, principles, practices and management and trouble-shooting skills and knowledge which are distinctive only to maintenance and upkeep of plants, equipments and machineries etc. Most of such aspects are discussed in this book. Separating maintenance engineering from other disciplines will enable mechanical, electrical, instrumentation and civil engineers etc to concentrate more on the development of their individual disciplines. Educational institutions should take a note of this.

QUESTIONS

1. What is difference between maintenance purpose and maintenance function?
2. Name 5 important basic functions of maintenance
3. How Does maintenance function differ from maintenance objective? Mention few typical maintenance objectives.
4. What is maintenance interdependency? Mention two functions on which maintenance department has more control and another two functions on which maintenance department has less control.

CHAPTER

2 PRODUCTIVITY, QUALITY, RELIABILITY AND MAINTAINABILITY (PQRM)

The problems of most final out-puts, in any industry, are the direct result of—

- Production systems and machines with inherent problems,
- Poor operating practices,
- Lack of in-between quality checking and corrections,
- Improper maintenance,
- Defective raw materials and tools.

Here consideration would be more on maintenance.

Again, **P**roductivity, **Q**uality, **R**eliability and **M**aintainability (PQRM) are very important aspects for profitability of any industry. Each one has different role and affect the performance of others Their interrelation is shown in Fig. 2.1 (PQRM Inter-relationship).

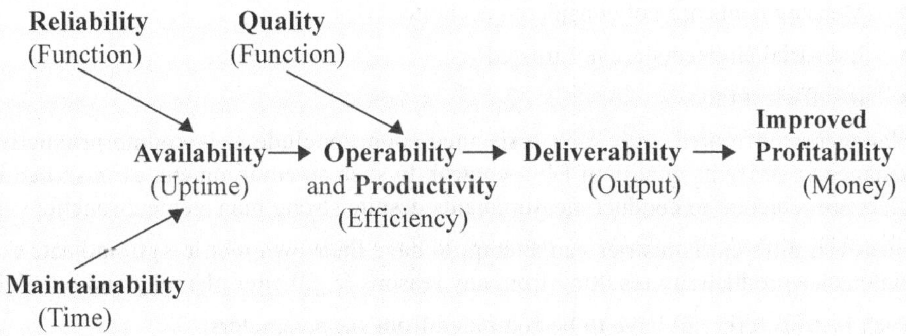

Fig. 2.1 PQRM Inter-relationship

2.1 PRODUCTIVITY

Productivity is a measure of economic efficiency which shows how effectively economic inputs are converted into output. Productivity is measured by comparing the amount of goods and services produced with the inputs, which were used in production. Labor productivity is the ratio of the output of goods and services to the labor hours devoted to the production of that output. Labor productivity is the most commonly used productivity measure as labor is an easily identified input to virtually every production process. Advances in productivity, that is the ability to produce more with the same or less input, are a significant source of increased potential national income. The productivity shows the changes from period to period in the amount of goods and services produced per hour. Although these measures relate output to hours at work of all persons engaged in a sector, they do not measure the specific contribution of labour, capital, or any other factor of production. Rather, they reflect the joint effects of many influences, including changes in technology; capital investment; level of output; utilization of capacity, energy, and materials; the organization of production; managerial skill; and the characteristics and effort of the work force.

In industries, employing many workmen, a term **Multifactor Productivity (MFP)** is used. MEP measures the changes in output per unit of combined inputs (or a set of combined inputs). Indexes of MFP are produced for the public/ private business, farm business, and manufacturing sectors/ industries, the railroad transportation industry, the air transportation industry, and the utility and gas industries ad other sectors of the economy. A change in MFP reflects the change in output that cannot be accounted for by the change in combined inputs. As a result, MFP measures reflect the joint effects of many factors including research and development (R&D), new technologies, economies of scale, managerial skill, and changes in the organization of production.

2.1.1 Maintenance Productivity

Maintenance productivity is little difficult to define and measure as maintenance is mostly a group activity. Whether the productivity should be measured for the whole department, or a fixed group who always remain together when on work or individual maintenance worker basis. If we consider the actual time of working (also called the wrench time) of maintenance workmen, we may find that, often, less than their operation counterpart; may, roughly turn around 35%. The maintenance department that checks worker productivity on a regular basis is rare - perhaps even nonexistent. Despite this, productivity checks remain the most effective method for verifying the quality of labor control. Numerous reasons are offered for not checking worker productivity—

- Measurements are misunderstood
- Workforce will suspect layoffs
- Personnel feel threatened
- Measurements are not considered accurate
- Industrial engineers are not trusted
- Insufficient time.

Managers confronted with such resistance soon conclude it is counterproductive to ask maintenance to verify its quality of labor control. In such an environment, even skilled industrial engineers are reluctant to conduct measurements despite strong management sanctions to do so.

However, different industries can attempt to have their own metric(s) to indicate the change in maintenance productivity resulting from any reasons (e.g. better planning and scheduling etc).

Such metrics (criteria) have to be combined from six parameters:

1. Downtime due to failure - highest expenses.
2. Downtime due to maintenance itself, including PM and PdM etc; adjustable expenses.
3. Cost of manpower.
4. Spare parts stock expenses (frozen money).
5. Tools, instruments, room, transport, soft for maintenance (investment with no return).
6. Equipment condition - very sophisticated parameter; e.g. - how long from now machine may work without any repair and all manufacturer parameters are OK. But other metrics may also be used.

All the above metrics may be converted in Rupees and a base value of combined metric to be worked out, either taking the long history or experience etc. Once the base metric is worked out, we can compare with that any changed situation.

Again, Maintainability is the key factor for Maintenance Productivity. The subsequent chapters in this book serves as **The Maintenance Productivity Enhancement Tool (M-PET).**

2.2 QUALITY AND QUALITY CIRCLE

Quality is a very general topic and is used in every aspect of our life; as such its definition depends on the purpose and for whom you are talking. Fundamentally 'Quality' can be defined as the ongoing process of building and sustaining relationships by assessing, anticipating, and fulfilling stated and implied needs. In industrial scenario, quality is the extent to which products, services, processes, and inter-relationships are free from defects, constraints, and items, which do not add value for customers and the commodity conforms to standards applied to the said commodity, be they mechanical standards or other standards. There are two forms of quality; objective and subjective. Objective quality is the degree of compliance of a process or its outcome with a predetermined set of criteria, which are presumed essential to the ultimate value it provides. Subjective quality is the level of perceived value reported by the person/ agency, who benefits from a process or its outcome. Often a component or service may encompass various intermediate quality measures, both objective and subjective. Quality can, again, be classified as static and dynamic. Static quality is what keeps us in the business whereas Dynamic quality allows us to excel or win it. They support each other - one cannot sustain without the other — but they are fundamentally opposed in philosophy. Static systems are dead - they control and maintain. Dynamic systems are creative, but may be anarchic. They do not endure. Static quality systems in the organization are like the unconscious muscles, which make the heartbeat and the lungs function. Unless they are in place, the organism cannot continue to live. However, in such system, the quality manager would be more busy to devise systems to assure aspects such as timeliness, product/ process matching the standard, cleanliness and safety etc; but in dynamic quality system, the manager would encourage spontaneous acts to excel and better standards or better satisfaction. Total Quality Management (TQM) or Total Quality Process (TQP) is, today, the culture in most of good organizations.

In industries, we often use "Quality of design", 'Quality of a product", "Quality of a process" and "Quality of a service" etc. Though most industries have a separate quality department responsible for testing, reporting and certifying the product/ process, quality is every one's responsibility. "Quality in Maintenance" or "Maintenance Quality" is a type of quality of a service. The quality concept helps maintenance managers and other concerned persons to develop, formulate and use quality indicators for maintenance management/ functions. By introducing quality aspect in maintenance function, we can be benefited in many ways such as —

1. Gain insight into how the maintenance function can be made more reliable, more cost-effective and more efficient,
2. Compare maintenance information systems with other good systems of your industry or other industries,
3. Align your maintenance program with your TQM initiatives
4. Monitor processing techniques,
5. Keep control check with replacement and life cycle costs;
6. Use of QAMM (Quality Assured Maintenance Management), discussed later.
7. Total Quality Preventive Maintenance Programmes that will— prolong equipment life, improve equipment safety and reliability, reduce the cost of major repairs and minimize the inconvenience of equipment downtime. Etc.

2.2.1 Quality Circle (QC) in Maintenance

Maintenance quality circle is also one of the employee participation methods. It implies the development and use of skills, capabilities, confidence and creativity of the people through cumulative process of education, work experience and participation etc. to solve their problems. Quality Circle is a small group of employees (normally 6 to 12) doing similar work, who voluntarily meet together on a regular basis to identify problems and improvements in their respective work areas, using proven techniques for analyzing and solving work related problems coming in the way of achieving and sustaining excellence leading to mutual upliftment of employees as well as the organization. QC is a method of capturing the creative and innovative power that lies within the work force. It provides self-motivation and happiness in improving environment without any compulsion or monetary benefits. The Quality Circle philosophy calls for a progressive attitude on the part of the management and their willingness to make adjustments, if necessary, in their style and culture. As such, management, normally, appoints a facilitator to each quality circle, to give the team members necessary managerial help.

The objectives of Quality Circles are multi-faced; e.g. Change in Attitude, Self Development (brings out 'Hidden Potential' of people), Development of Team Spirit and Improved Organizational Culture etc.

Organizational Structure for quality circles is generally as — Quality Circle members (from the worker level), with one leader of each QC, Facilitator, Coordinator and plant steering committee.

The process of operation, generally, is Problem identification, Problem selection, Problem Analysis, Generating alternative solutions, Select the most appropriate solution, Preparation of plan of action, Present solution to management and Implement the solution if it can be done by the team or help in implementation if it is to be done by others. Necessary problem solving techniques and tools are used.

Quality Circle concept is very well accepted and used in maintenance organizations for maintenance functions in most industries, with enormous benefit to those industries.

2.3 RELIABILITY AND RELIABILITY ENGINEERING

Reliability is a broad term that focuses on the ability of a product to perform its intended function. Mathematically speaking, assuming that an item is performing its intended function at time equals zero (i.e. assuming that the item is in a state to perform this required function at the beginning of the time interval), reliability can be defined as the probability that an item will continue to perform its intended function without failure for a specified period of time under stated conditions (in specified environments, and with a desired confidence that can be specified, designed in, predicted, tested and demonstrated). The term product, defined here, could be an electronic or mechanical product / component/ hardware, a software product, a manufacturing process, or a service etc. Usually, unless stated to the contrary, Reliability is deemed to include Durability.

The term reliability is also used as a reliability characteristic denoting a probability of success, or a success ratio of following two types—

- Mission or Assignment Reliability. The ability of an item to perform its required mission or assignment critical functions for the duration of a specified mission or life profile (e. g. missile or space ship etc),
- Basic Reliability. The ability of an item to perform its required functions without failure for the duration of a specified mission profile. (Products/ processes/ Services etc in industries).

Reliability Engineering is the application of reliability concept in engineering or industries and in more concerned with basic reliability i.e. life data analysis of inanimate objects, such as equipment, components, systems and processes etc. Reliability engineering covers all aspects of a product's life, from its conception, subsequent design and production processes, as well as through its practical use lifetime, with maintenance support and availability. Reliability engineering covers all three aspects of RAM i.e. reliability, availability and maintainability and all three of these areas can be numerically quantified with the use of reliability engineering principles and life data analysis. Reliability engineering helps in striking a proper balance between reliability, customer satisfaction, time to market, sales and features, and profitability.

We also often hear about "Human Reliability" and "Operational Reliability". Human reliability can be defined as the probability of accomplishing a job or task successfully by human at any required stage in system operation within a minimum specified time limit. Operational reliability is actually an amalgamation of 'Human Reliability", "Process Reliability", "Equipment Reliability" and "Equipment Maintainability", as shown in Fig. 2.2 .Operational Reliability introduces a systematic approach to eliminate the failure causes and bad reliability actors, affecting the critical processes and overall company profitability. For achieving business excellence, Operational Reliability has to take care of all the four components mentioned above.

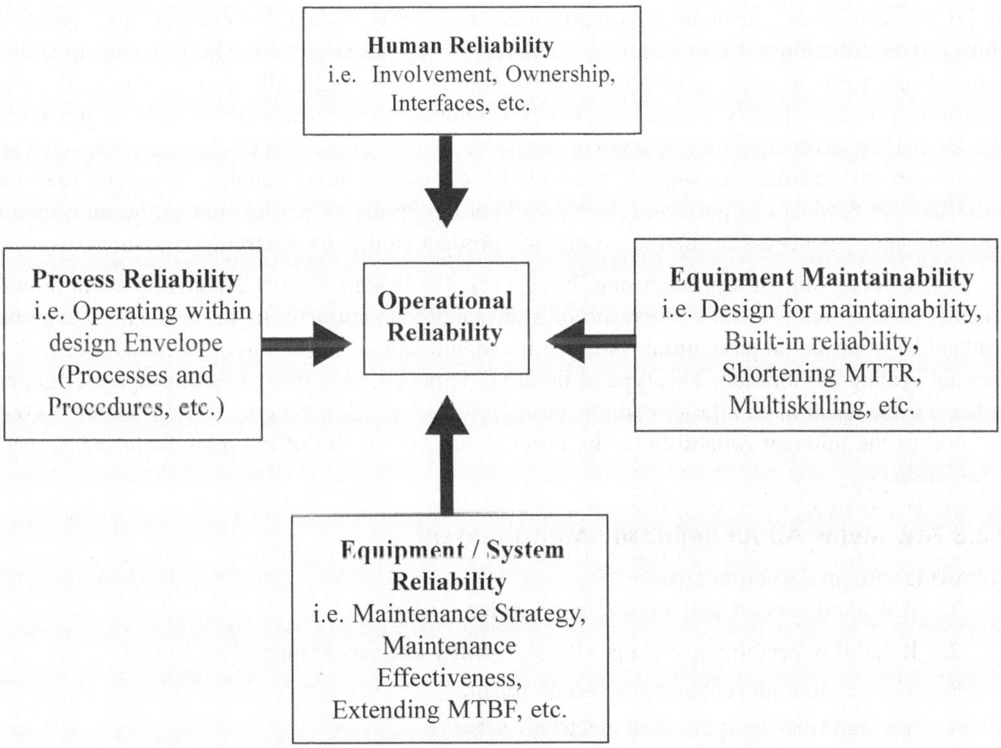

Fig. 2.2. Operational Reliability

2.3.1 Importance of Reliability

In today's highly competitive and technologically complex environment there are a number of reasons why reliability is an important product attribute, such as—

1. Reputation of the company/ industry
2. Customer Satisfaction,

3. Warranty Costs, If a product fails to perform its function within the warranty period, the replacement and repair costs will negatively affect profits,
4. Repeat Business; improved reliability has a positive impact on future business,
5. Cost Analysis; This life cycle cost analysis can prove that although the initial cost of their product might be higher, the overall lifetime cost is lower than a competitor's because their product requires fewer repairs or less maintenance,
6. Customer Requirements. Many customers in today's market demand that their suppliers have an effective reliability program.
7. Competitive Advantage. Many companies will publish their predicted reliability numbers to help gain an advantage over their competition who either does not publish their numbers or has lower numbers; etc.

2.3.2 Difference Between Reliability and QualitY

Although the terms reliability and quality are often used interchangeably, there is a difference between these two disciplines. While reliability is concerned with the performance of a product over its entire lifetime, quality control is concerned with the performance of a product at one point in time, usually during the manufacturing process. Reliability assures that components, equipment and systems function without failure for desired periods during their whole design life, from conception (birth) to junking (death). Quality control is a single, albeit vital, link in the total reliability process. Quality control assures conformance to specifications. Quality control also checks that the incoming parts and components meet specifications, that products are inspected and tested correctly and that the shipped products have a quality level equal to or greater than that specified. No product can perform reliably without the inputs of quality control, because quality parts and components are needed to go into the product so that its reliability is assured.

Again, even though a product may have a reliable design, it may effectively be unreliable when fielded, which is actually the result of a substandard manufacturing process. As an example, cold solder joints could pass initial testing at the manufacturer, but fail in the field as the result of thermal cycling or vibration. This type of failure did not occur because of an improper design, but rather it is the result of an inferior manufacturing process. As such, a highly reliable product is only as good as the inherent reliability of the product and the quality of the manufacturing process/ installation etc.

2.3.3 Few Myths About Reliability Management

Followings are myths and not true—

1. It worked once, it will work in any service,
2. Reliability performance is upto the Reliability Officer/ Manager,
3. The key to high reliability is speedy repair,
4. You can trust everything on electronic screen,
5. Management will always support the programme,
6. The craftsmen/ tradesmen men is only involved in repairs, not reliability,
7. Reliability is a project,
8. Mechanical engineers (or other engineers in specific cases) know every thing,
9. You can always trust fabrication shop (or other shops in specific cases), that have industry certification/ accreditation,
10. The above myth is not important, since you can always sue the concerned shop; etc.

2.3.4 Reliability Testing

Reliability Testing is a combination of Functional Testing and Performance Testing, focusing more on reliability aspects. Functional testing focuses solely on the outputs generated in response to selected inputs and execution conditions and verifies that an application works as intended from a user's perspective. Performance testing is a class of tests implemented and executed to characterize and evaluate the performance-related characteristics of the target-of-test, such as the timing profiles, execution flow, response times, and operational reliability and limits. Reliability testing does both these, although it breaks down reliability into Integrity, Structure, and Stress tests components etc.

Reliability Testing may include different tests, which may be categorized as given in Table 2.1,

Table 2.1 Few Reliability Testing Methods

Category of Tests	Few Specific Tests
Mechanical Integrity	• Mechanical Shock • Vibration, variable frequency • Thermal Shock • Solderability; Weldability and Weld Cross-section • Constant Acceleration/ Centrifuge • Particle Impact Noise Detection (PIND) • Sealability etc.
Endurance	• Temperature Cycle • Highly Accelerated Stress Test (HAST) • Highly Accelerated Life Testing (HALT) • Cyclic Moisture • Constant Acceleration, Accelerated Aging; etc.
Electrical Characterization	• E-beam testing and Microcircuit Isolation, • Navigation and CAD Tools for Circuit Layout Isolation, • Optical Beam Induced Resistance Change (OBIRCH) • Resistive Contrast Imaging ; etc.
Hermeticity	• Fine Leak Detect (Helium) • Gross Leak Detect (NID) negative ion detection; etc.
Special Tests	• ESD Threshold (HBM and MM) • Environmental Stress Testing (ESS); etc.

Often over-testing is done to determine the performance margin and to induce actual failure so that mode of failure can be examined and possible corrective action can be taken. The tests may be–

1. Abbreviated Life Tests; by predicting the characteristic failure patterns,
2. Failure-Repair Runs; by running till failure. Often the MTBF is considered as MTTF for end use,
3. Accelerated Test; by developing ways to have continuous actuations or revolutions etc what the equipment would have taken in two or more years. In these tests, units of MTBF or MTTF are not time but number of actuations or revolutions etc.
4. Test for Increased Severity; by increasing the required temperature, vibration, current strength or dust concentration etc,

5. Test for Large Sample Size; etc.

Like Quality Assurance, the term **Reliability Assurance** is the planned and systematic set of activities that ensure that process and products/systems conform to requirements, standards, and procedures. It normally includes activities involved in designing, manufacturing, installation, coding, testing and maintaining etc. Occasionally a reliability assurance agreement is also made with different agencies. Often, on high value contracts, contractors are required to submit Reliability Test Plans and Environmental Stress Screening reports if burn-in is a requirement of the item's technical data package.

2.3.5 Reliability Process, Tools and Techniques

"**Equipment Reliability Process**" represents the integration and coordination of a broad range of equipment reliability activities into one process for plant personnel to evaluate important station equipment, develop and implement long-term equipment health plans, monitor equipment performance and condition, and make continuing adjustments to preventive maintenance tasks and frequencies based on equipment operating experience. It includes activities like–

1. Performance Monitoring (System Performance, Component Performance, Predictive Trending Results, Operation Monitoring and monitoring other testing and inspection results)
2. Scoping and identification of critical components (Identify important functions, identify non-critical, critical and run-to-failure components),
3. Review existing PM and other maintenance strategies,
4. Necessary RBDs and Failure Analysis and prioritization of equipment failures (RCA and RCFA etc, discussed subsequently in this book),
5. Define reliability requirements to satisfy plant's/ industries needs; reliability prediction,
6. Device and implement PM and other maintenance strategies (including Reliability Centered Maintenance etc), introduce equipment condition feedback and post-maintenance tests etc,
7. Continue equipment reliability improvements, including development and use of PM Templates, adjustment of implement maintenance strategies and consideration of alternate maintenance strategies to ensure reliable equipment etc.
8. Life Cycle Management, including long term system and component health prioritization of improvement activities and integration of long term plans into management's long term activities.

Reliability Tools and Techniques cover system analysis (RBDs), FMEA and FMCEA, Life Data Analysis (Weibull Analysis), Accelerated Life Testing Data Analysis (ALTA), FTA and, Reliability Growth Analysis (RGA), Quality Tracking and Management System (QTMS), Failure reporting, analysis and corrective action activities (often called FRACA or FRACAS processes) and many similar techniques, many of which will be discussed subsequently in this book,

Reliability Block Diagram (RBD)- RBD techniques are useful to manage and interpret complex system architectures. RBDs enable splitting complex systems into manageable blocks, maintaining system-architecture integrity (i.e., logical connectivity that supports the mission profile functionality). When performing a Reliability Prediction analysis, failure rates for components, assemblies, and systems are calculated. The models used for Reliability Prediction assume that all components and assemblies are configured in series, and failure rate calculations are based on that assumption. However, often, systems will be designed with various redundancies of key assemblies in order to improve the overall reliability of that system. RBD is a complete, fully featured graphical reliability block diagram evaluator. The intuitive user interface enables you to draw a visual layout of your system on the screen. After entering your system parameters, calculation can be done to provide fast and accurate reliability and maintainability (RandM) results, including reliability, availability, MTBF, failure rate, expected number of failures, mean unavailability, total downtime, frequency, and hazard rate. Fault Tree (refer FTA in chapter 3) can be used in conjunction with RBD by linking fault trees directly to RBD blocks.

2.3.6 Reliability Predictions

Reliability predictions and Mean Time Between Failure (MTBF) calculations provide the basis for reliability evaluation and analysis. Prediction techniques have a worldwide acceptance and encompass a range of statistical models for various applications. By allowing you to assess the reliability of your design prior to production, reliability predictions enable you to build with confidence. You are able to pinpoint areas of concern, quickly analyze design changes, and improve overall product reliability. It contains mainly the following components-

1. Prediction Tool/ Method- It performs reliability allocation calculations, supports de-rating analyses, and provides a visual report designer to display your results in an informative and professional style,
2. Parts Libraries-It contain different components including electrical and mechanical part types with their associated data parameters. By using standard, proven component libraries you are able to accelerate the process of evaluating and modeling MTBF and product reliability calculations, gaining a significant time saving.
3. Reliability Prediction Analysis- It provides you with a comprehensive solution to meet all of your prediction needs.
4. Prediction Models- It enables you to mix models (from standard known models) within a single project, allowing you to select the model most appropriate to each section of your design.

Reliability Growth Analysis (RGA) - Reliability growth is the improvement in a reliability parameter over a period of time due to changes in product design or the manufacturing process. It occurs by surfacing failure modes and implementing effective corrective actions. Reliability growth management is the systematic planning for reliability achievement as a function of time and other resources, and controlling the ongoing rate of achievement by reallocation of these resources based on comparisons between planned and assessed reliability values. The growth projection is based upon information currently available from predictions or prior experience on identical or similar systems. Monitoring the reliability, the mean time between failures (MTBF), and the failure rate of the system, equipment or product also establishes a trend in the increase in the reliability, the increase in the MTBF, or the decrease in the failure rate. This is achieved with engineering, research, development, test-analyze-and-fix (TAAF), and/or test-analyze-and-redesign (TAAR) procedures until the product passes its acceptance tests and/or is delivered to the end-user. Reliability growth studies are necessary to insure that from information available at the beginning of a project, the reliability, MTBF, or failure rate goals are capable of being met.

2.3.7 Reliability by Redundancy

Redundancy, to a common person, means **superfluousness or unnecessary and is a form of dismissal** or the use of words in excess of those needed for clarity or precision, **but in engineering circle, it means duplication (or triplication or more) of** electronic, electrical or mechanical components or backup systems that are designed to come into use to keep equipment working if their counterparts fail or duplication (or triplication etc) of information in telecommunications in order to reduce the risk of error.

Again, in engineering circle, there are two types of redundancy; Load-sharing Redundancy and Standby Redundancy. **Load-sharing Redundancy** is a form of redundancy with dependent components, *i.e.* the failure of one component affects the likelihood of failure for the other component(s). Units in load sharing redundancy exhibit different failure characteristics when one or more fail. In Figure 2.3, blocks 1, 2 and 3 are in a load sharing arrangement and have their own failure characteristics. All three must fail for the container to fail. However, as individual items fail, the failure characteristics of the remaining units change since they now have to carry a higher load to compensate for the failed ones. If one component fails, then the component(s) that are still operating will have to assume the failed unit's portion of the load. Therefore, the reliabilities of the surviving unit(s) will change, normally decrease. Under **standby redundancy**, the redundant components do not share any of the load, and they start operating only when active components fail. In standby redundancy, the components are divided into two types, Active and Standby. The standby components have two failure distributions. One, for when they are in standby (quiescent distribution) and one for when they operate (active distribution). When the failure rate of the standby component is the same in quiescent mode as it is in active mode, then components are in a "hot standby" configuration and when the failure rate of the standby component is zero in quiescent mode (*i.e.* the component cannot fail when in standby), then you have a "cold standby" configuration. In-between configuration may be "warm standby". Normally reliability in cold-standby configuration is highest and hot-standby is lowest. Fig. 2.3 also shows standby configuration.

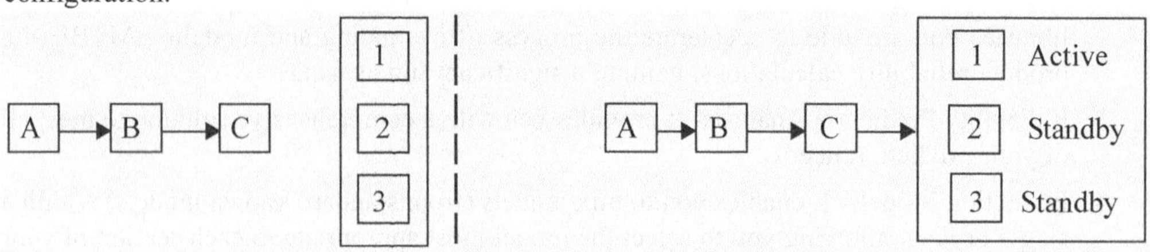

(*Load-sharing Arrangement*) (*Standby Arrangement— 1/3)*

Fig 2.3. Load-sharing and Standby Redundany

Though Load-sharing redundancy is also important consideration, mainly in electricity distribution systems etc, standby redundancy is more synonymous with reliability improvement by redundancy. Considering a system with "n" components, each with a probability failure "p", the probability of success of the system will be $(1\text{-}p)^n$ if all the components are serially dependent.

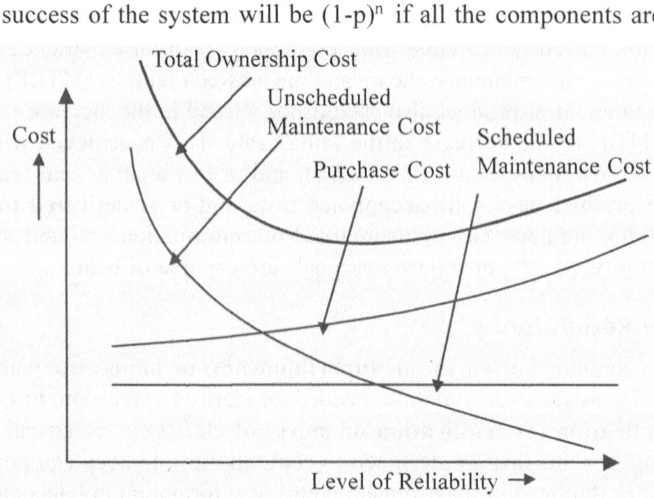

Fig. 2.4. Reduction in Life Cycle cost, based on Reliabiliy

However, if all the components are in redundancy position (normally arranged in parallel), the probability of success will be $1\text{-}p^n$.

However, there is cost for maintaining high reliability. Fig 2.4. (Reduction in life cycle cost, based on improved reliability) very roughly shows the cost reliability relationship for any system/ product. However, for basic reliability and mission reliability, different approaches of Total; Ownership Cost are considered for obvious reasons.

2.4 MAINTAINABILITY

It is an inherent characteristic of the system design pertaining to ease, accuracy, safety and economy in performing maintenance functions. It is the ease with which a system or component can be modified to correct faults, improve performance, or other attributes, or adapt to a changed environment. It can be defined as the ease in time and resources of retaining equipment in or restoring it to a specified operational condition. It can also be defined as the measures taken during the design, development and installation of a manufactured product that reduce required maintenance, man-hours, tools, logistic costs, skill levels and facilities etc. It directly affects the bottom line because it can impact operations, downtime, maintenance costs, and safety. It can also be defined as the measures taken during development, design and installation stages of a manufactured product, that reduce maintenance, man-hours, tools, logistic cost, skill level and facilities. Maintainability is an important aspect of any system's lifecycle, but process plant engineers typically give it little direct consideration. This is primarily the result of a short-term view of capital project costs that fails to consider lifecycle costs and downstream activities.

Maintainability Improvement - There may be many ways for improving maintainability. However, "design for maintainability is the most important way/ aspect, as if the equipment is not designed for good maintainability, we would feel problems in all our subsequent efforts. Human factors engineering can also be applied to systems design to minimize the time and effort required to perform periodic preventive maintenance as well as unscheduled maintenance. Field observation techniques can also be developed to ascertain the level of effort required to maintain existing systems and to identify opportunities for system improvement.

2.4.1 Design for Maintainability

The key to optimizing the reliability and maintainability of any component/ system is having a well-designed and effective programme/ arrangement. Before starting on a system or component design, two principles are to be keep in mind: Minimize maintenance from the beginning, and get your maintenance people involved. Reliability directly affects system or component maintainability. The more reliable a system is, the less maintenance it will require. As such, costs should be allocated carefully. For example, maintainability is best improved if the difficult-to-maintain parts of a system are made more reliable. Identifying reliable components, installation types and arrangements, and vendors is a key step to improving the reliability of instrument systems. Few of the considerations for design for maintainability are given here with–

1. Design Fundamentals—

- The simple word KISS (Keep it simple, stupid) applies well to maintainability. Even in the age of increasingly complex and sophisticated world, we should strive to keep it simple wherever we can
- Use modular design to divides the system into physical and functional modules, which can be arranged to facilitate design and maintenance. Modularization is preceded by unitization i.e. separating components into physically and functionally distinct units. Modularization refers to unitized components in which functional units making up a

module. Easily replaceable modules with logical organization reduce repair time, troubleshooting, training, and engineering. Interconnectability and interoperability should also be considered.

- Some thought must be given to future modifications or expansions. Failure to do so may make a system that is easily maintained today but difficult to work on tomorrow.

2. Standardize, Document, and Label—

- Using recognized national, industry, and company standards and codes is good engineering practice. It improves maintainability by reducing variations of design and installation for maintenance personnel. Standardization of components reduces inventory and improves reliability and maintainability.
- The use of commercial off-the-shelf components should also be considered, of-course meeting required standards,
- Documentation is extremely important in achieving good maintainability; specifically for the complexity and sophistication of the system increases. Make sure that maintenance documentation needs are being met.
- Labeling is essential in industries to make system components easily identifiable. Labeling must be consistent, standardized, clear, and accessible. While not a substitute for drawings, system identification that provide the capability of tracing wiring, power sources, and identification of components without use of drawings provides for more efficient and safe troubleshooting.
- Maintenance Displays, normally on or near operator's control desk/ panel, to give essential information etc.

3. Make wrong parts difficult to fit—As example, if many hydraulic hoses are to be located at nearby places, they should be different in size or length or otherwise so that a hose can not be fitted at wrong place.

Colour coding may also be used to fit components at correct place.

4. Access is Critical- Accessibility means having sufficient workspace and access to perform maintenance safely and efficiently and also for troubleshooting. Consideration should be made for opening of doors or removal of panels, removal of parts or modules, required manipulative tasks, needed body and tool positions and movements, duration of the access, and potential exposure to unsafe conditions during access. Low-reliability components should be the most accessible.

Configuration of equipments containing items requiring lubrication etc, should permit both, lubrication and checking of lubricant level etc without disassembly.

Lighting level and inbuilt handling facilities are also considered as far as possible.

5. Exposure of the system to weather, other environmental concerns, and stresses generated by other equipment such as heat, vibration, moving parts, etc., must be considered.

6. Protect and Accommodate Personnel—

- There are many maintenance activities that are less safe than others. As such provisionsshould be made to enable fixing proper safety measures before doing maintenance jobs. As example, for working on a transmitter 15 feet up in the pipe rack, proper location in the pipe rack to be provided to allow the use of a platform, installation of access platforms, or locating where a scissor lift or manlift can be used to improve the safety forsuch an installation.

- Mechanical and electrical hazards must also be considered. Pinches, sharp bends, edges andpoints, trip hazards, head knockers, and abrasive surfaces should be eliminatedorguarded. Adequate electrical and other mechanical guarding also must be provided.

7. Streamline Testing, Calibration, and Troubleshooting— A system or component should be easily tested for troubleshooting. Consider the failure modes and how they will be detected. If an equipment is difficult to be inspected or tested, it will be less likely to be inspected or tested. Modern equipment often comes with built-in test gadgets and self-test diagnostics. The required level of diagnostics is a function of the "invisibility," or level of abstraction, of the internal functionality of the instrumentation. However, diagnostics are not just the domain of the equipment supplier. Look for opportunities to add cost-efficient diagnostics to your instrument systems. Such diagnostics must have comprehensive coverage and provide speedy localization of the problem and help trouble shooting.

8. Plug-in connectors, quick disconnect devices, Lugs and crimp-on devices and other such maintenance friendly devices should be preferred.

9. Ensure Logistics are logical— Maintenance resource availability is also a key to good maintainability. Tools and training are fundamental. The proper tool for the job is a key to safe and efficient maintenance. Today's systems can be complex and sophisticated, making maintenance difficult for the untrained. However, training alone is not a good substitute for expertise, which includes knowledge + capability + experience + application.

10. Prevent and Predict— New technologies such as web-enabled monitoring, maintenance-based artificial intelligence, multimedia data access, wireless sensors, improved diagnostics, and self-validating devices will allow improved monitoring of equipment and systems, which will enhance instrument maintainability.

2.4.2 Massive Over-Design and De-rating to Improve Reliability

Though not economical and not a good practice, occasionally massive over-design is considered to improve the reliability of equipment/ component, though it may not improve maintainability. Again, occasionally, a standard designed higher capacity equipment is de-rated (down-rated) and supplied as lower capacity equipment. The author knows few cases where standard 50 T electric arc furnaces were de-rated by supplier and shipped as 40 T Furnace. However, these are not economically feasible in today's competitive market.

2.4.3 Maintainability Predictions

It enables you to analyze the repair metrics of your system or process. It provides a solid framework for performing maintainability analyses. By evaluating repair statistics, you can work to minimize system repair times, thereby minimizing downtime and increasing your system availability. It allows you to define your repair tasks and easily reuse this information throughout your design, using a host of maintenance parameters including the Mean Time to Repair (MTTR), Mean Corrective Maintenance Time, Mean Preventive Maintenance Time, and Maximum Corrective Maintenance Time. Maintainability Prediction can evaluate system maintenance parameters directly through repair tasks, FDandI (Fault Detection and Isolation) assignments, or a combination of the two for ultimate flexibility. Maintainability Prediction can be used independently or with Reliability Prediction and FMEA for failure rate data and system definition.

2.4.4 Maintenance Productivity Enhancement Tools (MPET)

Almost all the chapters dealt subsequently in this book will help in maintenance productivity improvement. A few of them are mentioned herewith as examples—

1. Standardize the maintenance jobs,
2. Documented standard maintenance practices with standard times to accomplish the Task,
3. Repetitive maintenance tasks should be tried for elimination, with RCA, FMCA etc,
4. Follow appropriate maintenance strategy—proactive, predictive etc,
5. Use modern tools like CMMS, TPM, Lean Maintenance etc,
6. Multi-skilling and continuous up-gradation of skill and knowledge of maintenance personnel, through various means as discussed later in this book,
7. Appropriate outsourcing of specialized jobs or monitoring, which can not be done properly in-house
8. Proper maintenance audit and effectiveness and KPI etc.

In this regards, following aspects of maintainability should also be considered for further maintainability improvement–

(i) Maintainability Planning
(ii) Maintainability Measures
(iii) Maintainability Analysis
(iv) Maintainability Design Check List, (refer Sec.2.4.1)
(v) Maintainability Testing
(vi) Maintainability Data Collection and Analysis

2.5 RELIABILITY, AVAILABILITY AND MAINTAINABILITY (RAM)

Reliability and maintainability jointly affect availability. A highly reliable equipment may fail rarely, but if its maintainability is poor, it would take very long time to repair, thus reducing availability. On the other hand, if maintainability is good but reliability is poor, it may fail very frequently. Again, maintainability is not considered in one-shot applications. Thus is the need for RAM study in details.

RAM is yet another concept being considered in recent maintenance management discussions. With proper consideration of reliability, availability and maintainability and (RAM) in the design, manufacturing, and installation phases, the number of failures can be reduced and their consequences minimized. To broaden the concept, occasionally "Supportability" is added to the concept to make that RAMS. RAM analysis is a powerful tool which utilizes failure information from a system's components in order to develop probability distributions that the system will be able to perform its intended function. Conducting an availability assessment of a system enables a project engineer to optimize a system's inherent performance. The assessment accomplishes this by identifying components which contribute the most to downtime and allowing the determination of the effect of design alternatives on system performance in a cost-effective manner (i.e., using computer modeling rather than prototype testing). It also considers FEMCA and Failure analysis (utilizing the appropriate tools, including block diagramming, fault trees etc) Analyzing the effects of maintenance activities is also necessary to realize the highest level of system performance. Establishing numerical RAM goals can provide the incentive for a vendor (of procured systems) to formulate a systematic performance improvement program and justify the design decisions. The term "availability" has been discussed in chapter

Another set of terms i.e. **Reliability, Availability and Serviceability (RAS)** are often used as key attributes in computing world. The term "RAS" is fairly common in the computing industry (particularly computers and storage) as computing becomes more fundamental. For example, a vehicle may depend on dozens of computers, and the consequences of the failure can be significant (e.g., an ambulance's engine won't start.

Equipment availability is discussed further in chapter 13.

QUESTIONS

1. Why is it not so easy to measure maintenance productivity?
2. What is PQRM in relation to Maintenance?
3. What is difference between quality and reliability in maintenance?
4. What is difference between maintainability and reliability of equipments?
5. What are the ways to improve maintainability of an equipment?
6. What is difference Reliability by redundancy? Give one example.
7. What is reliability? Explain how reliability data helps in performance of the maintenance functions.

CHAPTER

3 MAINTENANCE AND REPAIR JOBS AND TECHNOLOGIES

The jobs/activities and technologies required for various maintenance and repair activities/ functions may be classified in two categories–

1. Category-1. This includes the technologies required to determine if maintenance or repair action is required, and to quantify actual defects, residual life or operational expectancy and reliability; These would be discussed as maintenance management aspects in subsequent chapters. These technologies can also be termed as Health and Usage Monitoring System (HUMS).

2. Category-2. This includes the technologies required to actually repair/ improve the different parts/ components/ systems by more efficient techniques, shorter repair cycle time and lower cost etc.

The scope of this chapter is to discuss the category-2 technologies. It includes set of actions maintenance men or their associates have to do to in order to restore a failed or breakdown equipment /system to its original working condition or actions for upkeep and prevention of failures. Such actions and technologies are numerous and new technologies are being discovered every now and then. As such a comprehensive look of such actions and technologies is almost impossible. Only few of such actions and technologies would be covered in this chapter. First few basic actions would be discussed for beginners in little more details and later few composite / critical technologies would be discussed.

3.1 INSPECTION AND TESTING

This is normally the first step and then intermediate steps. Different types of inspection and testing have to be done by maintenance men, and a few of those are given herewith. Few of those are mentioned in some details in chapter 6.

1. Dimension and coordinates measuring,
2. Surface roughness and flatness checking,
3. Gauging and calibration—Use of gauges and fixtures for special checking,
4. Optical inspection of inaccessible space (as an engine cylinder) through Borescopeand Endoscope etc,
5. Chemical testing (e.g. in-situ testing gadgets for checking engine oil condition etc),
6. Tensile Testing,
7. Pressure testing and flow testing (visual or through instruments),
8. Few non destructive testing (NDT),
9. Hardness Testing (Vickers and Rockwell etc)

10. Grain Size Determination by Replication;
11. Electrical current and flux testing; etc.

3.2 DISMANTLING AND ASSEMBLING

These are the basic function/ job of maintenance, when a failed equipment or working equipment (during shutdown etc) is opened partially or completely)necessary repairs, adjustment or changing of elements and components are done and the equipment is fitted back in position, using common or special tools. However, enough importance is not given to these simple jobs/ actions, which, often, result in faulty repairs. Some of the common cares to be taken prior to and during dismantling and assembling are given below–

1. Check drawing before dismantling and mark on the drawing. Estimate the weight (if Not given in the drawing),
2. Check the job manual/ instruction and remember or list down the sequence of dismantling and assembling.
3. Put legible match mark, tags and labels on the two sides of couplings, flanges, pipe/ hose joints and electrical joints etc before dismantling and record if needed. If necessary, proper pipe schedule and electrical cable schedule may be made before dismantling as those may not be correctly reflected in drawings because of site modification etc.
4. Look for bench mark or level pads if base or foundation is to be repaired. This would help in maintaining the level during erection after repair. If no level bench mark is available, fresh bench mark is to be made by theodolite etc, before dismantling.
5. Use of proper tools is very essential, but often neglected. Torque wrenches should be used for uniform tightening of critical bolting. Use of slide wrench or pipe wrenchshould, normally, be banned for bolt joints.
6. Prepare joints and surfaces properly, by filing, machining, cutting and threading etc. Hand grinding and chipping of surfaces should, normally, be avoided as these may produce uneven surface.
7. Proper clamps and supports should be fitted, after assembling, to reduce vibration of pipes, tubes and hoses etc. Avoid twisting and undue bending of hoses. Use vibration absorbing pads with base of components or frames prone to some vibration.
8. Cover (or plug) all dismantled pipe end connections, hydraulic, pneumatic, electrical or other such items to prevent ingress of dust and moisture etc.
9. Proper depressurization of hydraulic and oil systems and putting oil collection trays before opening any such joints.
10. Proper alignment and balancing are to be checked after assembling, with the help of proper tools/ gadgets. Etc.

3.3 MAINTENANCE CLEANING

Cleaning of equipments, components, working tools, hands or working gloves and workplace etc, before taking repairs, during and after repairs is of prime importance, but is often not given due consideration. Cleaning is often considered "donkey's job" and is left to some unskilled worker to decide and do. But the type and extent of cleaning is purely a technical requirement, depending on the subsequent jobs to be done. Cleaning of components is, normally, assisted by kerosene, petrol, carbon-tetra-chloride (CTC) and many other solvents.

In hydraulic system, dirt and dust are the biggest enemy for sophisticated valves, pumps and control equipments/ items, which calls for proper covering and sealing of tanks, use of suitable filters during oil changing and repair of components to be done in dust-free room/ enclosure. For cleaning of rust, sediments and deposits from water-cooled components (furnace doors, heat shields), heat exchangers and tube-nests etc, suitable chemicals and solvents are used which soften and loosen the deposits so that the deposits come out by water or air force. Very often, high pressure water jets (pressure over 160 bar) are used for descaling of big tube-nests, heat-exchangers and condensers etc. Steam cleaning is also used often for many components.

Magnetizers have been developed (using permanent magnets or electromagnets), which can be clamped on the pipelines. The rust in water line or sludge in fuel lines get ionized and detached from the inner surface of the pipe lines and remain suspended or floating/ moving as long as magnetic field exists. By placing number of such magnetizers on the pipe length, the rust and sludge can be taken to the tank and removed.

Scrubbing, etching, skinning, burnishing, vacuum cleaning, extractors and many other such techniques are used for cleaning as per need. Many customized cleaning equipments are available and also many brands and types of cleaning fluids are available in market. You have to select the fluid, techniques and cleaning equipments as per specific requirement for job.

3.3.1 Types of cleaning

In short, cleaning can be of following type and selection has to be made as per need and item to be cleaned–

1. Aqueous Cleaning
2. Vapor Degreasing
3. Steam cleaning
4. Vacuum cleaning
5. Paint Stripping
6. One-Step Alkaline Cleaning (for Steel and Titanium)
7. Multi-Step Alkaline Cleaning and Acidic Descaling
8. Power Wash
9. Aluminum Oxide and Plastic Grit Blast
10. Light machining, etching, burnishing; etc

3.4 LUBRICATION

Lubrication is, one of the most important functions of maintenance. Lubrication for the machine plays the same role as blood for the human being. Improper or ineffective lubrication is the leading cause of mechanical failures and manufacturing/ production losses. For the production department, it is essential that machinery be correctly and adequately lubricated. Too often, machinery failures result from the wrong lubricants, or over/ under lubrication (too much or too little of the right lubricants). Over lubrication causes churning and heating and also wastes costly lubricants. Many problems are caused by the use of incorrect (or non-optimal) oil or grease and failure to change or top-off fluids regularly.

3.4.1 Function and Failure cause of Lubrication

The main functions of lubrication in bearings, gears and other rotating and sliding components are as follows–

1. Friction and Wear Control: To minimize friction at the point (including line or surface) of contact of mating components, normally by trying to introduce a separating lubricant layer. Minimizing surface-to-surface friction reduces wear.

2. Corrosion Control: To protect the precision finish of contact surfaces of components from being corroded,

3. Temperature Control: The lubricant absorbs thermal energy at the point of generation/ mating and transports it to a cooler portion or to the filters, dehydrators or tank where it can dissipate.

4. Contamination Control: To prevent ingress of foreign matters (including dust/ dirt, water etc) in the mating components; also to remove similar items from the components.

5. Power and Work Transfer: In hydraulic systems, the fluid (which doubles as the lubricant) is the medium by which work is accomplished. Etc.

To perform these functions, various types and grades of lubricants (oils, grease, synthetic lubricants, solid lubricants and mist lubricants etc) are available and various methods and systems (manual, semi-automatic, automatic, bath, feed, splash, centralized and decentralized systems etc) are used. Few surface treatments (coatings, plating and vapour deposition etc) are used for lubrication purposes also. Determining the quantity, quality and frequency is actually techno-economic judgment.

There can be various causes of lubrication or lubricant failure; e.g. Sudden Volumetric Loss, Failure to fill, Low level, Wrong lubricant, Base oil degradation, Particle contamination, Moisture contamination, Coolant contamination, Fuel/ chemical dilution, Additive depletion and viscosity depletion/ increase, Air entrapment, Foaming and oxidation, and/or thermal instability etc.

3.4.2 Few Tips on Lubrication

Few of the common tips on lubrication are given below–

1. If water in gearboxes has been a problem, use a water resistant gear lube and establish a drain interval to remove the water periodically,

2. Seal compatibility is a relative factor. A previous lubricant can make a difference in how a seal responds to the present product. Sometimes new seals are required when changing from one type of lubricant to another,

3. Store lubricants so that you use the first in (oldest), first out method (FIFO). This assures that you use fresh lubricants that have not been in storage for long periods of time. Stock rotation is part of a proactive maintenance programme. This does not bypass the checking of grease before use.

4. Check the operational or maintenance manuals of equipments regularly to assure that correct lubricants are being used at correct frequency.

5. Is your diesel engine putting out gray or black smoke? Check the air filter; it may be blocked and not allowing adequate airflow for complete fuel consumption.

6. Are you having excessive fuel consumption? Do you have incorrect idle speed? Possibly your low idle is misadjusted. Have a qualified technician check it.

7. When adding grease to your equipment, always make sure you purge all of the old grease. When adding the new grease, make sure that the area is clean of any foreign matter. Wipe off the grease fitting before and after relubrication.

8. You should consider using a color-coding system to ensure that the correct oil or grease is always used in each application by the operators. This also reduces the potential for contamination.

9. Though you should be diagnostic to select right lubricant for right application, you should also consider standardization, variety reduction and reclamation to the extent possible for reducing cost.

The lubricant supplier should have a strong relationship with the customer, providing focused, responsive services such as advice on preventive maintenance, lubrication and equipment diagnostics, consulting and training, conditioning services, and lubrication management systems, no matter where the plant is location.

For further study on lubrication, you may consult author's book "Oils, Lubricants and Lubrication" (Refer bibliography) or other similar books.

3.5 MAINTENANCE MACHINING (OR REPAIR MACHINING)

Machining or machine shop technology/ practice is a vast subject and many books are available on that. The main different between machining and maintenance machining is that maintenance machining is done to repair a damaged or degraded component and it lays more emphasis on in-situ machining. Though maintenance machining also involves in making few spares parts and components (to be used while maintaining an equipment) in conventional machine shop, it gives more importance in in-situ machining and repairs/ reconditioning.

Maintenance machining includes many jobs, such as–

1. Removing the jammed pins or shafts of a component, by drilling, boring or trepanning etc, in machine shop (if the component can be taken and mounted there) or in-situ.

2. Reconditioning of worn-out bearing journals or other worn-out holes or segments by boring to higher size, making weld deposits and machining back to original or desired size; either by taking the component to machine shop or in-situ, if the job can not be taken to machine shop or can not be mounted there (e.g. Rolling Mill housing journals, Mill Table frame journals etc).

3. Reconditioning of journal shafts and rollers etc for Sl. No. (2) above,

4. Re-drilling of damaged bolt holes of housing and frames to higher size (if use of higher size bolt possible) or by using **Threaded Sleeves** (drilling, boring and threading of damaged base frame hole to higher size and using a separate threaded sleeve of outer thread to match the new hole of frame and inner thread to original bolt size and then countersunk welding of sleeve with the frame for additional strength) or drilling and threading to higher size to match the next the **Thread Inserts**.

5. Replacement of damaged cast components, partly or fully, by making new parts, using machining and fabrication processes. This calls for special design efforts for converting casting drawings to to suit machining and fabrication processes.

6. In-situ drilling oh holes for oil injection or liquid nitrogen injection to help in dismantling of components,

7. Other different repair or reconditioning jobs of heavier and intricate components, replacement of which would be very costly and time consuming.

3.5.1 Thread Inserts

These are used for quicker repair and remaking of threads to original or desired size. Thread inserts are also used in few original equipments, where it can be removed in case of damage and renewed. These are precision formed screw thread coils of carbon steel or stainless steel wire having a diamond shaped cross section. Thread inserts increase the load-bearing capacity of screw joints, and they are very wear resistant. They are very useful in soft materials, and especially in frequently connected and disconnected assemblies. When they are installed into matching tapped holes, they provide permanent conventional 60' internal screw threads that accommodate any standard bolt or screw. Thread inserts are also available with special threads and also with tin sheet alloys, plastics and composites etc. Fig. 3.1 shows few shapes of such inserts, tapping of recommended holes and fixing (screwing) of inserts in the hole. Though self-locking external threads are used, adhesives like Loctite are also applied for strength and to keep "Thin-Walls" from backing out or vibrating loose. In addition, the adhesive also seals against liquids and gases ingress.

Fig. 3.1. Thread Inserts, Shape and Fitting

Though the blind thread inserts (splined body and low profile head) are mostly used, followings are some of the different types of thread inserts mainly basing on type of external fittings, available in the market–

1. Blind threaded insert with low-profile head,
2. Blind threaded insert with large flanged (splined body, flanged head),
3. Thread Insert with a hexagonal body style (Roll formed threads, hexagonal body),
4. Square bodied blind threaded insert (Roll formed threads, square bolt),
5. Blind threaded insert with improved locking mechanism, other than those mentioned above, etc.

Thread Inserts offer many advantages, such as–

- Reconditioning of damaged threads cheaper as compared to other methods,
- Offers high resistance to torque-to-turn and push-out to improve reliability,
- Speeds assembly,
- Increased thread strength; Also provides high-strength threads in thin materials starting at 0.5mm,
- If excessive torque is applied, the screw shank should fail before the insert, avoiding over-tightening problems such as thread stripping and expensive rework; etc.

3.5.2 In-situ Machining

In-situ Machining or On-site Machining or In-place Machining (the terms are interchangeable) is very beneficial for reconditioning of damages in heavy components or components which cannot be dismantled easily. It reduces the exorbitant cost of shutting down, removing and transporting parts for shop repair, which ultimately saves the customers time and money. In this, specialized portable machine tools are developed, which can be set up on the part requiring repair, using the latest techniques to ensure correct alignment and accuracy. The machining takes place often in a very short timescale and sometimes while adjacent machinery is still running. Often new custom-made (tailor-made) equipments are developed/ used to machine off the damaged material prior to using the conventional grinding method. Machining takes one tenth of the time of grinding so when a major failure occurs and a lot of material needs to come off we can reduce down time considerably.

Though most of the common machine tools can be used for in-situ machining, followings are more commonly used for in-situ machining—

- Facing and Boring Machines,
- Turning Equipment
- Pipe Machines
- Grinding and Drilling Machines
- Milling
- Honing (internal and external)
- Reaming
- Lapping
- Screw/thread cutting
- Metal Stitching
- l Selective plating ; etc.

Followings are some of the typical In-situ machining jobs—

1. Flange facing'
2. Tube sheet facing for heat exchangers,
3. Boring coupling and rudder palm bolt holes,
4. Line boring machine frames,
5. Line boring engine and compressor crankshaft main bearings,
6. Line boring gearbox bearing housings,
7. Turning damaged or built up areas on shafts,
8. Milling forging hammer and press bases,
9. Crankshaft machining and valve seat grinding of large engines,
10. Other big components of Steel, Power, Petrochemical, Marine and other industries.

Few advanced in-situ (on-site) repair techniques are briefly described below-

1. Electro-Disintegration Machine (EDM) Drill – Refer Sec.3.4.4.and 3.4.5.

2. Coupling Mirror Boring- It is a unique coupling boring process called Mirror Bore. This process effectively takes coupling bolt-hole boring out of critical path in an outage by allowing the individual coupling halves to be bored separately. When the coupling is later mated together, the bolt holes will line up exactly allowing easy installation of the coupling bolts. No final honing or other polishing is required. A key benefit of coupling Mirror Bore is the reduced cost resulting form shorter turbine outages. In addition, the Mirror Bore process allows complete interchangeability of the rotors so that spares can be quickly and easily changed out as needed. This process can be used for Bolt Hole Mirror Boring, Counter Bore Machining and Oversize Hole Repairs etc. One supplier of such machine is M/s G E, USA.

3. Fine-Line Welding- It can be performed on-site or in-shop. State-of-the-art, remotely operated equipment and highly trained personnel are dispatched to perform the highest quality weld repairs on various sophisticated equipment; such as Rotor Weld Repairs (Wheel/ Dovetail Replacements, Journal Restoration, Stub Shaft Replacements and Deep Groove Crack Repairs and Generator Repairs accomplished with windings in place etc0, Valves (Seat Repair and Replacements and Modifications and Upgrades etc) and Vessel Repairs and On-site Fabrication etc.

4.5.3 Metal Stitching

Also known as **Metal Locking** or Metlace, it is a mechanical approach to the repair of vital machine parts and was considered as the solution to the problems of welding and brazing etc.

Metal stitching has become a generic phrase that describes methods of repairing cracks in cast metals without welding. The use of metal stitching has grown in application from small water pump to huge turbine generating equipment, from small shop to largest steel plant. This cold process is extremely versatile and can be applied to the repair of a wide range of industrial equipments. No heat is applied and thus no stresses introduced and very little, if any, post-repair machining is required to correct misalignment or warpage. As no heating is needed, metal stitching repairs can be made in restricted areas, where flame or welding arc are prohibited. Often such repairs are made on site, without major dismantling.

The main component of metal stitching has been many things intended to replace the crack and surrounding material.; such as bolts (copper-and zinc-plated); tapered plugs (both threaded and non-threaded); threaded rods; screws; set screws; dowels; and just about anything else that could be stuffed, hammered or screwed into the space. Over time, these items have been called 'plugs', 'pins', 'stitching pins' and lock etc.

The process starts by drilling a line of holes at right angles to the crack in the casting and then converting them into a slot. Correct spacing for the holes can be achieved by special drilling jig with centres at 1/8", 5/32", 3/16" or 1/4" depending on the thickness of the material which is being repaired. Preformed locks are then fitted into the slots to create a bridge across the broken sections. The locks are made from some high nickel steel with the same co-efficient of expansion as the cast iron. This material to be chosen should be strong enough to take shear loads, but sufficiently ductile to provide the necessary elasticity. Most repairs will have a series of locks and stitches spaced at regular intervals along the crack. When this is completed, holes are drilled along the line of the fracture between each stitch. These are tapped to receive special screws which fill the crack and ensure that it is completely water tight. Lastly, the area which has been stitched is ground level to the surface of the original material to create a seamless repair. In this way, when the crack has been repaired the area of metal is often stronger than the original cast iron material, and when it has been primed and painted the repair can be invisible.

In some cases involving heavy loading stresses, such as the repair of cast iron columns, it is needed to provide extra reinforcement. In these circumstances, a 'master key' is inserted in addition to the normal 'stitches'. The 'master key' is a larger section of metal which can be varied in shape or size to suit the particular requirements of the repair, but basically it involves a larger bridging section which is fitted using the same technique as conventional stitching.

In cases where whole or a big sections of material are missing due to the effects of corrosion or mechanical damage, a patch of metal, known as an ' patch insert', can be made to fill the gap and stitched into position.

Followings are the advantages of metal stitching-

1. Repairs can be carried out in-situ removing need for dismantling and reassembly,
2. Can be used on any iron or steel casting of ¼" thickness or greater,
3. No heat is applied removing the risk of distortion and misalignment,
4. Minimizes down-time and lost production in industrial applications,
5. Capable of withstanding mechanical and thermal stresses,

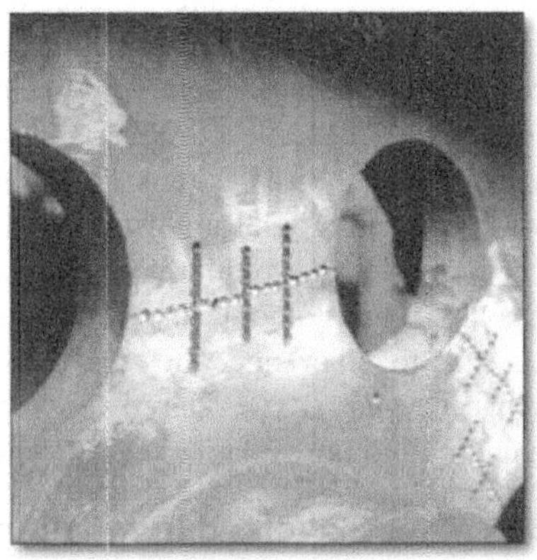

Fig. 3.2. Metal Stitching

6. Usually less expensive than replacement component cost or other possible repair; etc..

3.5.4 Metal Disintegrator

Metal disintegrator technology is a type of spark machining (or Spark Erosion Machining). It is primarily used for tap removal, stud removal, drill removal (with no damage to threads) and seized or broken bolts. The cutting action of a metal disintegrator is accomplished by creating a series of intermittent electric arcs that break down the hardest metals into minute particles. A branding electrode, held in the head of the disintegrator, vibrates as it cuts while a coolant is pumped through the electrode to wash away the powdered metal and to keep the roll cool to touch. Electro Arc Metal Disintegrators can use any material that conducts electricity as an electrode. However, different materials offer varying levels of efficiency in the process. It is important to use an electrode material that breaks down less than the work piece. Because their high melting temperature out performs other electrode materials molybdenum electrodes are often used, in a variety of shapes and lengths; round, square, and hex cross section; lengths in 6", 9", or 12". A 12" length can remove 35 to 40 taps.

The benefits of this technology are—Reclaim expensive parts quickly, easily, Reduce labor cost—jobs finished in seconds instead of hours, No waiting for parts to be returned from outside sources, Portable, flexible, and affordable, Eliminate production overruns and late deliveries and Scrap reclamation etc.

As this is a relatively new technology, very few manufactures are widely known. M/s Electro Arc Manufacturing Co, Michigan, USA is one of those.

3.5.5 Electro-Disintegration Machine (EDM) Drill

EDM Drills have been developed on similar principle of metal disintegrator. The primary application for EDM is the removal of finger dovetail pins on large steam turbines. Finger dovetail pins frequently become lodged or "stuck" in the pinholes due to corrosion and/or slight bending from centrifugal force. Using a brass electrode and tap water as a coolant and cutting fluid, the EDM will relieve a lodged pin much quicker than conventional drills. Computer control monitors the arc and all cutting to assure consistency and rapid material removal. The EDM can be used for Dovetail Finger Pin Removal, Cross Key Pin Removal, Balance Hole Drilling and Removal of Steam Strainer Rivets etc.

3.6 MAINTENANCE WELDING

Welding is the most economical and efficient way to join metals permanently. It is the only way of joining two or more pieces of metal to make them act as a single piece. Welding is vital to our economy. It is often said that over 50% of the gross national product of most countries is related to welding in one-way or another. Welding ranks high among industrial processes and involves more sciences and variables than those involved in any other industrial process. It is a vast subject and is used extensively for manufacturing various components. Though most of the welding technologies are used for maintenance welding also, maintenance welding poses greater challenge because of more uncertainties and positional difficulties often doing the welding in-situ. Repair welding and surfacing are both considered in the field of maintenance welding and are covered together since they are both done by the same welders. Often it is extremely difficult to separate what is considered repair welding, from maintenance welding, and surfacing can be included in both situations. The same basic factors apply to both weld repair and surfacing.

3.6.1 Steps involved in Maintenance Welding

The steps may vary, to some extent from case to case, few basic steps are given below–

1. The need for weld repair and surfacing– The earliest use of welding was for repair work. The most famous incident happened at the outbreak of World War I when German ships were interned in New York harbor. Their crews, hoping to make the ships inoperable, sabotaged the engines and machinery. However, by means of welding, repairs were quickly made and the ships were placed in transatlantic service to deliver material from the US to Europe.

Parts break and wear out continually. It is often possible and more economical to weld repair those, since the delay in obtaining the replacement part could be excessive and the cost of the new part would normally exceed the cost of repairing the damaged part. Weld repair is commonly used to improve, update, and rework parts so that they equal or exceed the usefulness of the original part. This is normally attained, with the possible exception of weld-repaired cast iron parts that are subjected to heating and cooling. Weld repairs on cast iron parts subjected to repetitive heating and cooling may or may not provide adequate service life. The problem is that cast iron parts subjected to high-temperature heating and cooling, such as machinery brakes, furnace sections, etc., fail originally from this type of service and due to metallurgical changes the weld may fail again without providing adequate service life. Again, the material of the part to be repaired has a great influence on the service life of the repaired parts. Parts made of low-carbon and low-alloy steels can be repaired without adversely affecting the service life of the part. On the other hand, high-carbon steels may be weld repaired but must be properly heat-treated if they are to provide adequate service life.

As such, it is absolutely essential that we know the type, specification, or composition of the metal that we are planning to weld. A part made of any metal that can be welded can be repair welded or surfaced. In fact, some of the metals that are not normally welded can be given special surfacing coatings by one process or another.

All the arc welding processes are used for repair and maintenance work. In addition, the brazing processes, the oxy-fuel gas welding processes, soldering, thermit welding, electro slag welding, electron beam welding, plasma welding and laser beam welding are also used. The thermal spraying processes are all widely used for surfacing applications. In addition, the various thermal cutting processes are used for preparing parts for repair welding. Normally, the selection of the welding process and appropriate surface preparation and other preparation processes depend on the same factors that are considered in selecting a welding or cutting process for the original manufacturing operation.

In the case of repair welding, there are usually limitations, such as the availability of equipment for a one-time job and the necessity of obtaining equipment quickly for emergency repair work etc. This limits the selection and it is for this reason that the shielded metal arc welding process, the gas metal arc welding process, the gas tungsten arc welding process, and oxyacetylene welding and torch brazing are most commonly used. However, for many routine and continuous types of repair work, some of the other welding processes may be more economical. For example, submerged arc welding is widely used for building up the surface of worn parts. The electro slag process has been used to repair and resurface parts for hammer mills, for construction equipment, and for rebuilding rolls for steel mills. Thus there is a difference in the selection of the welding process for the routine, continuing types of repair and surfacing work versus the one-of-a-type or breakdown emergency repair job.

2. Determination of Weldability– Before repairing any damaged materiel, it must be determined whether or not the materiel can be satisfactorily welded. Identification of metal of the part to be welded has to be done for identifying weldability. This determination may be based upon the following <u>factors</u>-

- Determine the nature and extent of the damage and the amount of straightening, fitting and stiffening of the part, that will be required,

- Determine the possibility of restoring the structure/ component to usable condition without the use of welding,
- Determine the type of metal/ material used in the damaged part, whether it was heat treated, and if so, what heat treatment was used; In emergency cases, some heat treated parts can be repaired (please refer next section). Determine if the welding heat would distort the shape or in any manner impair the physical properties of the part to be repaired,
- Determine if heat-treating or other equipment or materials will be required in order to make the repair by welding; The preferred metal of repairing heat treated steels, when practicable, requires the annealing of the broken part and welding with a high strength rod. This method produces a welded joint that can be heat-treated. Etc.

3. Analyze and prepare for repair- The success of a repair or surfacing job depends on the thought and preparation prior to doing any actual work on the repair procedure. One of the reasons for such an investigation is to establish the cause of the failure in the case of a broken part or the cause of wear or erosion in the case of a part to be surfaced. Few main points outlined are-

1. Make a detailed study of the actual parts that failed and its location/ associated parts,
2. Learn the background information concerning the specifications and design,
3. Make a listing of all of the facts so that at the conclusion the reason of failure would be as accurate as possible. This relates to the type of repair job since it also determines whether reinforcing may be required. Reasons for the part to fail or wear out can be either accident or misapplication or abuse or overload etc. If the part failed because of an accident or an overload or misapplication, it may be returned to service with the weld repair made to bring it back to its original strength and necessary guidelines can be issued to user. In the case of poor workmanship, poor design, or incorrect material the weld repair should eliminate the poor workmanship that was responsible for the failure. In a case of wrong material it would be assumed that the material was of a lower strength level, which contributed to the failure. In that case reinforcing would be required.
4. Take necessary approval from departmental and other concerned agencies; e.g.

- Inspector of Boilers and other such agencies for boiler components and pressure vessels etc, Such repairs need subsequent certification from concerned authorities,
- Containers and transportation equipment may be welded only with special permission and approval. These include railroad locomotive and car wheels, high- alloy high-strength truck frames, and compressed gas cylinders.
- Most pieces of power-generating machinery, including turbines, generators, and large engines, are covered by casualty insurance. Weld repair on such machinery can be done only with the prior approval of the welding procedure by the insurance underwriters etc,

5. Ascertain availability of welders with necessary certification (statutory and otherwise) for carrying out those critical repairs,

4. Working Plan and Procedure- The first step in this direction is to prepare a written repair procedure for doing either repair work or alterations. In the case of an alteration a record must be made and all alteration work must be approved. Alterations on bridges, large steel frames/ structures, and ships etc may be done only with special authorization. Such alteration work must be properly designed and approved. However, for minor repairs, such written procedures may not be needed.

The next step is to recheck the type of material being welded. If the material can not be found from drawings or specifications, we can do laboratory test, by sending filings or small piece. If that also is not possible, in few cases, maintenance men can use their knowledge and experience of similar parts; e.g. an automobile engine block is normally cast iron except for some, which might be cast aluminum, the spring of an automobile or truck would normally be high-carbon steel etc. The thickness of the metal and the position of welding etc are also to be kept in view as that also leads into the question of filler metal (thickness and material) to be used.

The third step is to select the welding procedure and welding tools/ equipments for the job. Followings are the main considerations-

1. The proper type and size of electrode, together with the current and polarity setting, must be determined if an arc welding process is used. If a gas welding process is used, the proper type of welding rod, correct gas pressure, tip size, flux, and flame adjustment must be determined. Determining on the thickness, quality and quality of weld deposit, MIG, TIG, Submerged, Plasma and other welding machines can be used.

2. In preparing the edges of plates or parts to be welded, the proper cleaning and beveling of the parts to be joined must be considered. The need for backing strips, quench plates, tack welding, and preheating must be determined.

3. Reducing warping and internal stresses requires the use of the proper sequence for welding, control and proper distribution of the welding heat, spacing of the parts to permit some movement, control of the size and location of the deposited weld metal beads, and proper cooling procedure.

4. Additional cooling (or other methods) for portion of the part under repair must be considered to eliminate or reduce heat transfer from the part under repair to other component and cause any problem. Seals etc are, normally, removed beforehand.

5. Automobile, aircraft or other such machines are designed for lightness and the safety factors are very strict. This necessitates consideration of reinforcement at the joint to compensate for the strength lost in the welded part due to the welding heat. A reinforcement must be designed that will provide the required strength without producing high local rigidity or excessive weight.

6. Safety Precaution- Before beginning any welding or cutting operations on the equipment, necessary safety precautions must be considered. Few of such precautions are—Draining fuel, oils or other inflammable materials and proper degreasing of the part; keeping fire extinguishers for small and common repairs and full fire engine standby for gas line, pressure vessel or boilers etc; Involvement of safety department personnel for critical repairs etc.

5. Actual Repair and Post Treatment- Actual repair and post treatment etc can be taken up in line with steps mentioned above.

3.6.2 Repairing Heat Treated Parts

Tough this has roughly been mentioned earlier, following considerations are made for repair of heat-treated parts-

1. In emergency cases, some heat treated parts can be repaired in the heat treated condition by welding with stainless steel electrodes containing 25 percent chromium and 20 percent nickel, or an 18 percent chromium-8 nickel electrode containing manganese or molybdenum. These electrodes will produce a satisfactory weld, although the heat of welding will affect a narrow zone in the base metal in the vicinity of the weld.
2. Minor defects on the surface of heat-treated parts may be repaired by either hard surfacing or brazing, depending on their application in service. In any of these repairs, the heat treated part will lose some of its strength, hardness, or toughness, even though the weld metal deposited has good properties,

3. The preferred metal of repairing heat-treated steels, when practicable, requires the annealing of the broken part and welding with a high strength rod. This method produces a welded joint that can be heat-treated. The entire part should be heat treated after welding to obtain the properties originally found in the welded parts. This method should not be attempted unless proper heat-treating equipment is available.

3.6.3 Metal Spraying/ Thermal Spraying

This technique is also commonly used for maintenance repairing, surfacing and coating and other such jobs. High Velocity Oxy-Fuel (HVOF) Coating, Combustion process, Dual Wire Arc process and Air Plasma process are some of the processes, which are used in maintenance repairs/ reconditioning.

3.7 SURFACING, HARD-FACING AND COATING

Surfacing and Micro-surfacing are used more for road maintenance; micro-surfacing being a thin polymer modified cold-mix paving system that can remedy a broad range of problems on today's streets, highways, and airfields. It is a low-cost preventive maintenance treatment, which retards deterioration of the pavement, maintains or improves the functional condition. However, in industrial maintenance, these terms are used for preparing, repairing or modifying surfaces of various mechanical, electrical and other components. In surfacing, the surface characteristics desired for the finished job depend entirely on the service to which the surface will be exposed. This is based on knowledge and experience and on the fact that the surface has deteriorated to the point that it needs to be reworked or resurfaced. When wear is involved, surfaces can be rebuilt many times without reducing the strength of the part and the service life will be greatly extended. Surfacing can be done by most of the welding technologies.

3.7.1 Friction Surfacing

It is a process derived from friction welding whereby a coating material, in rod form (often termed the Mechtrode) is rotated under pressure, generating a plasticized layer in the rod at the interface with the substrate. By moving a substrate across the face of the rotating rod a plasticized layer between 0.2-2.5mm thick is deposited (depending on electrode/ mechtrode diameter and coating material.). During the coating process, the applied layer of metal reaches a temperature near the melting point whilst simultaneously undergoing plastic deformation. The coating is thus the product of a hot forging action, as opposed to the casting mechanism inherent in welding and spraying processes.

Friction surfacing process allows special hard wearing alloys to be deposited directly at a knife's cutting edge. After finishing processes, the result is a blade with a hard, wear resistant edge and a tough backing material. With the extremely strong bond that is formed between the coated deposit (called TMC45) and the substrate, there is no risk of delamination.

(Wear Resistance = Hardness + Carbide)

3.7.2 Hard Facing

Hard-facing is also a surfacing process, which is done mainly to increase the hardness of the component. Hard facing can also called Hard-Surfacing. It is used for both, designing and manufacturing new product and also for repairing those. Hard-facing processes allow worn or damaged components to be resurfaced or rebuilt to the original condition without costly replacement and sometimes with increased wear resistance and extended life. It is mainly done by fusion process. Fusion hardfacing is a process, by which weld materials, with superior properties than the substrate, are applied to the substrate. Often, *two layers (*total 3 - 6 mm thick) are applied to reduce the surface hardness dilution of the relatively low-cost steel substrates in the expansive cobalt-base alloys. Hard-facing processes are very useful for improving wear and corrosion resistance to selected areas of machinery, such as cutting edges of earth-moving machinery. Common hard-facing techniques include arc, torch, and other processes, as shown in Table 3.3. In the table, Deposition efficiency is expressed as a percentage and is determined by dividing the weight of metal deposited by the weight of the consumable expressed as a percentage. Operator's efficiency is different from deposition efficiency. The operator's efficiency may vary from 20% for GTAW, 30% for manual welding to 65% for SAW.

Table 3.1. (Hard-facing Processes)

Category	Processes	Deposit Efficiency
Arc welding	Flux core arc welding (FCAW) Gas metal arc welding (GMAW) Gas tungsten arc welding (GTAW) Plasma arc welding (PAW) Shielded metal arc welding (SMAW) Submerged arc welding (SAW)	60~70% 95% 95% 85~90% 90~95%, wire only
Torch welding	Oxy/fuel gas welding (OFW)	95%
Other welding	Electron beam welding (EBW) Electroslag welding (ESW) Furnace braze (FB) Laser beam welding (LBW)	

Hard-facings are good for substrate materials that are suitable for welding, such as low-carbon steel. The most common hard-facing materials are nickel alloys and iron/chromium alloys used in wear resistance and high stress abrasion. Common materials for hard-facing, with their main purposes, are listed in Table 3.4.

Table 3.2 Hard-facing Alloys and Purposes

Metal Alloy	Main Purpose
Cobalt-base alloys	• Wear and corrosion resistance • Rebuilding worn machinery parts
Iron chromium alloys	• high stress abrasion
Manganese steel	• wear application
Nickel-base alloys	• metal-to-metal wear resistance
Tool steel	• tooling, wear application
Tungsten carbide	• high stress abrasion

Hard facing is extensively used for earth moving machines like dozers, excavators, loaders, shovels etc.

3.7.3 Coating and Cladding

Metal Coating and Cladding are similar technologies and are also used for metal surfacing. It provides high-tech coating applications for a variety of end uses to enhance functional performance. Metallic coatings provide a layer that changes the surface properties of the work-piece to those of the metal being applied. The work-piece becomes a composite material exhibiting properties generally not achievable by either material if used alone. Metallic coatings are deposited by galvanizing, electroplating, electroless plating, spraying, hot dipping, chemical vapor deposition and ion vapor deposition and thermal spraying etc. Whether it is a Medical, Industrial, Automotive, Military, or other industries, they often use this technique to enhance their products performance.

Cladding is, often, applied at the mill stage by the manufacturers of sheet, plate or tubing. Cladding by pressing, rolling or extrusion can produce a coating in which the thickness and distribution can be controlled over wide ranges and the coatings produced free of porosity. Although there is almost no practical limits to the thickness of coatings that can be produced by cladding, the application of the process is limited to simple shaped articles that do not require much subsequent mechanical deformation. The cladding process enables applying a cladding to metals that could not be clad using other processes. Cladding for roofing materials, wall and window panels are also quite common.

The basic principle of metal coating is localized melting of supplementary material into the surface. Direct diode lasers perform local or large-surface coating of metal components that suffer from corrosion or wear. The process can be easily automated and is typically performed by direct diode lasers. Hard particles or metals in the form of powder or wire are fed into the laser focus and fused with the base material. The latter is coated or locally repaired in a single process. The spot geometry of the diode laser can be adjusted to the size of the component and the type of application, thus achieving highest process efficiency.

Main applications are corrosion and wear protection in the automotive industry, repair welding/ reconditioning, tools and forms, coating with hard particles for wear protection and large area or local re-melting / heating etc. Typical applications of such coatings are for shafts, rods, valve parts, exhaust valves for engines, pump and turbine components, wear plates, tools, blades and moulds etc.

Coatings represent a variety of metallic mixtures and compound typically used to strengthen certain features of a product to make it better fit its purpose. Few such examples are titanium, aluminum, zinc, zirconium, chromates and phosphate for steel pre-treatment, TiN for machining iron-based materials, and CrC with its high temperature oxidation resistance, for use in die casting etc. Tungsten carbide/carbon (WC/C) is designed to coat and protect highly loaded precision components, gears and gear drives, engine components and hydraulic pumps and compressors.

As coating is a general word, many nonmetallic coatings are also used. These are mainly of two types, dispersion and powder. Dispersion coatings may be of baked phenolics, fluorocarbons, epoxies, plastisols, vinyls, urethanes, and others and powder coatings may be of nylon, epoxies, vinyls, polyester, PVDF, fluorocarbons etc.

The advantages of these processes are homogeneous alloying results, high process stability, optimized focus adaptation: square to line or circular spots and high process speed etc.

The procedure and steps for surfacing, hard-facing and coating etc are generally similar to those mentioned given in repair welding (Sec 3.5.1) The benefits of these technologies can be summed up as–

1. Building new parts or repairing damaged parts with assured longer life of elements subjected to wear and abrasion,
2. Often enhancement of life cycle of repaired part,
3. Rebuilding/ repair of worn parts at a fraction of replacement cost,
4. Savings in maintenance costs when rebuilding at the equipment operation place (in-situ),
5. Reducing breakdown time and increasing work efficiency,
6. Less Spare Parts Inventory: There is no need to keep numerous spare parts when worn parts can be rebuilt,

3.8 MICRO-WELDING AND MICRO-JOINING TECHNOLOGIES

Micro-joining is critical to the successful performance of many components for miniaturized products, with sophisticated joining processes needed to achieve low defect rates. It is considered a vital part of advanced packaging technology for the photonics, electronics and medical industries. As the name indicates, the welding application is of small and miniature nature, some times as small as a human hair. This state-of-the-art interconnection/ joining system is used for manufacture of micro-electronic mechanical systems (MEMS), which includes integration of mechanical elements, sensors, actuators and electronics on a common silicon substrate through the utilization of micro-fabrication technology) and other such items, and also for few packaging applications.

Micro-welding includes smaller/ miniature resistance welding, plasma arc welding, ultrasonic welding, wire bonding, laser-beam welding, micro-brazing/ soldering etc. Micro-joining, though may includes micro-welding, considers more of other hot technologies, such as Hermetic Electronic Packaging, Transient Liquid Phase (TLP) Bonding, Active Brazing of Ceramics, Flux-less Soldering, Fine Wire (<5 mil) Butt Joining, Glass-to-Metal Soldering, Plastic-to-Metal Soldering, Biocompatible Soldering/Brazing and Induction Brazing etc. However, these technologies are not much used in maintenance and, as such, would not be discussed further.

3.9 PLATING

Metal Plating is the method of applying a metallic coating to another material. Plating can be applied to almost any coarse material. The most common surfaces that are plated are metals and plastics. Plating can be used for different reasons. Some objects are plated to increase their sturdiness and provide a hard shell for whatever it is plate on. Some are plated to avoid corrosion, and a few are plated just to give off an attractive finish. Other benefits/ applications of plating are— it can conduct electricity, it has high resilience to wear and tear and it is long lasting. Nickel plating is one of the most common used in decoration. When combined with an under-layer of chromium, it provides a bright, shiny surface that is corrosion-free, and resistant to wear. It is used to coat over chemically metalized plastics, steel, brass, zinc, and aluminum.

There are a few ways of conducting the metal plating process. There is electroplating, selective immersion-free plating, electro-polishing, electroless plating, metal spraying, vacuum metallizing, tinning, galvanizing etc. Of these, first three are widely used for maintenance and reconditioning jobs.

3.9.1 Electroplating

It is the deposition of a metallic coating onto an object by putting a negative charge onto the object and immersing it into a solution which contains a salt of the metal to be deposited. The metallic ions of the salt carry a positive charge and are attracted to the part. When they reach it, the negatively charged part provides the electrons to reduce the positively charged ions to metallic form. Maintenance plating or plating for maintenance and repair of components involves coating of chromium, nickel, cadmium, zinc, copper and aluminum etc on components. In addition to vast arena of mechanical components like cylinders, plungers, piston-heads, valves, valve-seats, poppets, bumpers etc, such plating are used for maintenance of many electric and electronic components like PCB (printed circuit boards) etc.

3.9.2 Selective Immersion-free Plating

Basically, it is also an electroplating process, but the technique has been developed for selective in-situ repair/ reconditioning of big components. Instead of dismantling and immersing the component in electrolyte tank, the anodic device is developed into the surface profile of the component to be plated. For smaller deposits, the anodic device is suitably impregnated with required electrolyte and is brought over the surface to be plated. Under the effect of electric current, the metal is deposited over the surface of the component. This is referred as "**Brush Plating**". For more deposits, the electrolyte is made to circulate at high speed between anode and cathode. This process is referred as "**Circulation Plating**". Though the process calls for careful design and development of anodic device and control, it gives many benefits, such as –

- Huge cost saving by in-situ repair of big components (e.g. main columns of big Hydraulic Press etc) by saving downtime cost of dismantling and assembling and spare cost.
- Strength and properties of these deposits are, often, better than the conventional tank process. 10 to 20 micron thickness of such deposit may have equivalent strength of 50 micron thick deposit of conventional process.
- Improve solderability, EMI shielding, conductivity in select areas,
- This plating uses higher current densities - faster build-up of metal deposit,
- Such-plated deposit can be more dense, less porous, etc.

Electro-polishing is also a plating process and it produces a number of favorable changes in a metal part which are viewed as benefits to the buyer. All of these attributes translate into selling advantages depending upon the end use of the product. These include Brightening, Burr removal, Oxide and tarnish removal, Removal of surface occlusions, Increased corrosion resistance, Improved adhesion in subsequent plating, Reduced buffing and grinding costs, Removal of directional lines, Radiusing (making radius/rounding) of sharp edges, Reduced surface friction, Stress relieved surface and Removal of hydrogen etc.

3.9.3 Electro-Polishing

It can be considered also as a metal surfacing process as it produces a combination of properties which can be achieved by no other method of surface finishing. Mechanical grinding, belting, and buffing can produce beautiful mirror-like results on stainless steel, but the processes are labor intensive and leave the surface layer distorted, highly stressed, and contaminated with grinding media. However, electro-polishing is liable to chip or wear off. Electroplated surfaces may also exhibit hydrogen embrittlement which must be stress-relieved in a separate step. As such, its use is limited in maintenance.

3.10 REPAIR OF METALLIC MATERIALS

During service a component tends to degrade due to reasons mentioned in chapter 4. While in some cases the damage might have progressed to such an extent that nothing could be done to salvage the component, in many cases the extent of damage is within limit of being repairable. The repair limit depends upon the type of component, the operating environment and risk it is subjected to and the remaining life in it. For satisfactory repair, have to be following two main criteria have to be fulfilled–

1. The repair should not affect the continuous serviceability of the component,
2. The cost benefit of doing the repair must be clearly demonstrated. The repair to near new condition should, normally, cost less than 60% of cost of new part.

Of-course, repairs are, occasionally, taken to extend the life and to improve performance, maintainability and reliability. Conventional or standard repairs are of minor nature, employing simpler and fewer methods (normally mentioned earlier), whereas "Specialized repair" goes beyond the usual limits and employs a number of complex and advanced technologies. Modern specialized repair techniques strive to return the component to near new condition, if not improve upon that. For critical components, like repair of turbine blades or repair of components for aircraft, missile or deep-mining etc, following series of technologies would, normally, be used—

1. Methods of non-destructive inspection to detect damage (automated eddy current or ultrasonic inspection or real-time radiography etc); such inspection should take less time, should cost less and be quantitative. For turbine blades, it should be ascertained that minute defects are not present in critical areas.
2. State-of-art methods for cleaning the components, prior to repair; e.g. Hydrogen Fluoride cleaning and high pressure water jet cleaning/ striping etc. Inadequate cleaning would lead to poor braze or weld joints and poor adhesion of coatings.
3. Methods of repair/ rejuvenation and coating of components; it may include techniques like–Electronic Beam/ Laser/ Plasma Needle Arc Welding etc, Diffusion Brazing, Diffusion Bonding, inertia Welding, Transient Liquid-Phase (TLP) Bonding, Automated Welding and Grinding etc. Repair for newer design configuration (i.g. integrally bladed rotors) or rejuvenation by Hot Isolated Pressing and reheat treatments etc may also be needed. Coating technologies are very important for repair cases and, as such, this would again be discussed later in this section.
4. Analysis, testing and certification Technologies—We can use Standard Test Methods Damage Tolerance Assessment and Probabilistic methods. The repaired component may be approved jointly by repairer and user or along with third party certification, or by regulatory/ statutory authorities, depending on the case.

The assessment of cost benefit analysis/ stemming for the repair also takes into account the life extension that can be achieved, the performance improvement that can be obtained, the cost of repair vis-à-vis cot of new part, the availability of new part, the cost of field implementation and skill requirement etc.

3.11 COATING AND SURFACE MODIFICATION TECHNOLOGY FOR REPAIR

The main goals for these technologies are to improve repair durability, restore dimensional tolerances, reduce maintenance cost, reduce downtime and increase availability, safety and reliability. Various coating processes/ technologies, available in standard books, are used for designing and making new and original components, but repair coatings calls for more challenges. Most of the coating processes are equally applicable for repaired parts, with the understanding that the coating composition, that would be suitable for the repaired part may not necessarily be the same as the optimum choice for the new part. Fe example, a brazed joint in a repaired structure may have a totally different microstructure than that of a parent material and, as such, it is needed to match the coating composition to braze microstructure as well as the base metal microstructure. The available coating technologies/ processes can be summarized into following general categories—

- Flame and induction hardening,
- Hard-facing processes,
- Thermal spray processes,
- Diffusion processes
- Plating processes,
- Special surface treatments,
- Thin film coating by physical or chemical vapour deposition,

- High energy surface modification; etc.

For performing these coating technologies, many preparatory jobs have to be done in line with those mentioned in Sec.3. , such as, thorough cleaning/ stripping, matching coating composition to repair structure, associated repair jobs, recoating, resurfacing, refurbishing and other treatments.

Such repaired components are also to be tested and certified in line with those mentioned elsewhere in this book..

In addition to the goals mentioned above, the coating technologies have the potential to improve the performance of the repaired components even better than the new parts if a substantial improved coating is used. However, it should be kept in mind that repaired areas may be more prone to environmental attack than the base material and hence the need for environmental protection; particularly in the case of brazed repair areas in high temperature applications.

Cost benefit analysis for such repaired components involves cost of maintenance/ repair/ overhaul, estimation of life extension and life-cost saving and availability of new parts etc.

3.12 COMPOSITE STRUCTURAL REPAIR—MATERIAL SYSTEMS

Composite materials are extensively in aircraft industry, automobiles and many household and other equipments/ machines/ appliances. These materials range from high modulus carbon to lower modulus Kevlar and E-glass fibers and a variety of formulations of epoxy resin systems, etc. All of these material systems require elevated temperature cure or post cure processes to achieve sufficient material properties from the composite systems. In addition, many of the structures are of sandwich design, incorporating honeycomb core for high structural efficiency.

Repair of in-service damage to these structures have proved to be quite difficult, because of poor diffusion of composite material technology from the original equipment manufacturer (OEM) to the agency doing repair. The need to perform high temperature processes causes concern amongst repairmen, who are, generally, not equipped to do so. Transportation, storage and handling of such materials must be carefully controlled. Lack of skilled/ trained technicians for such jobs is also a problem. As such, maintenance men's or repair agency's capability is the deciding factor for such composite material's repair. Again the repair of Hazardous Materials (HAZMAT) is, generally, avoided.

Such resin based composite materials should have following characteristics—

1. Low temperature cure process, with minimum component surface preparation requirements for high strength and durability,
2. Fast cure cycle,
3. Good hot/ wet properties and damage tolerance characteristics,

The focus of this critical technology is to define the characteristics, identify or develop a resign system which will simplify the component repairs and which is very user friendly for even the lesser equipped repairmen. This resin system may be used for wet lay-up repairs or possibly with the required fibers/ pre-impregnated format. Activation of the cure process could be by heat or chemical or other means.

This resin technology has many advantages, such as-

1. Simplify, speed-up and reduce cost of repair and transportation,
2. Eliminate the need of expensive high temperature cure equipments,
3. Reduce technician training needs for basic repair,
4. Reduce machine downtime by reducing repair cycle time,

5. May provide temporary repair capability for large damages,
6. Will not degrade original structural characteristics,
7. Simplifies repairs for smaller and less sophisticated plant owners; Etc.

Testing and certification has to be done as mentioned in previous sections.

3.13 COMPOSITE PATCHES ON METALLIC COMPONENTS

The use of composite patches to effect repair of damaged metallic components, has been a recent phenomena. These techniques are used to repair damages, generally in the forms of dents or holes etc (caused by hits, impacts, collisions and lightening etc), in aircraft, road or sea transport vehicles or other such equipments. Main goals for these technologies are—

1. Improvements over traditional metallic repair technologies,
2. Reduction in repair cost and cycle time,
3. Simpler and less obtrusive repairs (weight, drag, no additional fasteners etc),

The existing traditional metallic repair technology consists of two stages—removal of damaged area and restoration of the strength and stiffness of the structure. This involves traditional sheet metal approach, with riveting/ bolting/ welding etc, which may introduce additional stress concentration and may affect the durability of the repaired structure.

The bonded composite patch technology has been developed to overcome the shortcomings of traditional repairs. Both, born and carbon fiber patches have been used because of their high stiffness and low density. In general, the boron patches have been preferred since its coefficient of thermal expansion is closer to the most common original substrate structure (aluminum etc). The main features of this technology are–

1. Design of patch to attract the desired load level; Load transfer into and out of the patch calls for complex stress analysis procedure, which must include residual thermal stresses, introduced during bonding,
2. Proper surface preparation of the metallic surface, where patch would be bonded,
3. The patch bonding process, including customized adhesion cycle; Etc.

Though this technology was developed for damage repair, it can be used for following types of applications–

- Crack patching- A patch is applied over the existing cracks, holes, dents or discrete damage to slow down or arrest its growth,
- Corrosion Repair- A patch is applied to the area whose section has been reduced to below minimum acceptable limit by the act of corrosion,
- Fatigue Life Enhancement- A patch is applied to an undamaged structure to enhance its fatigue life. The identification of such areas may be based on original equipment manufacturer's long-term durability test or equipment's user or maintenance men in-service experience.

3.14 COLD WELDING

It is the welding of two materials under high pressure or vacuum without the use of heat. Cold Welding may be of few types–Adhesion Cold Welding, Magnetic Pulse Welding, Ultra-sonic Cold Welding and Putty/ Epoxy based Cold Welding etc.

Academically, Cold-Welding means contact cold welding or **adhesion cold welding** or **vacuum welding**. It is a phenomenon that the metals weld together in room temperature and in ultra high vacuum environment where oxidized films are formed extremely slowly. Accordingly, two metals can be jointed without heating or pressurizing by using Cold-Welding. Cold welding, Cold welding, under vacuum, is also done by impact and fretting. However, such cold welding is used mainly in research organizations and sophisticated high-tech application like space technology. It is not much used in industries because of need of ultra high vacuum and difficulty in getting very smooth and very clean surface. Of-course powders in powder metallurgy use cold welding to best advantage because they present large surface areas over which vacuum contact can occur.

Magnetic pulse welding is another cold welding process, which uses a magnetic field to rapidly collapse one component onto another forming a metallurgical bond. The welding cycle is extremely short—typically less than 1 second. The technology is well-suited for joining dissimilar metals and cylindrical components, such as air conditioning tubing and tubular space frames. It offers numerous benefits to assemblers, such as no heat input, fast cycle Arial, ability to join dissimilar materials, and base metal strength in most materials. Magnetic pulse welding is a good alternative to brazing because it offers greater repeatability. The process is fast, clean, energy efficient and creates no heat affected zone to change the physical properties of either part. Additionally, it creates a true solid-state bond/ weld between either similar or dissimilar metals and no filler is needed. However, one disadvantage/ limitation is that the process is so fast that it does not lend itself to deep drawing of material since the material does not have time to stretch and, as such, the technology has not yet gained wide acceptance in common maintenance practice.

Ultrasonic metal welding is also called "cold welding" because it works without application of heat and without melting the metals. Ultrasonic cold welding most commonly joins nonferrous metals, but can also be used with other metals, such as when welding aluminum to ceramics. Ultra-Sonic welding technology allows the use of thin, strong and economical films. No special coatings are needed. It uses high-frequency vibrations. It works by placing the parts to be welded on the anvil. Next, force is applied to the parts. Then the upper part is vibrated back and forth with the ultrasonic horn. This causes the parts to rub together. The ultrasonic energy disperses oxide film layers (cleaning the surface) and results in the mixing of metal atoms, without melting the metals. The activated metal atoms join each other, causing a true metallurgical bond. However, this process is also not much used for repair and maintenance jobs.

Because of problems of hot welding and other cold welding processes, **Epoxy/ Resin/ Putty based Cold Welding** have been developed and many epoxy/ resin putty based cold welding compounds are being formulated and marketed for various repair and maintenance welding jobs. These are molecular compounds which are formulated using blends of reactive resins and hardeners. Epoxy resins refer to a family of molecules or oligomers containing more than one epoxide group (Oxirane). These products are solid or liquid with the consistency of honey and have the ability to react via the epoxy end-groups to generate three dimensional networks providing the final material with rigidity, hardness and the inability to flow. The final products generally exhibit excellent electrical properties, good adhesion due to presence of polar groups, low shrinkage, good impact resistance and moisture resistance.

The chemical chosen to react with these epoxides is referred to as the curing agent (or hardener), and it typically has active hydrogen attached to nitrogen, oxygen, or sulfur. The resins and hardeners have led to development of cold welding compounds, which are also called plastic steel putty, and are used extensively for easy, economical and dependable maintenance and repair work. Followings are the special features and advantages of such compounds—

1. Good adhesion,
2. Instant repair and in-situ repair possible,

3. Applies easily and needs no heat or pressure,
4. Repairs, patches and rebuilds in areas where other welding would be undesirable or impossible; No deformation or distortion of repaired components,
5. Conventional metal-working tools can be used to machine cured compounds,
6. Bonds to most metals, concrete and some plastic and wood materials,
7. Cures in 16 hours, or less, at room temperature in most applications,
8. When cured it can be drilled, tapped, machined or painted,
9. Excellent resistance to oil, petrol, water and many chemicals,
10. Its effectiveness against conventional causes of wear and friction,
11. Repair of pipe leakages possible without completely draining out of pipe; etc.

The limitations/ disadvantages for such application are the places where direct stress or pull is coming, the compound is little brittle in nature, shelf life of the compounds are low, the compounds are costly and the bulk repairs becomes very costly and selection of proper compound requires deeper study etc.

Different types of cold welding compounds are available, few of which are mentioned herewith—

- Multimetal— a metal repair system with excellent mechanical values and high chemical resistance; cures also in thick layers without tension or cracks,
- Plasticmetal— consists of 2 components metal powder + hardener. Special metal powders for an optimum metal adoption for Cast Iron, Steel, Bronze, Brass, Copper etc,
- Compounds for underwater repairs or coatings,
- Low Friction-way liners— for the production and overhaul of sliding parts of any kind for improved sliding properties and decrease in production costs with increase in productivity,
- Thin liquid— ready to use for metal impregnation with optimum capillary sealing action for all micro porous castings and machine parts. An easy, quick, very efficient and price-worthy impregnation repair liquid; etc.

There are different suppliers for such cold welding compounds, such as Diamant Metallplastic, GMBH, Germany, MultiMetall, Germany (The Myriad Techno Pvt. Ltd, Kolkata), Belzona Polymerics Ltd, UK, ITW Devcon, Germany (or ITW Signode), Durametal (M/s J T Denyer, UK), Thortex (USA), Denmech Felpro (usa) and Mecatec (M/s L and T) etc

3.15 VULCANIZING (HOT AND COLD)

The dictionary definition of vulcanizing is to improve the strength, resiliency, and freedom from stickiness and odour of (rubber, for example) by combining with sulfur or other additives in the presence of heat and pressure. The vulcanizing system is the combination of vulcanizing agent with activators, accelerators or co-agents needed to convert the rubber compound from an essentially plastic state to a shaped elastomeric thermoset. The system is added to the rubber polymer during mixing and remains inactive during subsequent processing and fabrication operations. Vulcanization then takes place, usually by the application of heat to activate the system. Crosslinks (e. sulphur, peroxides etc) are inserted between adjacent polymer chains to form a three-dimensional network which prevents further flow. The nature and composition of the vulcanizing system determines the type and concentration of crosslinks. Stiffness and resilience (elasticity) increase with crosslink density, whereas set, stress-relaxation and creep all decrease. For most rubbers, tensile strength reaches a peak and then falls as crosslink density is raised.

However, these days, cold vulcanization is becoming more popular for many jobs, especially repair jobs. In this, new splice materials are chemically activated by contact with other chemically active materials, without requiring the use of a vulcanizing press. In general, the splice geometry is the same for hot and cold vulcanized splices, but the materials differ, of course. Materials and procedures for cold vulcanizing usually must be obtained from the cold vulcanizing material manufacturers or their authorized agents, and seldom from the belt or other component manufacturers.

Vulcanizing is used for making new components, and also for conversion, repairs and rebuilding of old components. Use of vulcanizing in maintenance and repair include tyres and tubes (including tubeless tyres) of different automobiles and earth moving equipments (including punctures/ damages to rubber tire tread crown, shoulders and even sidewall areas in some cases), Repair of inner lining of metallic tube or inner tube by vulcanizing patches, Pulley lagging, Reconditioning of abrasion resistant liners for transfer points, chutes, hoppers, bins, tanks, and storage silos other impact areas and conveyor rolls etc, using rubber or urethane or UHMV or Ceramic etc, Conveyor belts and other beltings; Etc. As an example, repair of conveyor belts will be described very briefly in subsequent paragraph.

Conveyor Belt Repair- Conveyor belts are extensively used in different industries to transport different types of materials. These belts account for substantial spares and inventory cost. Repair of these belts is extremely cost saving and time saving. Most of these belts are rubber based and cold repair/ vulcanizing is mostly the method used. Different firms have developed different adhesive and hardening solutions/ compounds (separate for rubber and supporting fabric layers), to be used as per their recommendations. Continual conveyor belt maintenance and repair ensures long conveyor belt life. Repairing holes, rips and edge wear as soon as they happen prevents costly downtime. Some of the typical damages and rough repair methods on these belts are given below—

1. Repair of rubber /cover damages (for both, single layer and multiple layers)—

- Patch damages of rubber covers- Repair patches, slightly bigger than damage size, are used (in single coat or double coat) after proper cleaning and buffing the existing surface (good and un-removed), with solutions; patches are generally of round, square, rectangle or diamond shapes.
- Repair of worn longitudinal of the rubber covers of the belt- Repair strips of suitable width are used with preparations as mentioned above.
- Repair of rubber sheeting—covers are in the shape of bigger patches,

2. Repair of longitudinal tears—These are small longitudinal damages/tears of both, rubber cover and fabric ply; repair pieces of both rubber and fabric layers are used, generally 100 mm longer at both ends of tear, with different solutions etc.

3. Repair of perforations—Generally used for upto 20% f belt width and includes both rubber and fabric layers,

4. Repair of torn edges and frayed edges—These include both rubber and fabric layer damages; Normally limited upto 20% of belt width.

Instead of fabric based, High Tension Steel Cord or Kevlar or PVC cord based conveyor belts are also used in industries and can be repaired, as mentioned above or otherwise. Mechanical belt fasteners are also often used to join/ splice the cut/ damaged belt ends.

Mechanical Belt fasteners offer an economical, reliable and long lasting belt splice/ joining method, for both, light duty and heavy duty applications and are commonly used in industry for repairs. The majority of belts being manufactured today are made of synthetic materials, which lend themselves to mechanical attachment. All belt fastener processes, fastener installation, countersinking, etc. are designed and made in short notice rather than waiting for specialists or outside contractors. This translates into significant cost savings as downtime is minimized and production may continue without substantial interruption. Most belt fasteners are manufactured in two styles, solid plate and hinged. Each has its particular advantages and the selection of fasteners should begin with this basic understanding. Solid plate fasteners span opposing belt ends that have been butted together, forming a tight, sift-free splice, such as bulk conveying applications. Hinged belt fasteners are applied as individual segments to each belt end and then brought together and connected by means of a connecting hinge pin. Mechanical fasteners can be of many types, such as Wire hooks, Common bar wire hook, Common bar lacing, Stapled plate, Plastic spiral loop, Plastic hinged plate, Bolted hinge, Riveted hinge plate, Heavy stapled plate, Bolted solid plate and Riveted solid plate etc. Selection of mechanical fasteners depends on working tension, pulley size and construction of the belting and other physical and environmental factors.

Trouble shooting for mechanical fasteners should be examined in line with Fastener "comb-out" through end of belt without opening, Fasteners open up and release from belt, Belt breaks behind fastener, Splice failing at edges, Fastener parts fracture and fail, Belt fails under splice and Fastener wears out prematurely etc.

3.16 MATERIAL IMPROVEMENT

Selection of material is done at the design stage, after due considerations of specification, duty requirement, factor of safety needed and, occasionally, after consultation with maintenance engineers. However, during maintenance and repair of any equipment, occasionally change of material or use of improved or better/ improved material is done to arrest the repeated failures or to give better performance. Such material improvement may be needed because of following reasons–

1. Some omission or inadequacy at design stage,
2. Change in duty requirement, e.g. demand of higher output or better product or cheaper inputs etc.
3. Change in environment, e.g. ingress of moisture, dust, various fumes, magnetization and other environmental changes; Also change in statutory regulations,
4. Lubrication and other cares not being maintained properly because of frequent damages or difficult approach/ access,
5. Equipments not manufactured as per design or not erected/ commissioned as per specification/ need,
6. Non-availability of some spares or components because of obsolescence or otherwise,
7. Recent availability of better and cheaper consumables, components or alternatives, etc.

Technological innovations are leading to development of new and improved materials every now and then and maintenance men have to keep themselves aware of such developments so that they can select the best alternative known and economically available at the time when they decide to change the material. Development of new materials, for possible use in repair and maintenance, are mostly in fields of metallic alloys, non-metallic or alloys of metals and non-metals. They are mostly composites of following nature—

- Polymer Matrix Composites (including resins, fibres and glass-fibres etc),
- Metal Matrix Composites (metal with another metal or ceramic or boron nitride etc) ,

- Ceramic Matrix Composites (fiber reinforced ceramic composites, e.g. aluminum oxide matrix reinforced with silicon carbide fibers; carbide tools etc)
- Structural Sandwich Composites (a layered composite formed by bonding two thin facings to a thick core or otherwise); to reduce the weight of sandwich core, often honeycomb or corrugated construction used.

Essential properties (e.g. tensile strength, tensile modulus corrosion and abrasion resistance etc) of these composites are much higher and, also, densities are much lower than other materials like aluminum alloys, steel, titanium and most non-metallic etc. Different companies have produced and are still developing various types of improved materials/ composites, with different brand names. Information about such materials and their properties can be found from manufacturer's/ dealer's information sheets or from other books. A few older version of such materials, commonly available in India, are mentioned below—

1. Sintered self-lubricating bushes and pads (oilite bushes etc)
2. Ultra Bronze, a type of non-ferrous alloys, cheaper and superior substitute to brass, phosphor bronze, Al-bronze and nylon etc
3. Ultra High Molecular Weight Polyethylene
4. Cast Nylon (especially Cast Nylon-6), e.g. Arkaymid and Garflon brand names
5. FRP (Fibre Reinforced Polymer)
6. Fibre-glass moulding, etc.

3.17 ARMATURE WINDING AND MOTOR REPAIRING

In maintenance, electrical repairman's (electricians, winders etc) job is very specialized one and may include jobs like—

1. Synchronous Motors (and other motors generators etc) fault diagnoses
2. Rewinding, overhaul and remanufacture of AC/ DC motors
3. Rewinding, overhaul and remanufacture of generators, relay, circuit-breakers, switchgears, PLCs etc
4. Testing and repair of oil and air-cooled transformers
5. Aligning, dynamic balancing and vibration analysis — in the field or in our shop
6. Other electrical equipments checking, repairs and calibration etc.

However, this is a vast subject and will not be discussed in details in this book.

3.18 OTHER TECHNOLOGIES AND COMPONENTS

As per the definition of maintenance in chapter 1, maintenance encompasses all sciences. Accordingly, many topics/ subjects of mechanical engineering, electrical engineering, instrumentation engineering and civil engineering etc also form the required technologies and components for appropriate maintenance engineering. Though such list would be very vast, in addition to those already mentioned in this chapter, few more are just enumerated below as more examples—

1. Tribology (wear, friction, lubrication and their interaction)
2. Lubricants and lubrication systems and lubricating equipments
3. Hydraulic and pneumatic systems, circuits and components
4. Instrumentation and monitoring tools

5. Pumps, Compressors, Blowers, Fans, Exhaust Systems etc,
6. Boilers, Heat-exchangers, Turbines and steam and electrical power generation systems,
7. Bearings and bearing systems,
8. Mechanical Drives- Gears, chains and sprockets, couplings etc
9. Electrical energy and power,
10. RCL circuits and programmable control systems etc,
11. Electronics and electronic component's maintenance/ replacement,
12. Automobiles and other traction and transportation equipment's maintenance,
13. Building and structural maintenance
14. Handling and lifting gadgets/ equipments, scaffoldings and ladders etc
15. Maintaining Plant Sanitation and Housekeeping, etc.

As further discussions about these technologies will make the book unnecessarily voluminous, details about these can be obtained from the books of respective disciplines.

Table 3.3 (Good Maintenance Practices)

Mainten-ance Tasks	Standard	Required Good Practices	Consequences of not following good practice	Probability of future failures
Lubricate Bearing	Lubrication Interval; TimeBased;+10%	• Clean Fittings • Clean end of grease guns • Lubricate with proper amount and right type of grease • Lubricate within variance of frequency	• Early bearing failure, • Reduced life by about 20% ~ 80%	20:1
Coupling Alignment	Align motorcoupling, using dial indicator or laser alignment procedure; Straightedge method unacceptable	• Check runout on shaft and coupling, • Check for soft foot, • Do angular alignment • Do horizontal alignment • Align total equipment specification, not coupling specification,	• Premature coupling failure • Premature bearing and seal failure in motor and drive unit, • Excessive energy Loss	7:1
V Belts	Measure the tension of V Belts through tension and deflection, utilizing a belt tension gauge,	• Identify the proper tension and deflection for the belt, • Set tension as per specification, faliure of drive or driven unit,	• Premature belt failure through rapid 'belt wear or 'total belt failure, • Premature bearing failure of drive or driven unit, • Belt creeping or slipping, causing speed variation, • Motor shaft breakage	20:1

3.19 GOOD MAINTENANCE REPAIR PRACTICES

As a general estimation, about 70% of equipment failures are self-induced. Between 30% ~50% of self-induced failures are the result of maintenance personnel not having enough knowledge of basics of maintenance. Again about 20% ~ 30% of self-induced failures are because maintenance personnel, though having enough skill, do not follow good maintenance repair practices. Thus Good /Best Maintenance Practices" are necessary for maintenance personnel to keep equipments operating at peak reliability and cost-effectively.

The previous paragraphs of this chapter help in evolving good/ best maintenance practice for any organization. Training of maintenance personnel (chapter 15) and other chapters of this book also help in evolving best maintenance practices. Table3. shows sample good/ best maintenance practices, which maintenance personnel must know. In the table, probability of future failure has been taken as number of self induced failures versus those following the bes practices.

More such discussions later in this book.

QUESTIONS

1. What is in-situ machining and why is it more challenging than conventional machining?
2. What is maintenance welding and why is it more challenging than conventional welding?
3. What is metal-stitching or metal locking? What are its applications in maintenance?
4. Explain briefly two cases of In-situ repair.
5. What are thread inserts and its uses in repair of components?
6. What is Immersion-free plating and its importance in maintenance?
7. What is machine lubrication and what are functions of lubrication?
8. What is cold welding? Explain briefly few advantages and one application of epoxybased cold welding.
9. What is Vulcanizing and what are its uses in repair of conveyor system?
10. Give few ways to repair and modify the surface of a worn-out shaft.
11. What is difference between repair welding and manufacture welding? What additional cares are to be taken in repair welding?

CHAPTER

4 DEFECT/FAILURE LIST GENERATION AND FAILURE ANALYSIS

4.1 DEFECT/ FAILURE DEFINITION

Defect and Failure may be synonymous and are normally taken to mean the same, though one may try to find some difference between the two. Defect is a more general term, normally meaning fault, mistake, flaw, vice, deficiency or shortcoming of any type or situation. Failure normally means not-succeeding, deteriorating, decaying, fatigue or stoppage etc of any item for any reason. Defect may be used in broader perspective. In an industry, there are following five major sources of defects—

1. Failure- Unintentional damage i.e. bearing seizure, shaft or gear teeth breakage,
2. Workmanship- Not enough skill or motivation, e.g. misalignment, poor welding etc,
3. Design- Design not fitting actual requirement an condition etc,
4. Material- Defect at manufacturing, sourcing, storage or handling stages etc,
5. Operation- (a) Mal-operation; Incorrect use by operating personnel,
   (b) Operational problems, i.e. pump cavitating a pump or ignoring vibration etc,

However, in industrial situations, the words defect and failure are normally taken to mean the same. **Let us take a look at some examples of common industrial failure definitions—**

- Failure is any loss that interrupts the continuity of production,
- Failure is a loss of asset availability,
- Failure is the unavailability of equipment,
- Failure is a deviation from the status quo,
- Failure is not meeting target expectations,
- Failure is any secondary defect; Etc,

It may be noted that there are no perfect failure definitions. For instance, "Failure is any loss that interrupts the continuity of production" has to include planned shutdowns, staggering production, rate reductions for decreased sales, etc. It would not pick up failures on equipment that is spared since it does not interrupt the continuity of production. As such, failure definition must be concise and easily understandable and should focus on the priority issues.

4.1.1 Failure Rate

It is the frequency with which an engineered system or component fails, expressed, for example, in failures per hour or so. It is often denoted by the ë (lambda). The failure rate of a system usually depends on time, with the rate varying over the life cycle of the system. In practice, the 'Mean Time Between Failures' (MTBF) is often used instead of the failure rate. Thus the Failure rate is simply the multiplicative inverse of the MTBF (1/ë). The term 'failure rate' is more used for reliability calculations. Here, we would use MTBF.

4.2 FAILURE MODE

It is the manner in which a system or component fails. Failure mode generally means the category, type, style, way, mood or status of a failure event. The term "Failure mode" is often used in all failure analysis etc. It fosters good communications since everyone knows what is important and it also provides a basis for a common understanding of what the facility's needs are. Its terminology is very specific and it must only address one topic. A broader way of differentiating failure modes may be 'conceptual', 'technology' or 'organizational' etc. Premature operation, failure to operate at the prescribed time, failure to cease operation at the prescribed time, failure during operation, and degraded or excessive operational capability may be another way of classifying failure modes. Different category of components (bearings, gears etc) may also be considered as different failure modes and we would be considering more of this classification. More about Failure Mode in Sec.4.11 (FMEA).

4.3 FAILURE CODE

It is simply a code that illustrates why an asset / equipment failed or the reason that the asset/ equipment failed. Code can be a number, which is cross referenced to a list of actual code (failure codes etc) descriptions or more conveniently a series of alphanumeric characters or simply alphabetic abbreviation, that are a logical abbreviation of their descriptions. They are generally used in maintenance systems or more commonly computerized maintenance management systems. They can be used on a work order for an asset or equipment failure. Failure codes provide a convenient method of getting statistics about equipment failures or breakdowns. CMMS systems will generally have a reporting function that allows reports to be run on specific failure codes for your equipment. Whatever method you chose, your operators and maintenance personnel will soon become familiar with your own codes providing that the list is limited to a manageable number. However, with modern database technology and available disk space, the full descriptions (failure modes) are increasingly being used instead of alphanumeric codes.

Some of the recent and most commonly used failure analysis techniques are mentioned in this chapter. Of-course, few may appear little similar in approach to others, but with different terminology.

4.4 FEW BASICS OF FAILURES

Few experts say that the causes for more than 90% of all plant failures can be detected with a careful physical examination using low power magnification and some basic physical testing. Before explaining how to diagnose a failure, we should review the effects of stress on a part. When a load is put on a part, it distorts. In a sound design the load isn't excessive, the stress doesn't exceed the "yield point", and the part deforms elastically, i.e., when the load is released the part returns to its original shape. Occasionally, with repeated loading in this range, the part may be permanently deformed. Greater increases in load will cause the part to actually break.

Few of the basic reasons of failures are as given below—

1. Fatigue Failures: When fluctuating loads (often also called as Fatigue Loads), are applied, such as those in a hydraulic cylinder or in an automotive connecting rod, the fatigue strength of he material comes into play. When the fatigue strength is exceeded, a crack can develop. This fatigue crack can slowly work its way across a part until a fracture occurs. The face of a fatigue failure tells us both the type (bending, tension, torsion or a combination) and the magnitude of the load. To understand the type of load, look at the direction of crack propagation.

True Brinelling Failures occurs when loads exceed the elastic limit of the ring or bush material. Brinell marks are indentations at ball/roller/bushes frequency caused by any static overload or severe impact.

2. Corrosion: Corrosion can greatly affect /reduce the fatigue strength and increase chances of failure. Corrosion results from the chemical attack on parts/ materials by hostile fluids or atmospheres. Symptoms include red/brown areas on rolling elements, raceways, cages and other components. Corrosion usually results in increased vibration followed by wear, with subsequent increase in radial clearance or loss of preload etc.

3. Ductile Overload Failure and Brittle Overload Failure: A "ductile failure" is one where there is a great deal of distortion of the failed part. Commonly, a ductile part fails when it distorts and can no longer carry the needed load, like an overloaded steel coat hanger. Generally, some ductile parts break into two pieces. A "brittle fracture" is one when a part is overloaded and breaks with no visible distortion. This can happen because the material is very brittle, such as gray cast iron or hardened steel, or when a load is applied extremely rapidly (and heavy shock) to a normally ductile part. In a brittle overload failure, separation of the two halves (or more pieces) isn't quite instantaneous, but proceeds at a tremendous rate, nearly at the speed of sound in the material.

4. Stress concentration: A stress concentration is a physical or metallurgical condition that increases the local stress in the part by some factor; such as a small or no radius in a machined component, where the dimension changes. It has a great effect on crack initiation because of their effect on increasing the local stress. If a part is relatively lightly stressed, the cracking will start at only one point. However, if a shaft is more heavily loaded, then cracks can start in several places and work their way across the part.

5. Contamination: In addition to other cases, contamination is one of the leading causes of premature bearing failure. Symptoms of contamination are dents or scratches embedded in the bearing raceways and balls/rollers, resulting in undue bearing vibration and wear. Contaminants may include airborne dust, dirt or any abrasive substance that gets into the bearing. Principal sources are dirty tools, contaminated work areas, dirty hands and foreign matter in lubricant or cleaning solutions.

6. Wear: A variety of mechanisms result in loss of material by mechanical removal, leading to failures through above mentioned reasons or otherwise.

7. Hydrogen Embrittlement: When tensile stresses are applied to a hydrogen embrittled component it may fail prematurely. Hydrogen embrittlement failures are frequently unexpected and sometimes catastrophic. Hydrogen can be introduced into the material in service or during materials processing. Hydrogen embrittlement is an insidious type of failure as it can occur without an externally applied load or at loads significantly below yield stress.

Liquid metal embrittlement is the decrease in ductility of a metal caused by contact with liquid metal. Very small amounts of liquid metal are sufficient to result in embrittlement.

4.5 FAILURE GENERATION

As briefly mentioned earlier, failures/ defects may be generated or induced at various stages of equipment/ component life-cycle, such as—

1. ***Design Stage***: Some failure may be induced because of design efficiencies; e.g. someomissions or incorrect assumptions of duty condition or factor of safety etc.

2. ***Manufacturing Stage***: For making a product cheaper, the manufacturer, sometimes, takes lower safety factor and design properties and deviates from required close tolerances. Occasionally, due to non-availability of some material, the manufacturer inferior material (and not better/ costlier material) so that the cost does not increase and delivery is not delayed. This causes failure at operation stage.
3. ***Erection Stage***: Equipments are, sometimes, not erected as per laid-down procedures and precautions, in order to reduce cost or time, Few of the common omissions are—inadequate stress relieving after major welding, no pre-stressing of big bolts and non use of facilities like torque wrenches for uniform tightening etc. Such omissions, though not easily noticeable during inspection and commissioning stage, ay induce failure at operation stage.
4. ***Commissioning Stage***: Often, during commissioning stage, proper procedure is not followed and full load test is not done, either to save time or due to non-availability of full load at that time. Sometime the equipment is accepted conditionally, even without few inspection and monitoring appliance, which later becomes a permanent feature. Such inadequate commissioning prevent surfacing of inbuilt defects timely.
5. ***Operation Stage***: Many faults generate at this stage due to wrong operating practices and inadequate care and upkeep. Overloading of machine, no or improper cleaning and lubrication of machine, bypassing limit-switches and built-in test equipments (BITE), not immediately reporting and correction of minor defects and continuing production in that condition etc are few examples to generate fault at this stage.
6. ***Maintenance Stage***: Defects are also generated at maintenance and repair stage by using incorrect dismantling procedure, use of improper tools, improper cleaning and lubrication, non-uniform tightening and improper torques on bolts and use of inferior consumables and pares etc. Because of these, faults will surface out in subsequent operation and maintenance stage.
7. ***Environmental Degradation***: Many defects are generated because of change in environmental conditions, such as more corrosive and abrasive fumes and dusts, more moisture, more radiation and magnetic fluxes and more heated environment etc.

Basing on above mentioned stages and sec. 4.3 and others, the failures can be classified into following categories/ types–

- Failures based on inherent reliability characteristics (design failure),
- Dependent failure- these are chain effect or secondary failure as a result of design defect or otherwise,
- Manufacturing or burn-in failures; these may include failures due to poor manufacturing, erection and commissioning etc,
- Wear-out failures, including fatigue failures etc,
- Operation induced failures,
- Maintenance induced failures,
- Equipment damage due to accident, environment degradation or otherwise, etc.

4.6 FAILURE REPORTING AND COLLECTION

The faults, so generated at various stages in different equipments and assemblies in an industry, are to be recorded, reported and collected at proper, often centralized planning section, for analysis and planning further action. In different industries different ways of reporting and collection of failures may be followed. In general, following sources/ agencies are used for such jobs—

1. **From Maintenance Customers**—Maintenance customers are production departments, facilities, operating departments and administrative groups. Each maintenance customer may have unique expectations and requirements regarding maintenance. (Check lists and work request from these customers),
2. Maintenance Personnel (Maintenance check list and maintenance work order),
3. Centralized Inspection Agencies and Inspection Schedules- The agencies may be centralized inspectors, centralized lubrication cell, centralized hydraulic cell, centralized building and structural inspection cell, safety, house-keeping and environmental inspection cells and other such agencies. They may have different inspection schedules.

Table 4.1 shows a typical inspection schedule.

System/ Plant - Vapour Recovery System Code No.		**Date of Inspection**		**Inspected By**
Equipment or Code	**Failure Number**	**Defect/ Failure**	**Verification Procedure**	**Defect Category**
System		• Absence or disconnection of any component, required to be used, • Installation or use of any uncertified component • System not in compliance with static pressure decay test criteria	• Direct observation • Direct observation • TP201.3 or equivalent	
Nozzles		• Nozzle automatic liquid shutoff mechanism malfunctions in any way	• Direct observation	
Hoses		• Any coaxial hose with slit, tear or perforation in excess of 5mm in dia.	• Direct measurement	
Processing units		• Vapour processing unit inoperative • Pressure drop through vapour return path exceeds units the given limit,	• Direct observation and measurement	

4. Reports of Built-in Test Equipments (BITE),
5. Reports of other on-line and off-line monitoring equipments. This may be coordinated through centralized condition monitoring cell or through regular maintenance workmen/ staff.
6. Left-over jobs/ problems from previous PM or major repairs, deferred for want of time or spares or facilities; etc.

The failures, so received from various agencies, are to be compiled equipment-wise and area-wise and analyzed for taking corrective and preventive actions. These jobs may be grouped into following categories—

- Fault (Failure) Detection,
- Fault Diagnosis and Failure Analysis.

4.7 FAILURE ANALYSIS (FA)

Defect Analysis or Fault Analysis or Failure Analysis are generally similar approach in industrial maintenance scenario. Though one may argue that Defect Analysis covers wider areas than Failure Analysis, but, henceforth, we would use the term "Failure Analysis" to include all defects, faults and failures etc in industries.

Failure analysis is the process of collecting and analyzing data to determine the cause of a failure and how to prevent it from recurring. It involves investigation of reasons as mentioned in the basics above. We may use different types of visual inspection, electrical testing, non-destructive evaluation, destructive evaluation and other examinations.

4.7.1 Failure Detection/ Analysis tools/ Methods

In a big industry, many types of tools and methods are used for failure identification and analysis. Few are already mentioned in para 3.0 of chapter 3. Though there may be some repetition, few more are mentioned herewith–

1. Visual examination of dimensions and general physical features,
2. Representative specimen/sample selection for truer analysis,
3. Proper Failure Mode Determination,
4. Chemical identification/ analysis,
5. Crack detection (Dye-penetrant, ultrasonic etc),
6. Fracture analysis, including fractography and micro-fractography,(SEM and TEM)
7. Microscopes (Scanning Electron Microscopy SEM, TEM and Laser Signal Injection Microscopy (LSIM),
8. Fatigue analysis,
9. Micro-structural analysis and metallographic analysis,
10. Electron dispersive analysis by X-ray (EDAX),
11. Corrosion and contamination analysis and corrosion meter
12. Stress Corrosion Cracking,
13. Particle Analysis/Identification,
14. Filter Residue Analysis,
15. Welding, Soldering and Brazing Analysis,
16. Vibration Analysis,
17. Process/Manufacturing Problem Analysis,
18. Reverse Engineering of Components,
19. Evaluation of life cycle test failures,
20. Plastic and Rubber Investigation,
21. Finite Element Analysis (FEA) and Engineering Mechanics Analysis,
22. Operational Dynamics Analysis
23. Engineering Calculations,
24. Atmospheric Testing,
25. Semiconductor probing,
26. Simulation tests,
27. Software Based Fault Location Techniques (CAD Navigation, Automatic Test Pattern Generation (ATPG); Etc.

More details about above mentioned testing/ analysis methods are beyond the scope of this book. While analyzing the failures, the types of failures, mentioned in Sec.4.4 and others can be grouped in categories as below–

- Degradation Failure— by these, the equipment gradually looses its performance, while in use and causes loss due to degrades efficiency, even though the equipment does not stop or breakdown,
- Quality Deterioration Failure—by these the equipment gradually deteriorates to the extent that the quality of the processed products are lowered,
- Catastrophic Failure—These occur when the equipment suddenly breaks down due to partial or total damage of some parts, but the equipment can be restored back by replacement or repair etc.

The analysis includes defining the failures properly, finding out function of defective units, finding alternate means to achieve the function if needed, shutdown losses and costs etc. After analyzing the failures and its frequency/ pattern, different maintenance strategies can be adopted. Fig.4.1 shows the Failure Analysis vis-à-vis different maintenance strategies.

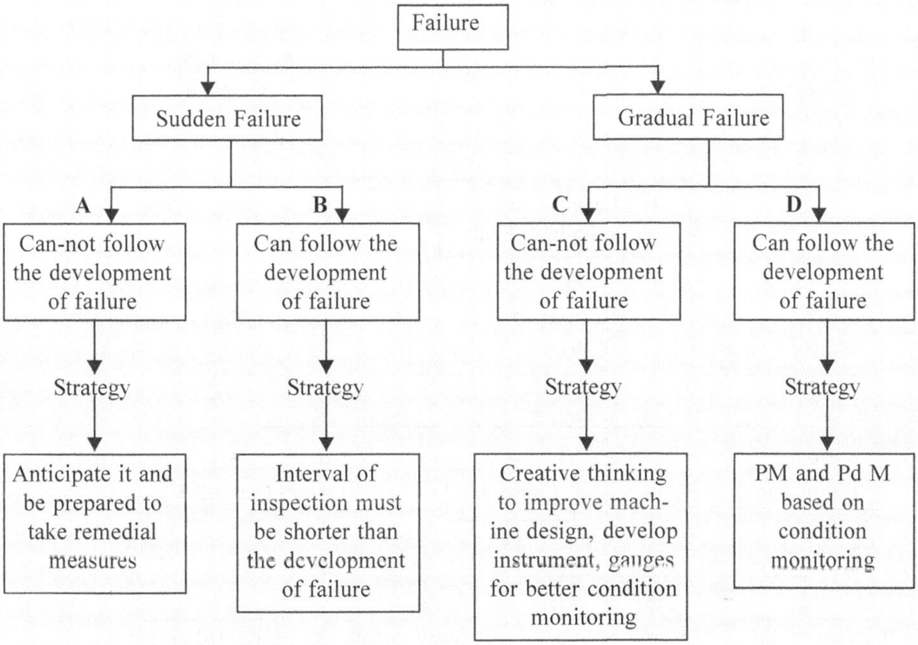

Fig.4.1 Failure Analysis vis-à-vis Maintenance Strategies

The objective of such failure analysis should be to bring most of the failures from left side to right side of Fig. 4.1, i.e. from sudden to gradual and from "can-not follow the development" to "can follow the development". The ideal situation of maintenance in an industry ill be indicated when most of the failures fall in "D" category and least in "A" category in Fig 4.1.

Many specialized failure analysis methods/ techniques are being practiced in industries for improving the maintenance and, thus availability and reliability etc. Few of those will be discussed in subsequent portion of this chapter.

4.8 FAULT TREE ANALYSIS (FTA)

Fault Tree Analysis (FTA) is a technique for reliability and safety analysis. Bell Telephone Laboratories developed the concept in 1962 for the U.S. Air Force for use with the Minuteman system. Later, it was gradually adopted by many industries. FTA diagrams, also called as Fault Tree Diagrams (FTDs) or negative analytical trees, are logic block diagrams that display the state of a system (top event) in terms of the states of its components and sub-components (basic events).

FTA is a logical, structured process that can help identify potential causes of system failure before the failures actually occur. The most serious fault/ failure, such as explosion, toxic release, breakdown, damage, malfunction, etc. is selected as the Top Level Event/ Fault. A fault tree is then constructed downward by relating the sequences of events, which individually or in combination, could lead to the Top Level Event/ Failure. Ishikawa's cause and effect diagram (fish-bone diagram) or other techniques may be used for that. FTA may also rely on solid theories such as Boolean logic Theory (which reduces the Fault Tree structure into the combinations of events leading to failure of the system, generally referred to as Minimal Cut Sets) and Probability Theory. As shown in Fig. 4.2, the successive causes are analyzed, both quantitatively and qualitatively, and necessary remedial/ corrective actions (including possible design changes and other risk-mitigating actions) are planned to eliminate those causes, which, in turn, aims to eliminate the top level event/ fault.

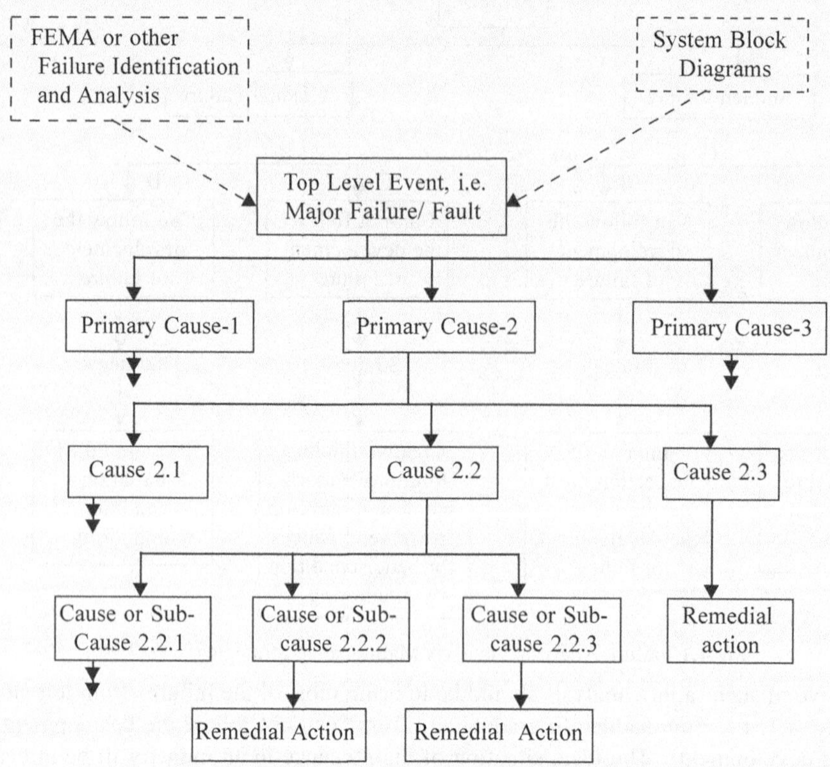

Fig. 4.2 Preparation of a Fault Tree

In modern FTA diagrams, special symbols like "And-Gate", "Or-Gate", Rectangle, Circle, Diamond, Oval and Triangle etc are used. However, in normal industrial situations, the types of symbol does not matter much and the spirit and methodology, as shown above, are considered. The primary advantages of FTA are the meaningful data they produce which allow evaluation and improvement of the overall reliability of the system. It also evaluates the effectiveness of and need for redundancy. The limitation or disadvantage of the FTA is that the undesired event are also evaluated and all significant contributors to the failure must be anticipated. This effort may be very time consuming and expensive.

4.8.1 The main difference between FTDs and RBDs is that in an RBD, one is working in the "success space", and thus looks at system successes combinations, while in a fault tree one works in the "failure space" and looks at system failure combinations. Traditionally, fault trees have been used to access fixed probabilities (*i.e.* each event that comprises the tree has a fixed probability of occurring) while RBDs may have included time-varying distributions for the success (reliability equation) and other properties, such as repair/ restoration distributions.

4.9 EVENT TREE ANALYSIS (ETA)

It is a visual representation of all the events, which can occur in a system. As the number of events increases, the picture fans out like the branches of a tree. Event trees can be used to analyze systems in which all components are continuously operating, or for systems in which some or all of the components are in standby mode - those that involve sequential operation logic and switching. The starting point (referred to as the initiating event) disrupts normal system operation. The event tree displays the sequences of events involving success and/or failure of the system components. The goal of an event tree is to determine the probability of an event based on the outcomes of each event in the chronological sequence of events leading up to it. It helps in FTA and can be considered as part of FTA.

4.10 ROUTE CAUSE ANALYSIS (RCA)

Root cause analysis is a generic term used commonly to refer to structured problem solving within organizations. RCA is normally done in those unwanted situations, which consume resources and causes problems/ failures in a repeated fashion. It is finding the real cause of the problem and dealing with it rather than simply continuing to deal with the symptoms. It considers mainly following four points–

1. How does one determine which situations are items/ candidates for root cause analysis?
2. How does one figure out what the root cause is? (Finding the route cause),
3. Does the removal of the cause entail less resource expenditure than it takes to continue to deal with the symptom?
4. Removal of route cause.

Determining or finding out item/ candidate for root cause analysis is often quite difficult in normal chaotic organizational environments because the situations which repeat are either distributed over time so one doesn't realize they are actually recurring, or the situation happens to different people so there isn't an awareness of the recurring nature of the situation. When an industry/ organization is using an automated problem resolution support system, it is easy to determine which situations are recurring with what frequency. All one has to do is to run reports against the system to determine which solutions are being used with what frequency. Those situations, which are recurring with the greatest frequency and consume the greatest amount of resource to rectify are the items/ candidates for root cause analysis.

Most problems/ failures, which arise within an industry, have multiple approaches to solution. These different approaches generally require different levels of resource expenditure to execute. Due to immediacy or urgency, most of past solutions might have been based on symptoms rather than real or root cause, which might lead the problem/ failure to reoccur. One has to learn from those situations and solutions o find out the root cause.

Once the root cause is determined, it has to be determined whether it costs more to remove the root cause or simply continue to treat the symptoms. This is often not an easy determination. Even though it may be relatively easy to estimate the cost to remove the root cause it is generally very difficult to assess the cost of treating the symptom and repeated minor failures.

Once removal of root cause has been found economical, one has to go for that. Though RCA is beneficial for maintenance, one may argue that "Root Cause Analysis" is really not an appropriate phrase. In this apparently endlessly interconnected industrial scenarios, everything seems to influence so many other things. Seeking the "Root Cause" is an endless exercise because no matter how deep you go there's always at least one more cause you can look for.

RCA has to be done with total business process in perspective, otherwise it may be a confusing label. As per the categories of causes, there may be many root cause analysis, e.g. human factors-styled root cause analyses, barrier root cause analyses, change, cause-and-effect, energy flow, and fault root cause analyses etc, but all these can be unified with total business perspective of the company.

4.11 ROOT CAUSE FAILURE ANALYSIS (RCFA)

Root cause failure analysis (RCFA) is another name for similar technique, specifically for maintenance. RCFA focuses on eliminating the risk of recurrence of the failures by identifying the physical, human and latent (organizational) system roots that lead to the failure. RCFA is a simple, yet disciplined process used to investigate, rectify and eliminate equipment failure, and it's most effective when directed at chronic breakdowns. RCFA takes corrective action past the component stage and into the system deficiency or latent root stage. Here, you solve not just one, but a multiplicity of problems and generally those problems will not recur because their deepest root causes have been corrected. The methodology of RCFA is also generally the same as mentioned for RCA. Often RCFA combined with RCM (Reliability Centered Maintenance) gives a successful formula for maintenance.

4.11.1 Difference Between FA (Failure Analysis), RCA and RCFA

Though all three are useful and related techniques, there are some differences between the three. FA is general analysis and normally stops at the Physical Root Causes. The Physical Root is at a tangible level, usually a component level. We find that it has failed and we simply replace it. We do not go much in detail about how and why of failure.

RCFA indicates conducting a comprehensive analysis down to all the root causes (physical, human and latent), but connotes analysis on mechanical/ electrical items only. RCFA goes more in details about failures but here "Failure" has a mechanical/ electrical connotation to most people.

RCA is applicable to more than just mechanical/electrical situations. It conducting of a full-blown analysis that identifies the Physical, Human and Latent Root Causes of "how" any undesirable event occurred. It also includes safety incidents, quality defects, customer complaints, administrative problems (i.e. - delayed shutdowns) and the similar events.

4.12 FAILURE MODES AND EFFECTS ANALYSIS (FMEA)

FMEA is a technique first utilized in the aerospace industry to find problems with an aircraft before it ever left the ground. In short, it is a way of looking into the future and determining where potential failures might be located. While using this in industries, instead of looking into the future, we look at our past failures. It is a disciplined approach used to identify possible failures of a product/ equipment or service and then determine the frequency and impact of the failure and also its causes associated with the designing, manufacturing and maintaining of the product/ equipment.

Using a simple example of a bearing failure, the steps can be as below—

1. Define Failure mode– Bearing failure,
2. Identify cause of failure–Lack of grease,
3. Identify effect of failure– Bearing jamming, leading to pump stoppage and production loss. In past, it failed 12 times in a year.

4. Corrective action– Change bearing and improve lubricant or lube system.
5. Strategy to improve– Installing some built-in-test equipment for low grease pressure or no grease warning.

It has become the basic requirement for many reliability specifications. The basic aim is to identify potential design weaknesses and other weaknesses through a systematic analysis, considering all possible ways in which any equipment can fail (i.e. mode of failure), the possible causes and effects of each failure, the probability of occurrence of each failure and initiating suitable measures/ intentions for reducing those.

Taking the above example of bearing failure, Table 4.2 (Sample Failure Event/ Mode) demonstrates what failures are occurring, in our facility/ industry, and what their impact and frequencies are.

Table 4.2 Sample Failure Event/Mode

Failure Event	Failure Mode	Frequency	Impact	Total Loss/year
Failure of pump p-102	Bearing Failures	12 failure / yr.	Rs. 2,000/ failure	Rs.60,000
....	....	....	....	....

We may do such calculation for each failure event, but we should consider the more significant ones for FMEA (generally about 20%). Table 4.3 gives the commonly used steps involved in performing a FMEA–

Table 4.3 Common Steps involved in FMEA

Steps	Descriptions
1 Perform preparatory work	Develop failure definition and mode, make flow diagram, gap analysis and preliminary worksheet and interview schedule.
2 Collect data	Interview/ discuss concerned personnel to determine what the failures are, their frequencies and their impacts
3 Summarize and encode Results redundancies.	Input into an electronic spreadsheet and determine any
4 Calculate losses	Multiply frequency x impact for every failure event in the analysis
5 Determine "Significant Few"	Determine the 20% or less of the failures that result in 80% of the losses.
6 Validate results	Verify that the results are valid
7 Issue a report	Communicate results to plan actions

The above steps are mostly self-explanatory. The steps of FMEA can be explained further by Fig.4.2 (FMEA Process Flow Diagram). In this diagram, RPN (Risk Priority Number is a mathematical product of the numerical Severity, Occurrence, and Detection ratings. RPN = (S) x (O) x (D). This number is used to place priority on items that require additional quality/ job planning. "Determine Detection" in the diagram means assessment of the likelihood that the existing Controls (design and process) will detect the Cause of the Failure Mode or the Failure Mode itself, thus preventing it from reaching the machine /system.

One of the most confusing issues for new practitioners of FMEA is that any Cause that itself has a Cause might be a Failure Mode. Any Effect that itself has an Effect might also be a Failure Mode. In different contexts, a single event may be a Cause, an Effect, and a Failure Mode. Most real-world systems do not follow the simple Cause-Effect model. A single Cause may have multiple

Effects. A combination of Causes may lead to an Effect, or they may lead to multiple Effects. The Failure Mode must also be considered in all of these models. Figure 4.3 illustrates the relationship between a Function, a Failure Mode, Potential Causes, and Effects. In the FMEA model Causes do not automatically result in the Failure Mode.

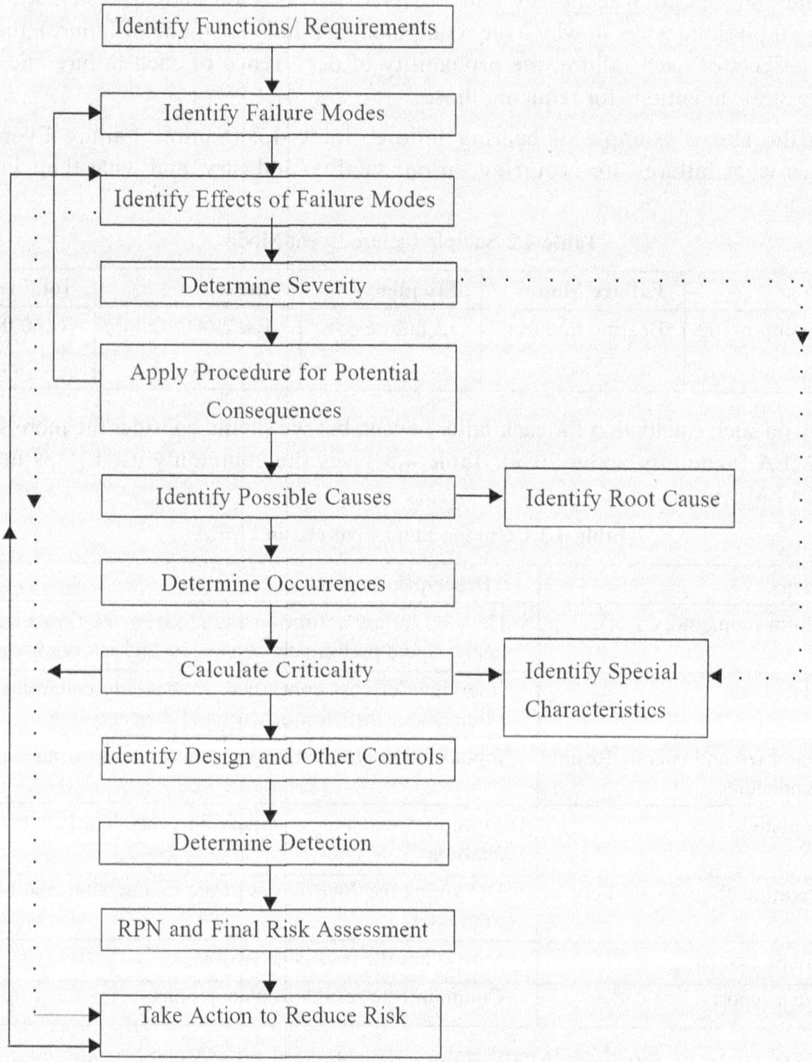

Fig. 4.2 FMEA Process Flow Diagram

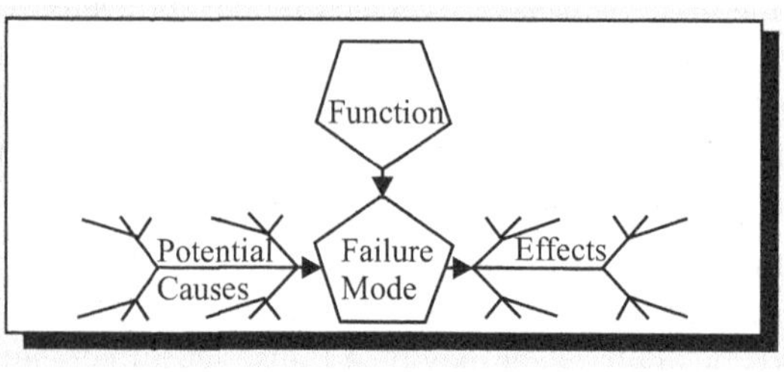

Fig. 4.3 Failure Mode Relationship in FMEA

Table 4.4 gives few illustrative examples of Failure Modes, Effects of Failure Modes and Causes of Failure Modes (without interlinking). Table 4.5 shows a sample FMEA spread sheet.

Table 4.4 Few Failure Modes, Effects and Causes

Few Failure Modes	Few Effects of Failure Modes (Rank Severity of Effect)	Few Causes of Failure Modes
Corrosion Hydrogen embrittlement Electrical Short or Open Torque Fatigue Deformation Cracking/ Breakage Incorrect Speed Etc.	Injury to the user Inoperability of the product, Processor system Improper appearance of the product or process Odors Degraded performance Noise Damage, Etc.	Improper torque applied Improper operating conditions Contamination Erroneous algorithms Improper alignment Excessive loading Excessive voltage Wear and Tear Etc.

Table 4.5 FMEA Spread sheet

Subsystem Name- DC Motor	P- Probability (chance) of occurring	FMEA Date ....
Equipment Code No.	S- Seriousness of failure of item	Prepared by..........
	R- Risk Priority Measure (R=PxS)	Reviewed by ....

(1-Very low, 2- Low or Minor, 3-Moderate, 4- High, 5-Very High or Catastrophic)

Part Name and Number	Function	Failure Mode	Causes of Failure	Effects of Failure	Risk Priority P	S	R	Recommended Action	Action Taken
Drive	Receive speed demand	Incorrect speed demand being received	Fault in Position Controller output	Extensive damage to the machine	2	4	8	Indicator and Audile Warning	
	Measures Actual Speed	Incorrect Speed Reading Procedure	Wear and Tear Check	Extensive damage	4	4	16	Voltmeter, Improve	
....	....	....	....	....				....	

FMEA is an integral part of any quality QS 9000 compliant quality system.

4.12.1 Benefits of FMEA

FMEA is designed to assist the engineer improve the availability and reliability of the machines/systems and its design. Few of the benefits are given below—

1. Improve product/process/equipment reliability, quality and availability,

2. Early identification and elimination of potential failure modes,
3. Prioritize system/process deficiencies,
4. Documents risk and actions taken to reduce further risk,
5. Emphasizes problem prevention,
6. Minimizes late changes and associated cost, Etc.

4.13 FMECA (FAILURE MODE EFFECT CRITICALITY ANALYSIS)

This technique is an extension of FMEA, which also examines criticality, in addition to mode, effect and analysis of failures. Normally, FMEA also considers "critically" of the failures, without specifically mentioning that. Table 4.5 shows that FMEA Spread Sheet has considered criticality (Probability and Seriousness). However, FMECA is, often, getting more prominence in many industries or consultancy firms as it considers two distinct analyses; failure mode and effect analysis (FMEA) and criticality analysis (CA), where the failure modes are ranked according to a combination of severity and the probability of that failure mode actually occurring. The aim, again, is to identify potential design and other weaknesses through a systematic analysis/ investigation of all possible ways in which a component or functional block can fail, possible causes of each failure, the probability of its occurrences, the criticality of failure the effect of each failure on the component and system operation and to plan/ initiate corrective action to prevent or reduce probability of potential problems from occurrence in future.

Few special features of FMECA are as follows—

1. Criticality analysis considers mainly severity and probability of failure modes and rates them in scale of 1~5 or 1~10 or 1~15, as given in Table 4.6 (FMECA Criticality Analysis Matrix),

Table 4.7 FMECA Criticality Analysis Matrix

Severity (Rank in Scale 1~5 or 1~10 or 1~15)		**Probability** (Rank in Scale 1~5 or 1~10 or 1~15)	
1	Catastrophic	A	Frequent
2	Critical	B	Reasonably Probable
3	Marginal	C	Occasional
4	Minor	D	Remote
		E	Extremely Unlikely

2. It has option to assign Risk Priority Numbers (RPN),
3. Failure Effects and Severities traced through to system hierarch,
4. Multiple failure effects permitted for a single failure mode,
5. Automatic criticality calculations, Etc.

The steps and methodology for FMECA is generally similar as mentioned for FMEA.

4.14 FRACAS-(FAILURE REPORTING, ANALYSIS, AND CORRECTIVE ACTION SYSTEM)

It may again be considered as an extension of FMEA. Many companies employ a multi-purpose, closed-loop corrective action process to continually monitor products, systems, or services to continually improve quality and product reliability. These processes are generally known as a FRACAS. The process typically provides the means for collecting field information, analyzing the data, then implementing and tracking corrective actions. The FRACAS process is carried out in a wide range of industries, in a variety of ways, though some of them may not use the term FRACAS.

4.15 FINITE ELEMENT ANALYSIS (FEA)

FEA was first developed for use in the aerospace and nuclear industries where the safety of structures is critical. Today, with the growth of computer technology, FEA is capable of solving the most sophisticated problems, not just in structural analysis, but for a wide range of phenomena such as steady state and dynamic temperature distributions, fluid flow and manufacturing processes such as injection molding, metal forming and other industries. FEA consists of a computer model of a material or design that is loaded and analyzed for specific results. It is used in new product design, and existing product refinement/ reconditioning. Modifying an existing product or structure is utilised to qualify the product or structure for a new service condition. In case of structural failure, FEA may be used to help determine the design modifications to meet the new condition.

Mathematically, the structure to be analyzed is subdivided into a mesh of finite sized elements of simple shape. Within each element, the variation of displacement is assumed to be determined by simple polynomial shape functions and nodal displacements. Equations for the strains and stresses are developed in terms of the unknown nodal displacements. From this, the equations of equilibrium are assembled in a matrix form which can be easily be programmed and solved on a computer. After applying the appropriate boundary conditions, the nodal displacements are found by solving the matrix stiffness equation. Once the nodal displacements are known, element stresses and strains can be calculated. It is essential to have an understanding of the technique and physical processes involved in the analysis. Only then can an appropriate and accurate analysis model be selected, correctly defined and subsequently interpreted.

In recent years, FEA has been used almost universally to solve structural engineering problems. Due to the extreme demands for faster, stronger, lighter and more efficient aircrafts, automobiles and other such big machines, manufacturers have to rely on this technique to stay competitive and ensure safety. FEA has often been used in high volume production and manufacturing industries for many years, as to get a product design wrong would be detrimental. FEA is a very important tool for those involved in engineering design, it is now used routinely to solve problems in the many areas, such as—

- Structural strength design,
- Analysis of Shock (underwater and in materials),
- Structural interaction with fluid flows,
- Crash simulations,
- Electrical analyses
- Dynamic analyses
- Metal Forming,
- Thermal analysis
- Acoustics
- Vibrations
- luid flows
- Buckling problems
- Electromagnetic evaluations

etc.

The terms 'finite element method' (FEM) and 'finite element analysis' (FEA) seem to be used interchangeably in most documentation, though FEM is more used academically where as FEA may be used more practically.

As FEA is more concerned with design, reliability and maintainability and is less used for routine repairs, FEA would not be discussed further here. Interested readers may consult Internet site" http://www.dermotmonaghan.com/index.php" for further reading.

4.16 WEIBULL ANALYSIS (OR LIFE DATA ANALYSIS)

It is the process of discovering the trends in product or system failure data, by statistical means, and using them to predict future failures in similar situations. By learning these trends, you can attempt to correct or compensate for them, thereby improving product reliability. It can be used to study a variety of fields, practices, and disciplines. Weibull distribution is one of the most widely used distributions for failure data analysis, especially useful for mechanical, chemical, electrical, electronic, materials, and human failure analysis.

- Weibull analysis can help us in finding the followings—
- What type of failure mechanism is the root cause?
- How many failures are expected?
- How reliable is the existing part compared to a possible new design?
- When should I replace an existing part with a new one to minimize maintenance costs?

Weibull analyses study the relationship between product reliability and product lifespan. They provide insight into the decrease in reliability as the usage of a product or system increases. The primary advantage of Weibull analysis is that it can provide reasonably accurate failure analyses and failure forecasts with extremely small data samples. This facilitates cost-effective and efficient component testing.

The steps of Weibull analyses is generally the followings—

1. Gather life data for the product,
2. Select a lifetime distribution that will fit the data and model the life of the product,
3. Estimate the parameters that will fit the distribution to the data,
4. Generate plots and results that estimate the life characteristics, like reliability or mean life, of the product; Etc.

For further reading about this analysis, refer to standard books on Statistics or Reliability.

4.17 ELECTRICAL STRESS ANALYSIS

It looks for the components that are likely to fail due to operating conditions. The analysis considers component application, input and operating voltages and currents, voltages and current peaks and surges, junction temperatures and temperature rise. Analysis results can e compiled in a table that will shows for each components, actual operating parameters and safe limit for those parameters. Such table can highlight the components that are overstressed. Such analysis is common in maintenance practice.

QUESTIONS

1. What is failure in maintenance? What are basic reasons of failure in mechanical items?
2. What do you understand by Failure Mode and Failure Code and what are their respective uses?
3. What is Failure Analysis, name the various techniques of failure analysis and explain briefly any two?
4. What is RCA in maintenance and what are its uses? How does it differ from RCFA?
5. What is FMEA and what are its uses. How does FMEA fifer from FMECA?
6. What is Fault Tree Analysis? Please make one FTA for a car break not starting, going at-least up-to two levels and remedial measures.
7. What are the failure generation stages and how failure lists are obtained in industries?
8. What are the ways of failure collection and reporting in industries?
9. Give some idea of Finite Element Analysis and Weibull analysis in relation to maintenance.

CHAPTER

5 MAINTENANCE STRATEGIES/ TYPES/SYSTEM

5.1 DEFINITION

Maintenance Strategy may be defined as a long-term plan, covering all aspects of maintenance management which sets the direction for maintenance management, and contains firm action plans for achieving a desired future state for the maintenance function.

Maintenance system though meaning almost the same as maintenance strategy, assumes little broader approach. **Maintenance system** can be defined as a methodology and strategy for establishing a good maintenance management principles and providing a framework for identifying, mapping and controlling the elements of maintenance functions.

Maintenance type, though meaning somewhat same as other two terms, may be the more appropriate term for this chapter, as the other two terms cover relatively broader approach. **Maintenance types** may be defined as what kind of maintenance functions are to be performed and what type of maintenance methodology is to be chosen to plan and execute those functions in most efficient way.

However, for further discussions, we would mean the three terms the same.

Almost every business/ industry has some sort of maintenance programme for its physical assets. Even in domestic sector, people having car or many house hold appliances, knows the principle and advantage of preventive maintenance programmes, whether that may mean routinely changing of engine oil etc For few machines it may be easy to keep track what need to be done and when, but when scope includes thousands of equipments/ assets, involving millions of Rupees, it becomes significantly complex and expensive. As such, over period of time, many maintenance strategies or methodologies have been developed, mainly basing on time for doing maintenance, frequency of maintenance, quality of maintenance, complexity and sophistication of equipments and value of total assets etc.

5.2 BASIS OF SELECTING MAINTENANCE STRATEGIES

Though, basically there are four distinct maintenance strategies (Reactive, Preventive, Predictive and RCM), we would consider many others in this chapter, to cover the topic broadly. By optimizing equipment maintenance strategies against both target availability and the penalty of failure, you can optimize your asset life-cycle costs.

There are various reasons for evaluating/ considering/ selecting maintenance strategies such as –

- Which set of tasks that should be performed and their frequency,
- With aging of the plant a different mix of tasks is needed to maintain reliability,

- Perhaps with reduced staff it is not possible to perform all the tasks that used to be done and only the most effective tasks are selected, or
- The plant operators may want to reduce maintenance expenditures without elevating risk or reducing reliability; etc.

Another set of conditions to help determining the strategy may be as followings–

1. Is the equipment regularly available to perform planned/routine maintenance activities?
2. Will the cost to shut down the equipment be less than the cost of failure (break- own)?
3. Is equipment downtime a problem?

The resulting strategy must strike a balance between maintenance cost and plant reliability. This evaluation develops tools and techniques for determining the best mix of maintenance tasks. There may not be a sacrosanct maintenance strategy for the total plant/ industry for all times; you may have to continue to improve maintenance task selection techniques, maintenance strategy, develop industry databases of equipment maintenance and reliability information such as the task selection templates, and improved software tools to support your strategy and other approaches. An element of the maintenance strategy is to prioritize maintenance tasks. One approach is called Risk-based Evaluation and Prioritizing (REAP). REAP has been successfully used to prioritize routine and outage maintenance tasks, based on improved reliability and cost of the task.

Any good maintenance strategy must have an integrated information system for interacting with all concerned areas such as shown in Fig 5.1

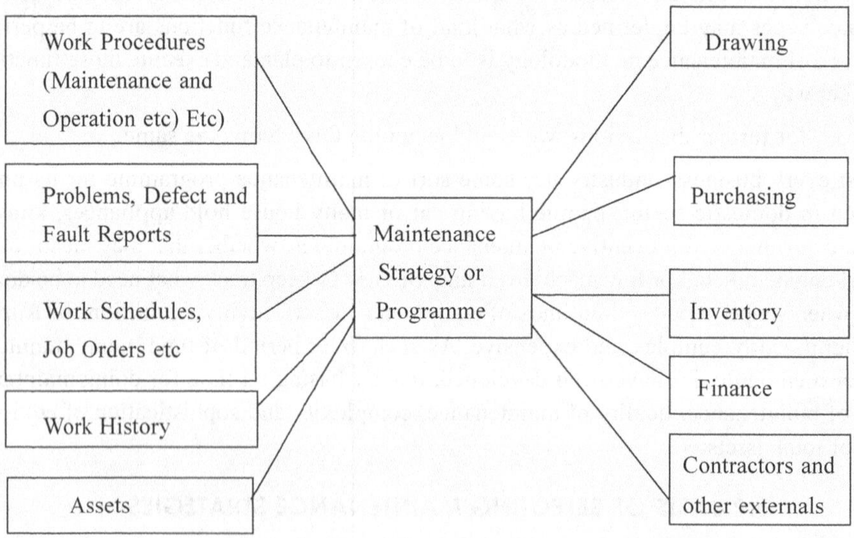

Fig. 5.1 Integrated Information System for Maintenance Strategy

Presently many maintenance systems/ strategies are in use in industries and selection has to be made depending the need, complexity and reliability needed and also considering techno-economics. Most of the commonly used strategies can be grouped as shown in Fig. 5.2.Reactive maintenance, as the name suggests, are maintenance jobs in reaction to breakdowns or problems or deficiencies already occurred. This may include breakdown, emergency or corrective maintenance etc.

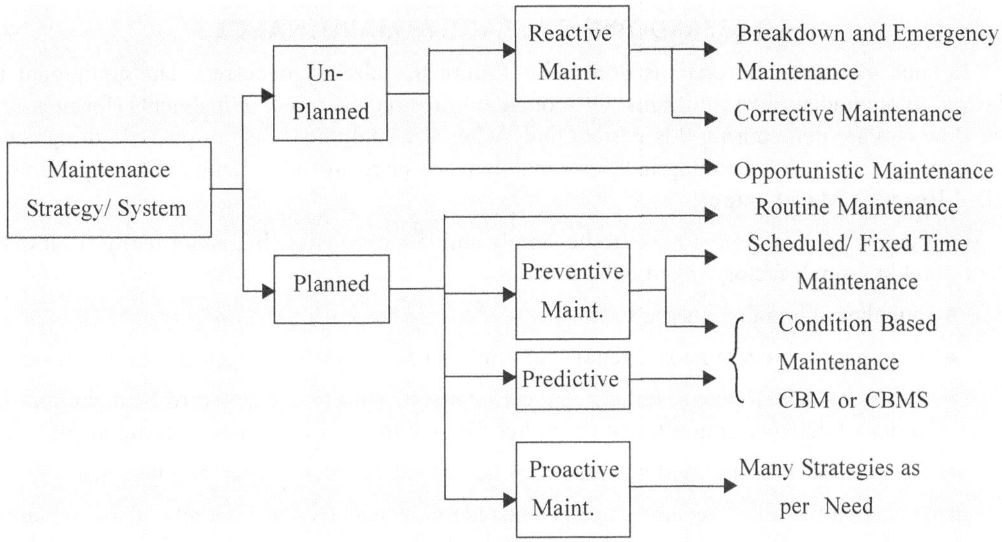

Fig. 5.2 Types of Maintenance Strategies

5.2.1 Maintenance Decision Diagram

Fig. 5.2.1 shows a maintenance decision diagram, giving another basis for selecting maintenance type/ strategy.

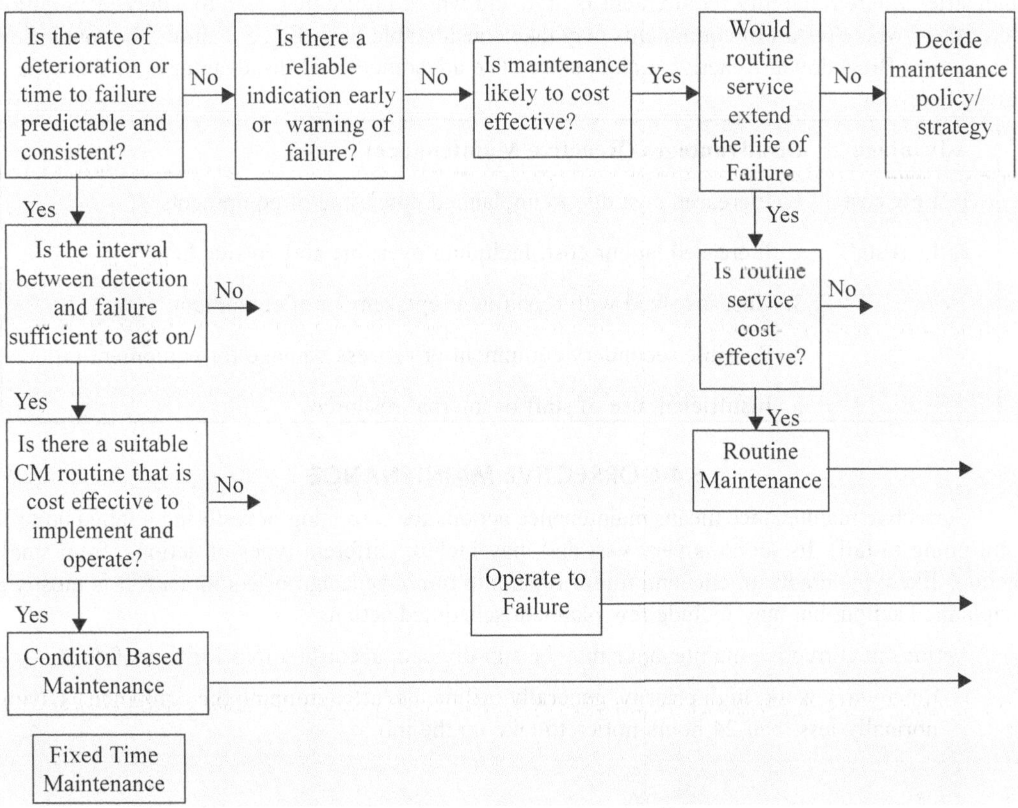

Fig. 5.2.1 Maintenance Type Decision Diagram

5.3 BREAKDOWN OR REACTIVE MAINTENANCE

In such maintenance repair is done after failure has already occurred. The equipment is allowed to run undisturbed till it fails. Off-course lubrication and minor adjustments (for pressure and flow etc) are done during this period. Only when the equipment fails to perform designated functions or comes to a grinding halt, any maintenance or repair job is taken. This type is also called **Reactive Maintenance.**

On face value, this system looks to be simple and less expensive, but is not really so. It may work good in a small factory/ plant, where—

- Number of equipments are few,
- Equipments are simple and repair does not call for specialists or special tools/ tackles,
- When sudden stoppage/ failure of the equipment will not cause severe financial loss in terms of delivery commitment or further damage to other equipments/ components,
- When sudden failure will not cause any severe safety or environmental hazards.

In such small factories, generally no specialized maintenance crew are kept and maintenance is normally done by persons operating the machine and other concerned persons. Occasionally help of outside agency may also be taken. Maintenance is generally done to put back the breakdown machine into operation and not much attention is given to investigate and to prevent recurrence of such breakdown.

However, such maintenance system can not work in big industries, having large number of equipments, some of which may be quite intricate. This is not used for chemical or process industries where reliability requirement is high and where failure may lead to safety or pollution hazards or where restart of equipments may take considerable time. Fig. 5.3 shows the rough flow diagram of Breakdown/ Reactive maintenance. The advantages and disadvantages of this type is given herewith.

Advantage	Disadvantages (Reactive Maintenance)
1. Low cost	1. Increased cost due to unplanned downtime of equipments
2. Less staff	2. Increased labour cost, including overtime and outside help
	3. Cost involved with repair and replacement of equipment,
	4. Possible secondary equipment or process damage for equipment failure
	5. Insufficient use of staff or internal resources.

5.4 CORRECTIVE MAINTENANCE

Corrective maintenance means maintenance actions for correcting or restoring a failed unit (or unit going to fail). Its scope is very vast and may include different types of actions, from small actions like typical adjustments and minor repairs to minor redesign of equipment. It is mostly a unplanned action, but may include few planned/ scheduled actions.

Actions in corrective maintenance may be sub-divided, according to priority, as follows—

1. Emergency work, high priority, generally offline, i.e. after stopping the equipment, Giving normally less than 24 hours notice to take up the job

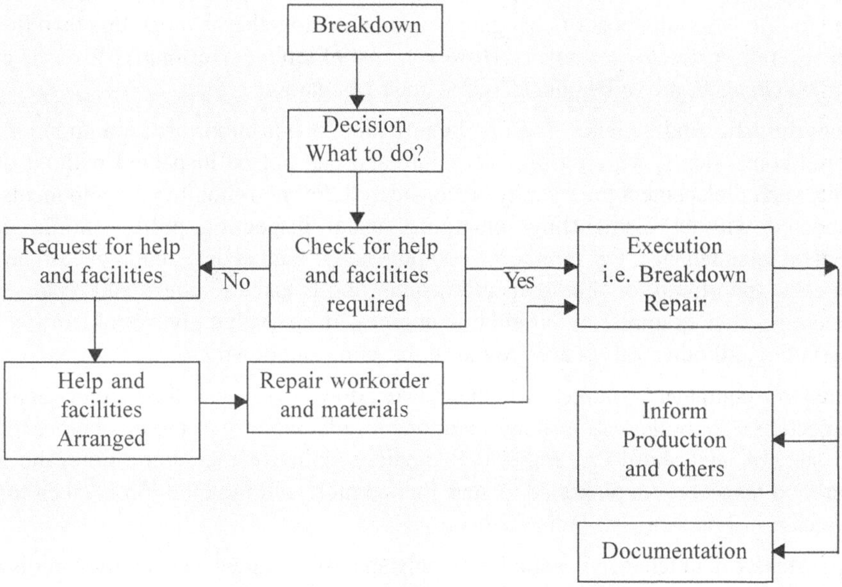

Fig. 5.3 Flow Diagram of Breakdown/ Reactive Maintenance

2. Deferred works–jobs of lower order priority; generally off-line,
3. To eliminate /reduce repetitive breakdowns,
4. Specific reconditioning or redesign jobs, generally small or medium in volume.

Corrective maintenance is generally one time task; once taken up, completed fully. Each corrective job may differ from other. Some bigger corrective maintenance jobs may have the following stages—

- Collection of data/ information and analysis,
- Identify likely causes,
- Find out the best possible solution to eliminate those likely causes,
- Implement those solutions etc.

Some of the differences between corrective maintenance and preventive maintenance are as follows—

1. PM jobs are generally taken before the equipment has stopped working but corrective may be taken before or after the equipment has stopped working,
2. Level and type of PM jobs are normally decided within maintenance department whereas In corrective maintenance, help of other departments may be taken.
3. PM jobs are planned well in advance whereas corrective maintenance jobs may be taken at shorter notice.

Corrective maintenance may include some of design-out maintenance jobs.

5.5 OPPORTUNISTIC MAINTENANCE

In multi-component system, with several failing components/ equipments, often it is advantageous to follow opportunistic maintenance also. Hen an equipment/ system is taken down for maintenance/ changing of one or few worn-out component, opportunity can be utilized for maintaining /changing other worn-out components, even though they have not failed. This would be, probably, economical in the long run rather than taking shutdown when other components fail or going to fail. Normally cost of changing several parts together is much less than the sum of the costs of several separate replacements. However, cost of leftover (residual) life of the components, going to be changed, are to be considered in such calculations.

Opportunistic maintenance is very beneficial for non-monitored components. For non-monitored components, which are inaccessible and can not be inspected without dismantling/ changing, such replacement policy may be considered. For non-monitored components, which can be inspected without dismantling/ changing, usual inspection policy should guide about replacement/ maintenance. For monitored components, if fault is detected in one of the few similar components, opportunity of shutdown and dismantling is taken to check and repair other similar components. As an example, in automobile engines, if one valve gives problem (worn-out) and needs grinding, all other valves are ground in the same shutdown.

Often in a equipment complex, which are taken down every year for statutory annual overhaul and inspection (like boilers etc), if any component fails one or two months before the scheduled date of start of next shutdown and if that repair is going to take some time, the next annual overhaul and inspection is preponed to start immediately and the total job is taken together. This is also a case of opportunistic maintenance.

Opportunity maintenance is actually not any specific maintenance strategy but is a system of utilizing an opportunity, which may come any time. To carry actual jobs, we may use other maintenance systems/ strategies.

5.6 ROUTINE MAINTENANCE

Routine maintenance is the simplest but very essential form of maintenance system. Earlier the routine maintenance was considered about preventing failures. Today routine maintenance is being considered about avoiding, reducing or eliminating the consequences of failures. As the name implies, it means carrying out minor maintenance jobs at regular intervals. It involves jobs such as cleaning, lubrication, inspection and minor adjustments pressure, flow, tightness etc and tightening of loose parts etc. It also includes inspection of bearings, V-belts, couplings, jointings, foundation bolts, earthings and protective covers etc. The small and critical defects, observed during such inspection, are rectified immediately and bigger jobs are planned for rectification during next available shutdown. Such maintenance is essential for effective scheduled and preventive maintenance.

Routine maintenance is not necessarily need-based. In a equipment /plant some motors may be running for 4 hours a day and some others may be running 20 hours a day but, in routine maintenance, all may be inspected at the same frequency. This may lead to some amount of over maintenance in some components, but the system pays up handsomely in the long run. "Regularity", .e. carrying out planned jobs regularly in simple cyclic schedule, is very essential in routine maintenance. Such schedules are simple (like check, clean, lubricate, tighten, adjust etc) and repetitive. Routine maintenance may also be considered as a small portion of preventive maintenance.

Frequency of routine maintenance is generally once every shift or every day (normally at the start). Of-course in sophisticated and automatic working equipments or in equipments having enough condition monitoring gadgets to indicate failures or deviations, the period of routine maintenance may change. Again, if such jobs are more and time availability is less, one group of job may be planned for Monday, another group of jobs for Tuesday and so on.

Routine maintenance needs very little investment in time and money. The duration of routine maintenance is generally so small that it does not affect the output of machine appreciably.

However, the cost of not-doing routine maintenance may be very high as a small defect may develop in big and catastrophic failure.

As one example of routine maintenance, in few railway suburban electric trains system, whenever the train stops at few bigger stations, a group of maintenance people immediately starts checking and doing minor jobs like identifying loose parts and tightening, cleaning moisture traps and checking brakes etc. The whole job may take 10 to 12 minutes by the time the train is due to start for onward journey. In industries, during shift change periods, a small group of maintenance personnel carry out necessary inspection, lubrication, adjustments and tightening etc, which may take about 15 minutes.

5.7 PREVENTIVE MAINTENANCE

Preventive maintenance (pm) has a vast scope. As the name applies, it may include any action to prevent equipment downtime/ failure. An enhanced PM may include Predictive maintenance, CBM, CMMS and Proactive maintenance etc also. Fig. 5.4 gives the Process Flow for enhanced PM. The steps include review of existing PM or other maintenance programmes, Analysis, assessing new reliability and other requirements and planning of new PM programmes, integrating programmes of new or other equipments with existing system if any, establishing system support requirements and configuration, define reimplementation plan and fixing performance monitoring metrics etc.

However, the conventional PM, using less of predictive and proactive maintenance, is the oldest, easily understood and extensively used form of maintenance. PM is the planned maintenance of plants and equipments (including and resulting from routine maintenance inspections, other inspections/ reports and condition monitoring etc), in order to prevent or minimize breakdowns and deterioration/ depreciation rates. It covers a vast area and occasionally people get misled by its coverage. After PM repairs, the equipment's health is, restored back nearly to its original condition. However, it does not include much improvement and up-gradation jobs.

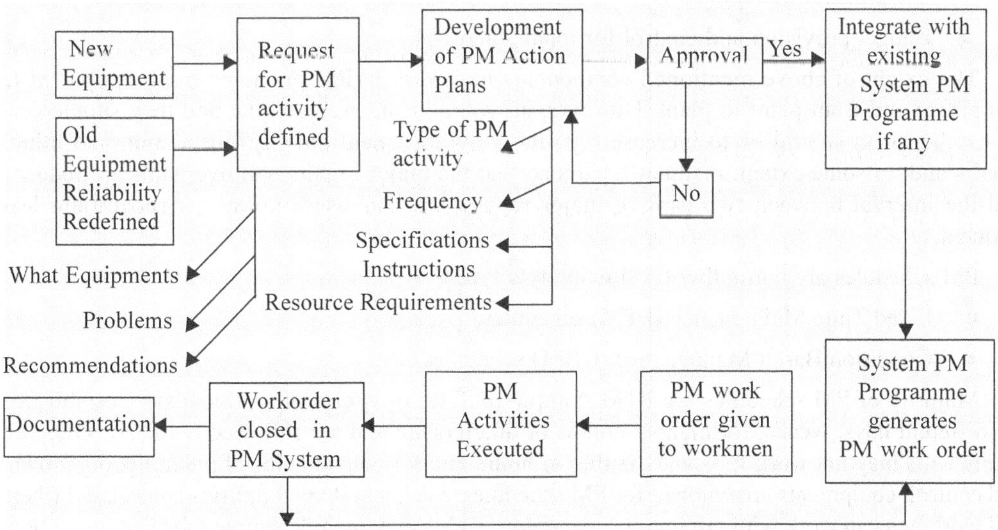

Fig. 5.4. Enhanced PM Process Flow

In general, the various components of PM are as follows—

1. Check drawing, design and installation of equipments including subsequent redesign and minor modifications, depending on specific nature of problems,

2. Proper identification of all items, proper documentation and codification; check—
   - History cards/ records,
   - Spares catalogues, equipment catalogues and inventory lists,
   - Job manuals and standard maintenance practices (SMP),
   - Maintenance workorders and other pending workorders,
3. Periodic inspection of plant and equipments—
   - Use of checklists by inspectors and its frequency; shift-wise, daily, weakly and monthly Etc,
   - Well qualified and experienced inspectors,
   - Use of necessary aids; test equipments, Vibration meters, Ultrasonic and X-ray Equipments etc,
   - Preparing total defect/ failure list and its categorization,
4. Repetitive servicing, repairs, upkeep and overhaul—
   - Minor repairs,
   - Medium repairs— roughly around 50% of jobs of major overhauls,
   - Major overhauls or capital repairs,
   - Emergency repairs or corrective repairs,
   - Recovery or salvaging; when equipment has undergone several major repairs, etc
5. Adequate lubrication, cleaning and painting of equipments; Changing of oils and lubricants of systems as per inspection reports,
6. Typical failure analysis and plan for their elimination,
7. Organization for PM,
8. Budget provision and control for repairs and PM.

The extent of above mentioned components may vary in PM practices from equipment to equipment and from plant to plant. However, all components are essential and they supplement each other. Aim should be to increase (qualitatively and quantitatively) inspection and minor repairs and, to some extent, medium repairs so that the major repairs and overhauls are reduced and the interval between two capital/ major repairs are increased; i.e. major repairs are less frequent.

PM schedules are normally of following two types—

- Fixed Time Maintenance (FTM) schedules,
- Condition Based Maintenance (CBM) schedules.

Majority of PM schedules are FTM. Simple routines or cycles of workloads are established for different days, weeks, fortnights, months or quarters etc and are followed regularly. Often all equipments may not work for same number of hours but, for convenience of planning, organizing and control, equipments are clubbed for PM schedules. Such a system simplifies control and filters and workmen can work with minimum supervision. FTMs may include simple jobs like inspection, calibration, testing and adjustment etc or bigger jobs like replacement and rebuilding etc.

CBM mostly comes under predictive maintenance category but it overlaps to some extent with PM also and reasonable amount of condition monitoring equipments are used to determine the jobs of PM. A good PM programme uses modern diagnostic tools and condition monitoring tools as part of inspection tools and also uses some amount of routine and corrective maintenance. Again planning and scheduling of all activities of PM should be done beforehand so that common delays of following types are avoided—

- Waiting for job orders at the start of shift and also after finishing one job,
- Visiting the site or other place to find out what to do and how to do,
- Unnecessary trips to stores as complete list of tools/ spares are not available at the start of shift and tools are brought from stores as and when needed,
- Operating personnel not clearly aware of time of sparing the machine and doing their preparatory jobs (e.g. opening of dies/ jaws etc), before handing the machine to maintenance for PM.
- Loosing time because of non-availability of safety clearance/ permit in time.

Crews carrying PM jobs should, normally, be different from crews carrying breakdown maintenance. If the two jobs are given to the same crew, the PM jobs often gets neglected or gets less importance and less supervision as not doing PM jobs does not immediately reflects in breakdowns.

5.7.1 Frequency of Major Repair in PM

Scheduled periodic/ cyclic major repair is an important component pf PM. The interval between the two PM major repairs for the same equipment, i.e. frequency, may not be the same throughout the lifecycle lf the equipment. Frequency is dictated by the well known bathtub curve; the initial small de-bugging or wear-in phase (just after commissioning the equipment after major repair), chance failure phase (long normal working period) and wear-out phase (just before the equipment is taken for major repair. It is also governed by the inspection results through different diagnostic and condition monitoring tools. Our past experience and manufacturers recommendations, if any, also help in deciding the frequency. Fig.5.5 shows the rough curve for major repairs of PM. In the curve "x" is the working period of the equipment, when its condition deteriorates from good (near 100% condition) to bad when the failure rate is expected to be un-acceptable and "y" is the period of PM major repair, when the equipment is restored to nearly 100%. In actual practice, the periods x_1, x_2, x_3 etc and y_1, y_2, y_3 may not be the same, but is often nearby.

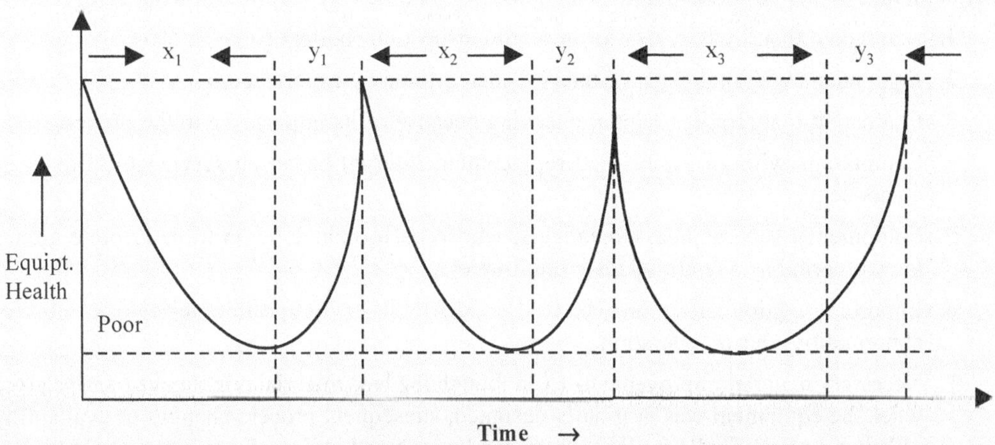

Fig. 5.5. Frequency of PM Major Repairs

5.7.2 Sources for PM Procedure Development

Following sources may help in the development of PM procedures—

1. Vendor recommended PM: The maintenance manuals provided by equipment manufacturers are invaluable when trying to build a PM pro-cedure. However, the procedure in the manual should not be used in its original form. Additional tests, repairs or inspections should be added which reflect the special use of the machine in your facility.

2. Plant Developed PM: Mechanics who work on equipment and the operators who run it can help define PM that might be specific to the equipment's environment. Working together, they often discover PM procedures that would not otherwise be obvious.

3. Generic PM: Generic PM procedures, developed for general classes of equipment, can be modified to fit a specific facility requirement. Some typical procedures of this type are developed by professional organizations such as the Institute of Electrical and Electronic Engineers (I.E.E.E.) USA, etc.

The advantages and disadvantages of PM can be summed up as following–

Advantages (Preventive Maintenance)	Disadvantages
1. Cost effective in many , capital intensive processes	1. Catastrophic failures still likely to occur,
2. Flexibility allows for the adjustment of maintenance periodicity,	2. Labour intensive,
3. Increased component life cycle,	3. Includes performance of unneeded maintenance,
4. Energy savings,	
5. Reduced equipment or process failure.,	4. Potential for incidental damage to components in conducting unneeded maintenance.

5.7.3 Selecting equipments for Periodic Maintenance

Periodic maintenance is commonly referred to as somewhat combination of Preventive and Routine maintenance. Periodic maintenance jobs may also be taken from following categories–

1. Equipment that, by law, requires periodic inspection (boilers etc)
2. Equipments with maintenance intervals determined by experience,
3. Equipment that requires regular checking because of its importance to the process,
4. Equipments with an established replacement interval based on serviceable life of its components,
5. Equipments, such as heat exchangers, whose performance begin to deteriorate after a known period as a result of scale build-up etc,
6. Important equipments, for which it is difficult or impossible to detect/ monitor abnormalities during operation.
7. Necessary material improvement- Even though the best material might have been selected when the equipment was originally designed, subsequent process changes or availability of better materials calls for jobs which can be included in periodic maintenance. Original drawings and other records should also be corrected accordingly.

8. Changes in safety or housekeeping standards- Such changes, as compared to original standards may also call for jobs which can be included in periodic maintenance.

This section will supplement the sections on routine maintenance and preventive maintenance of this chapter.

5.8 PREDICTIVE MAINTENANCE (PD M)

As a broader definition, PdM is a technique that uses regular evaluation of the actual operating condition of plant equipment, production system and plant management function to optimize total plant operation. Predictive maintenance (PdM) compares the trend of measured physical parameters against known engineering limits for the purpose of detecting, analyzing, and correcting problems before failure occurs. A predictive maintenance approach strives to detect the onset of equipment degradation and to address the problems as they are identified. This allows casual stressors to be eliminated or controlled, prior to any significant deterioration in the physical state of the component or equipment. In Pd M, maintenance needs are based on the actual condition (normally measured quantitatively and qualitatively) of the equipment/ asset, rather than on some predetermined schedule. It involves predicting the failure before it occurs, identifying the root causes for those failure symptoms and eliminating those before those result in extensive damage to equipments. Thus there are three distinct stages of PdM— **Detection, Analysis** and **Correction**.

Though PM and PdM have some overlapped areas and often supplement each other, the main differences between them are as follows–

- In PM, overhauling of equipments are, generally, done at fixed time interval basis, but, in PdM, overhauling of equipments are taken on the basis of condition monitoring, only before the equipment is going to fail or deteriorate.
- In PM, the basis of jobs and frequency are the planner's judgment, past experience and equipment manufacturer's recommendation etc and not much sophisticated monitoring tools are used; but in PdM, good amount of simple and high-tech monitoring tools are used regularly (at shorter intervals or continuous),
- PM programmes are more suitable for industries having large number of similar or near similar equipments, such as weaving machines and powerlooms of textile mills, machine tools of machine shop, pumps, compressors and electric motors etc. But PdM is more suitable for heavy, costly and more critical equipments where overhauling/ repairs requires more downtimes, such as turbines, heavy gen-sets, big mill stands and furnaces.

Pd M calls for regular and continuous inspection and monitoring of equipment's condition and health and collection of data and analysis to find out the root causes of failure symptoms. Based on these or otherwise, several minor repair jobs and adjustments are planned and executed and necessary major jobs are also planned. Often such repairs are planned in some cyclic pattern, which amounts to PM jobs. Thus is the overlapping of PM and PdM.

Condition Monitoring and Condition Based Maintenance (CBM) are the two most essential components of PdM, in addition to some PM. Condition Monitoring and tools/ gadgets used for that will be discussed in chapter 6.

5.8.1 Components of a Well-Organized PdM Programme

In line with Fig. 5.1, there may be many components of PdM, few of which are given below–

- Equipment inventory,
- Equipmentnumbering system
- Spare parts inventory

- Spare parts forecast (reduction of inventory)
- Sequence of tasks (both, PM and PdM routes); Combining of PM and PdM jobs,
- Equipment and maintenance performance indicators and trends,
- Planning and scheduling for PdM; the preparatory steps,
- Organizational requirements
- Qualification of personnel, training,

5.8.2 Measuring Results of PdM

Following parameters measure results of PdM–

- PdM database/data collection,
- Costs of PdM (equipment/instruments, labor and services),
- By determining PdM benefits and return on investment (ROI),
- Decision factors for in-house vs. contracted PdM or PM jobs.

5.8.3 Advantages and Disadvantages of PdM

Those can be summed-up as followings—

Advantages (PdM)	Disadvantages
1. Increased component operational life/availability	1. Increased investment in diagnostic equipment.
2. Allows for preemptive corrective actions,	2. Increased investment in staff / workmen training,
3. Results in decrease in equipment and/or process downtime and prolongs the time intervals between shutdowns/ overhauls,	
4. Decrease in costs for parts and labour,	3. Savings potential not readily seen by management.
5. Improved worker and environmental safety,	
6. Better product quality,	
7. Energy savings	
8. Estimated 8% to 12% cost savings over preventive maintenance program,	
9. Improved workmen moral.	

5.9 CONDITION BASED MAINTENANCE SYSTEM (CBMS OR CBM)

In crude way, CBM is an age-old practice. In earlier days also, whenever a bearing was giving noise or getting hot, the equipment was taken down, as soon as possible, to check and rectify the problem. But in those days, early detection of fault was not possible and, hence, and hence the system did not give much benefit. Recently, CBM has become one of the most important maintenance system/ strategy, because of—

- Development of good and cheap condition monitoring tools (both, portable and on-line) and superior monitoring techniques, which detect the fault generation very early,
- Increased awareness and acceptance of CM tools and techniques by common maintenance personnel and other concerned personnel,

- Increased cot of equipment downtime because of reactive or preventive maintenance (in terms of cost of maintenance and resources used and cost of loss of production) etc.

The impact of CBM or any maintenance initiatives, including CM, should be predictable and measurable, and indeed somewhere around the performance and dependability of the operating unit. Among the most sensible measures of performance is the production rate or throughput of the unit. However, in performance and dependability analysis, the "random" nature of failures should not be forgotten.

5.9.1 CM vis-à-vis CBM

The objective of condition monitoring (CM) is to predict upcoming maintenance need by monitoring condition indicators. CM is simply attempting to predict an imminent failure. But CBM (and other PdM techniques) takes the appropriate measures to delay the failure as long as possible through the obvious like appropriate lubrication, but also through other jobs like dusting down cooling fins on motors, vacuuming the distribution board and cleaning the pool of oil under the machine so an increase in 'drip rate' is noticeable etc. A CM programme cannot become a successful CBM practice unless the links to the operation are made and the financial metrics determined.

5.9.2 Frequency of CM

The frequency of CM is a difficult question and assessment. The frequencies of CM tasks are a primary function of the rate of decay of the failure rather than the MTBF. MTBF only comes into the equation if the inspection confidence is known and is less than 100%. The main question that you need to ask yourself in determining the appropriate frequency for a CM task is "How quickly does it fail, once an incipient failure is detected?" If it fails more quickly, then inspect more quickly. Again, for CM frequency determination, you must first decide what results you want from your programme. If it is simply to predict imminent failure, follow the conventional intervals and gadgets. But if you want to optimize performance, minimize costs and extend the useful life of your critical production systems, you must base the interval and methods used on a design review of the installed system. This review must include the designed operating envelope of the system (i.e. what was it designed to do?), the actual installation and how it is really being operated.

The earlier saying was "The frequency of condition-based maintenance tasks should be based on the frequency of the failure and/or failure criticality of the item". The new saying is "The frequency of condition-based maintenance tasks should be based on the failure period (also known as the "lead time to failure" or "P-F interval)".

5.9.3 Few Technical Barriers for widespread Application of CBM

- The difficulty to accurately and reliably predict the remaining useful life of the equipment,
- The difficulty to continually monitor a equipment, and
- The difficulty of maintenance system to learn and identify impending failures for an equipment and then recommend what action should be taken.

However, these barriers can potentially be overcome through innovations in technical areas, such as Prognostication capabilities, Cost-effective sensors and monitoring systems and Reasoning or Expert systems. Prognostication includes areas such as estimation of remaining useful life (or time to failure or degraded performance), development of confidence levels (uncertainty estimates) and recording an audit trail of how the estimate and confidence level were obtained.

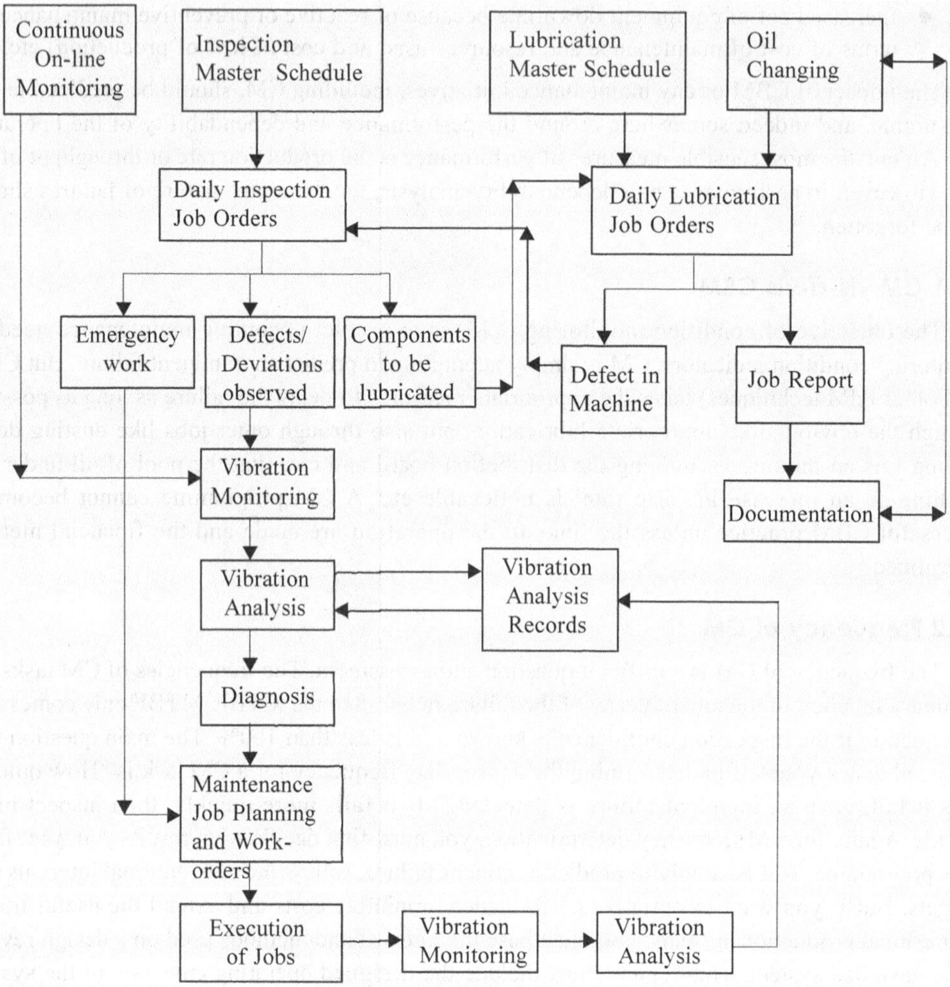

Fig. 5.6 CBMS Flow Sheet

5.9.4 Methodology of CBM

A simplified method of CBM System has been shown in Fig. 5.6 (CBMS Flow Sheet), taking only few monitoring tools.

A simplified CBMS normally uses following steps–

1. Listing and codification of all equipments/ components: for proper identification and location,

2. Selecting critical equipments/ machines and systems: identify very critical, critical and less critical, basing on importance for production, cost of repair and safety and environmental hazards etc. CBMS is more beneficial for critical and very critical machines. If the machines are classified as Vital, Essential and Desirable (VED analysis), continuous monitoring, periodic monitoring and fixed time maintenance may be preferred accordingly.

3. Identify components/ items: Very critical, critical and less critical equipments/ machines/ systems are subdivided into process components, mechanical components and control components etc. From these, individual items/ components are identified for monitoring; e.g. bearings, seals etc.

4. Fixing condition parameters: For identified items/ components, condition parameters are to be assigned which can be actually measured and monitored; e.g. vibration, temperature, pressure, flow, noise, strains, magnetic flux and electrical insulation etc.

5. Fixing Monitoring Techniques: For each condition parameters, relevant measuring and monitoring techniques are selected and suitable tools/ instruments/ implements are identified and obtained/ installed. Some of the techniques and tools/ instruments are discussed in chapter 6. All techniques can not be used and also not required for all machines/ components. After this, the mode of monitoring is ascertained, such as displacement mode, velocity mode and acceleration mode for vibration monitoring etc. Then comes fixing of monitoring points or sampling points.

6. Monitoring Schedule and frequency: Next comes the decision for frequency and schedule of monitoring (daily, weekly, monthly, continuous etc). As indicated in section 5.8.2, frequency assessment is a tricky judgment and has to be regularly updated. Some condition parameters can be monitored "on-line" i.e. when machine is running (e.g. vibration, temperature, strain etc) while some others can be monitored only "off-line" i.e. when machine is stopped (e.g. checking internal clearance and back-lash etc).

7. Trend Monitoring: The inspection records, thus obtained, are decoded, analyzed and compared with maximum allowable limit or earlier data, if available. Today, equipment manufacturers, often, supply vibration signature/ record and records of other condition parameters of new machines, along with their test certificates. However, if no records are available, successive inspection/ monitoring records of same machine are analyzed and a trend is established, which helps in finding out the extents of deviations and defects. In case of increasing trend, more frequent inspection are made. Failure statistic and MTBF/ MTTF, if available, can also be considered.

8. Repair schedule and execution: Based on monitoring, necessary repair actions are scheduled and executed for correcting the deteriorations to avoid severe degradations.

9. Follow-up: After repairs, the monitoring continues, as shown in Fig 5.6.

5.9.5 Objectives and Advantages of CBM

The advantages and disadvantages are generally similar in line with those mentioned for PdM in section 5.7.3 and 5.7.4. The main aims/ objectives are—to intervene before the failure occurs, to do maintenance only when needed, to reduce number of failures and number of shutdowns, to reduce cost of maintenance and production loss and to increase the operating life of the replacement items, etc.

The earlier thinking was that "If both are technically appropriate, fixed interval overhauls / replacements are usually both cheaper and more effective than CBM. The new thinking is that "If both are technically appropriate, CBM is nearly always both cheaper and more effective than fixed interval overhauls /replacements throughout the life of the asset".

5.9.6 PD.M vis-à-vis CBM

Apparently PdM and CBM may appear to be somewhat similar. Though they are actually quite different, there are some overlapping areas. CBM may include a good amount of PdM, in addition to other maintenance strategies. As an example, the following is the spectrum of PdM, which is also used in CBM—

Equipment Category	Equipment Type	Failure Mode	Failure Causes	Detection Methods
Rotating	Pumps, Motors,	Premature	Excessive	Vibration Analysis
Machinery	Compressors,	Bearing Loss	Force	Lube Oil Analysis
	Blowers	Lubrication Failure	Over, under or improper Lube.	Spectrography and Ferrography
Electrical Equipments	Motors, Cables, Starters, Transformers	Insulation Failure	Heat, Moisture	Time/ Resistance test, I.R. Scans, Oil
		Corona Discharge	Moisture Splice methods	Analysis Ultrasound
Heat Transfer Equipments	Heat Exchangers, Condensers	Fouling	Sediments, Material Build-ups	Heat Transfer Calculations
Contaminants and Transfer equipments	Tanks, reservoirs, Piping, Reactors	Corrosion	Chemical attack	Corrosion Meter, thickness checks
		Stress Cracks	Metal Fatigue	Acoustic Emission

The advantages and disadvantages of CBM is somewhat similar as for PdM (Sec.5.7.3).

5.10 PROACTIVE MAINTENANCE

Dissimilar to preventive maintenance that based on time intervals or predictive maintenance that based on condition monitoring, proactive maintenance concentrate on the monitoring and correction of root causes to equipment failures. The proactive maintenance strategy is also designed to extend the useful age of the equipment to reach the wear-out stage by adaptation a high mastery level of operating precision. It takes a micro view on machine damage — concentrating on the causes instead of the symptoms of wear etc. Proactive maintenance commissions corrective actions aimed at he sources of failure. It is designed to extend the life of mechanical machinery as opposed to (1) making repairs when often nothing is broken, (2) accommodating failure as routine and normal, and (3) preempting crisis failure maintenance - all of which are characteristics of the predictive/ preventive strategies/ disciplines.

Proactive maintenance relies on the use of electronic instrumentation and computerized programs to effectively pinpoint problems in rotating (also sliding) equipment. Because most problems are mechanical, **vibration analysis** has evolved as the leading diagnostic tool in proactive maintenance to determine problems and their sources. Other diagnostic systems have made their debut and each has, in its own way, contributed to enlarging the scope of predicting potential problems in rotating equipment and structures. It also uses many other tools/ techniques, few of which are given below–

- Infrared Thermography,
- Ultrasonic leak detection,
- Lube oil analysis, Contamination Analysis, Wear stability,
- Phase angle analysis for alignment and looseness checking and nodal and anti-nodal points on piping and structure etc,
- Laser shaft alignment, Precision balancing,

- Resonance correction,
- Fluid chemical stability, Fluid physical stability,
- Cavitation stability;
- Soft-foot conditions checking and correction, etc.

At this stage, techniques like RCA, RCFA, TPM (Total Productive Maintenance) and RCM (Reliability centered maintenance) etc are effectively used. It adopts a continuous improvement model in moving to a "improved precision" and/or "World Class" domain and involvement of people and integration of cross-functional groups becomes essential because all groups contribute to the creation of defects and they must contribute to their elimination. The knowledge base for doing Proactive Maintenance activities is then accessible to the entire workforce instead of only a few experts.

All of these programs have brought maintenance activities out of the dark and into a new light. One critical factor for this is that of the "hands-on" aspect of machinery maintenance. If there is to be a complete cohesion between those who analyze and those who correct the problem, it becomes imperative that all maintenance personnel receive training about all of the proactive maintenance programs in their plant. Some of the techniques, mentioned in chapter 11 are also used in proactive maintenance.

5.11 RISK BASED MAINTENANCE

This is not any specific maintenance strategy/ type but may be any or combination of above mentioned strategies, but devoted to the risk factor of the plant/ equipment. Maintenance plays a pivotal role in managing risks at many sites (mines etc) and it is important that right risk assessment tools/ techniques are applied to capture and evaluate the hazards at hand to allow a functional risk based approach. You have to examine how risk concept can be used in establishing preventive safety and management strategies, what types of tools are available and what gains can be made.

Risk is inherent in all human activities. We manage risk continuously and subconsciously make decision, based on previous experiences. In industries, risk management, often captured in a Safety Management System (SMS), is a complex issue with many facets and its success is based on our ability to balance frequent low consequence events as well as rare high consequence events, such as fires, explosions, flooding etc.

Fig.5.7 shows the a risk analysis method Normally qualitative method are not much effective in assessing system hazards. Quantitative methods are better for maintenance application where some data is available and decisions on system safety and criticality can be made.

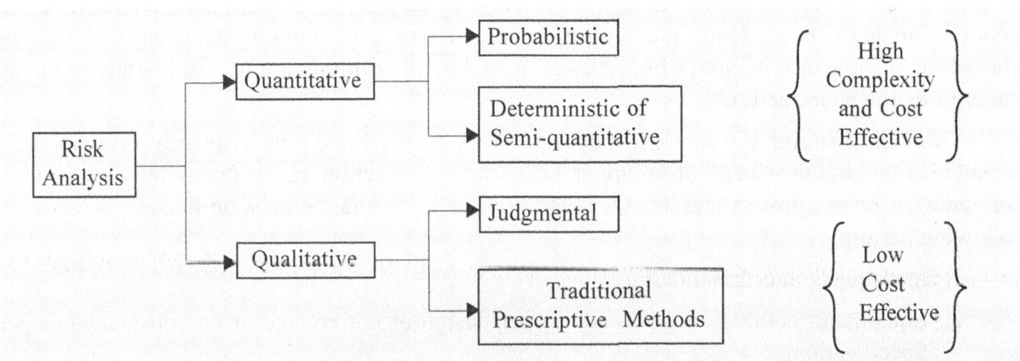

Fig. 5.7 Risk Analysis

From Risk Analysis, we can finalize "Risk Assessment" which also includes Hazard Identification, Risk Estimation, Risk Evaluation, Risk Tolerability and analysis of various options. Then after, we can decide on maintenance strategy and actions.

5.12 QUALITY ASSURED MAINTENANCE MANAGEMENT (QAMM)

QAMM is another set of terminology, which uses some of the earlier maintenance strategies, with quality aspects specially in view. The objectives of this strategy may be the followings—

1. Prediction of impending failures of critical plant components (like rotors, shafts, pulleys etc) in real time, resulting in enhanced safety, operation reliability, availability and maintainability,
2. Establishing an alarm level, based on the variations of physical parameters, like vibration, temperature, power consumption etc,
3. Reduction of the life cycle cost by optimizing plant operation and maintenance scheduling and proper implementation,
4. Facilitation of design revision to permissible level to incorporate new technologies for active control, including built-in test equipments and structural materials; Etc.

Identification of actual maintenance work and minimum necessary improvements, are to be properly assessed. Maintenance strategies, used in QAMM, are, normally, combination of preventive, predictive and proactive etc.

5.13 DESIGN-OUT MAINTENANCE

It is design oriented curative measure aimed at rectifying a design defect or defects originated from improper method of installation or poor material choice etc. Design-out maintenance requires a strong maintenance-design interface so that maintenance engineer works in close cooperation with design engineer. Whereas most maintenance systems/ strategies aim to minimize number of failures or effect of failures, the design-out maintenance aims to "eliminate the cause of maintenance". It is more suitable for equipments of high maintenance cost. The choice is to be made between the cost of redesign and cost of recurring maintenance. Recently, design-out maintenance has gained more importance for maintenance engineers because of its potential in reducing breakdowns, minimizing maintenance efforts and costs and, at times, totally eliminating maintenance activities for those items.

Though the corrective maintenance includes some amount of design-out activities, it differs from design-out maintenance. In corrective maintenance the design and redesign activities are carried-out by the maintenance engineer or his department, but in design-out maintenance, the maintenance engineer involves design engineer and design department and works in close cooperation with them. Thus, the design-out maintenance includes both, small redesign of corrective maintenance nature or big redesign jobs, but all with the aim of eliminating/ reducing cause of maintenance activities.

However, whenever any equipment does not work properly and fails frequently, efforts should be made to find out the causes of failure and take corrective action, rather than immediately jump for redesign or design-out maintenance. Design-out maintenance may be needed because of following reasons—

1. Inadequacy in original design,
2. Equipment/ component not made as per design or not erected or commissioned as per Specification,

3. Change in performance parameters and needs from original specification,
4. Non-availability of spares or components, because of obsolescence or otherwise,
5. Availability of much cheaper and better spares, consumables and components etc,
6. Change in statutory obligations, especially with respect to safety, pollution or environmental needs; etc.

Such design-out maintenance (also called as maintenance modification) can be done in one installment (such as total replacement or revamping) or n the form of small continuous improvements (generally preferred by Japanese—Kaizen). Again, the design-out maintenance differs with plant modification and up-gradation jobs to the extent that the aim of design-out maintenance is limited to eliminate the cause of maintenance,

5.14 REVIEW/ IMPROVEMENT OF MAINTENANCE STRATEGIES

Many large organization may have sophisticated preventive and predictive maintenance departments that are staffed with trained technicians. Other companies may have their PM programmes embedded within general maintenance workmen, with the necessary technical and specialized supports provided by contractor vendors or other external help. Again, some PM programmes are maintained by a CMMS software application and interfaced or integrated with predictive or performance based systems that support an extensive reliability programmes, while some PM programmes exists as a combination efforts and control methods or more manual in nature as simple card files and periodic rotations. However, we have to review occasionally if the ongoing maintenance programmes/ strategies are really what we think of and is it giving the result what we want. For such review, we have to dig into the effectiveness of our maintenance programme/ strategy. We may take the following steps for that—

1. Review the existing programmes/ strategies (quality, accuracy and compliance etc)
2. Analyze and device new programmes and planning,
3. Establish system support requirements and configuration,
4. Define reimplementation plans with rough timeline,
5. Redesign performance monitoring metrics,
6. Define human resource development plan if any needed,
7. Develop benchmark, Audit programmes etc;

Each of the above mentioned steps calls for substantial investigation and thought. Step (2) also calls for defining criteria for maintenance activities, check lists and cost model etc. Step (3) also includes development of system codes (how data can be selected and sorted), Maintenance triggers (what and how your system generate maintenance work orders), System Default code (auto populates data fields) and Work completion codes (supports the ability to analyze and trend) etc. Step (4) also includes the amount and type of equipments needed, resource availability and project/ programme's constraints. Step (5) calls for examining existing metrics, if any, researching of industry benchmarks and developing applicable metrics. Step (7) also calls for defining a changed management programme defining maintenance audit and accounting responsibilities.

Fig. 5.8 further shows how to arrive at appropriate maintenance system/ strategy from failure pattern (Sl. Nos. 1, 2, 3 shows the 1st, 2nd, 3rd preference). In this, RCM and Proactive Maintenance and other special maintenance systems are not shown as that would complicate the diagram.

Though the percentage may vary as per the type of industry and the reliability required, in one type of industry, the actual percentage of different maintenance work-orders are as follows—

1. Preventative Maintenance .... 50 %
2. Predictive maintenance and CBM.................30%
3. Corrective Maintenance................. 15%
4. Breakdown...<5%

Reliability Centered Maintenance (RCM) will be discussed in chapter 11 of this book.

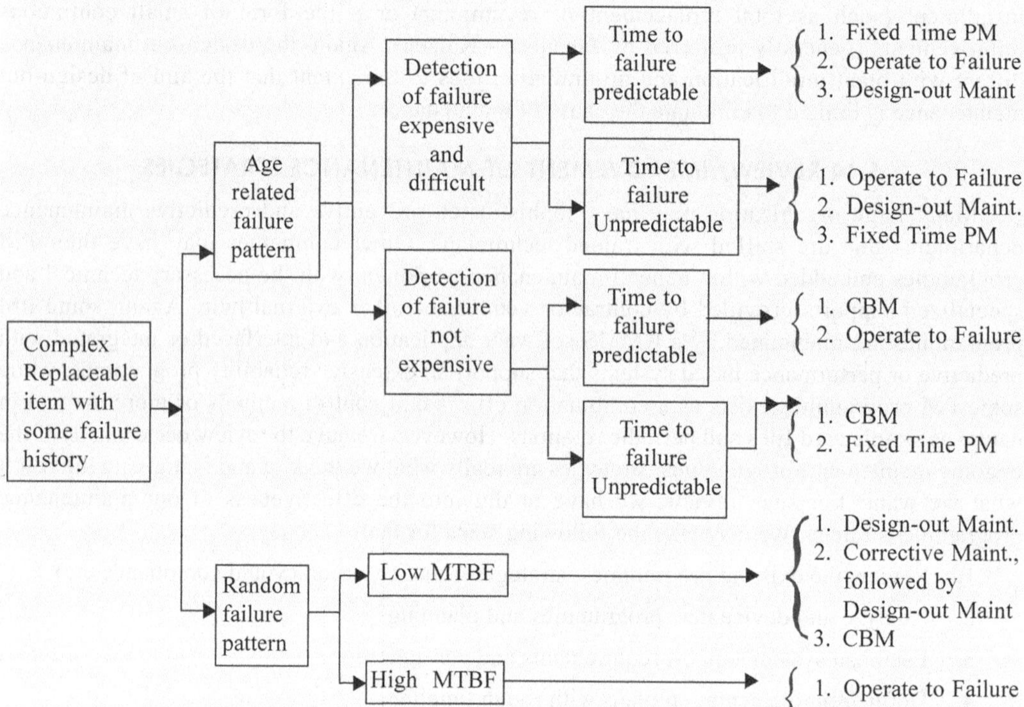

Fig. 5.8. Decision Diagram to select appropriate Maintenance System

5.14 COMPARISON OF MAINTENANCE STRATEGIES

Considering the Breakdown Maintenance, Corrective Maintenance and Opportunistic Maintenance as "Breakdown Maintenance" or "Reactive Maintenance" , including Routine Maintenance in Preventive Maintenance and including CBM or CBMS in Predictive Maintenance, we can divide maintenance strategies in four broad groups—Breakdown, Preventive, Predictive and Proactive maintenances. Rough comparison of these four strategies are given in Table 5. 1.

Table 5.1 Comparison of Maintenance Strategies

	Maintenance Strategy	Maintenance Approach	Significance and Techniques Needed	Human Body Parallel
1	Breakdown (Reactive)	Fix-it when breaks	Large maintenance budget	Heart attack or stroke
2	Preventive	Scheduled Maintenance	Periodic component replacement/ Repair	By-pass or transplant surgery
3	Predictive	Condition-based Monitoring	Monitoring condition by vibration, heat, alignment, wear	Detection of heart disease using EKG or ultrasonic

4 Proactive	Detection of Sources of Failures	debris etc Monitoring and correction of failing root causes	Cholesterol and blood pressure monitoring with diet control

The differences can also be explained in following way—

1. Scattered Acts/ actions—Breakdown and Reactive Stage,
2. Getting your acts together—Reactive to Planned Stage,
3. Getting Beyond the Boundaries—Planned to Predictive/ Proactive Stage; Maintenance must e recognized as an integral part of plant's production strategy, by which products are delivered to customers. Though improvement in maintenance organization is necessary, "going beyond the boundary" is also considered to get other departments to adjust their responsibilities related to plant's upkeep.
4. Fix the Process, not just the Problem—To Proactive and World Class; maintenance, Production and other concerned departments should share few responsibilities for improving plant's productivity and profitability etc.

5.16 EVOLUTION OF MAINTENANCE STRATEGIES

In past, a simple expectation from maintenance was keeping an equipment running or restoring it to the desired operating condition, but most management now see maintenance efficiency as a factor that can affect the all business effectiveness and risk-safety, environmental integrity, energy efficiency, product quality and customer service and not contained only to plant availability and cost. Thus the evolution of new and better maintenance strategies became necessary as the business became more competitive and complexity in industries increased. In general, the evolution of maintenance changes usually is categorized into 3 different generation, the period of 1930's-1940's which usually referred as the First Generation (early days of industrialization), between 1950's to 1970's often recognized as the second generation (period of increasing mechanization and complexity in industries), and from 1980's till recent times which is commonly accepted as the third generation (period of higher complexity, automation, competition and safety hazards etc). This has been shown in Table 5.2 (Evolution on Maintenance Strategies).

Under the third generation maintenance strategy/ principles, many organizations have shifted more focus on reliability than availability and have stated zero breakdowns/zero in-service failures as their maintenance goals. However, since no amount of maintenance can guarantee the total elimination of failures, it is no longer a really achievable objective and a more realistic approach is to avoid, reduce or eliminating the consequences of failures.

Table 5.2 Evolution of Maintenance Strategies

First Generation	Second Generation	Third Generation	Fourth Generation
• Fit when breaks	• Scheduled Maintenance/ Overhauls • Preventive Maintenance and Controls • Low-tec Computerization	• Predictive Maintenance • Condition Monitoring, CBM, CBMS etc, • Design for Maintenance and Reliability (Design-out Maintenance) • Hazard Studies • High-tech Computerization • FMEA and FMECA etc	Under Development (More use of IT, Computer modeling, Expert System and Artificial Intelligence to predict and prevent failures)

		• Expert Systems • Multi-skilling and Teamwork • Proactive to limited extent	
1930	1950 1970	1990	2000

Due to the rapid changes in the development of equipment and process, accelerated with the help of faster computers, it is only a matter of time when the maintenance scenario entered its fourth generation. The third generation maintenance has undergone a shift of focus in maintenance to highlight those areas where the inherent design of the assets yields probabilities of failures that are unacceptable, and provide some guidance and motivation for improving those assets. Hence, the basic principles of the fourth generation of maintenance, although are expected to be based on the previous three generations, will have some signified features, few of which may be as below–

1. Definite deliberation of risk, notably at higher levels of organizations, when dealing with equipment design and maintenance strategies,
2. Coherence between functional demand, equipment design and maintenance will be greater than the currently existing integration,
3. More use of developments in information technology to detect, predict, diagnose and prevent equipment failures; Increased usage of computer modeling in maintenance strategies, more use of artificial intelligent and expert systems, computer simulations and modeling etc to provide the predictive tools of the future; Etc.

5.17 TOTAL MAINTENANCE MANAGEMENT (TMM OR TM2)

TMM has been coined by some specialists/ consultants. It consists of multiple modules (using the strategies already mentioned earlier in this chapter) that, when put together, make up a robust management system. These modules bring together all aspects of managing a maintenance organization such as work order tracking; scheduling of preventive maintenance; inventory control, purchasing, and receiving; labor scheduling and costing; and project management and estimating. TMM keeps track of all of the labor and inventory costs for all of your work, equipment, and facilities so that you can run detailed cost analysis reports. This would not be discussed further as most of the principles of TMM have already been discussed in this book.

QUESTIONS

1. What do you understand by maintenance strategy/ system? Name few common maintenance strategies and explain the basis of their selection.
2. What is reactive maintenance, why is it called so and what does it include?
3. What is preventive maintenance and what are its components? Why is PM better than reactive maintenance?
4. Explain any example when PM is not preferred and machine is allowed to run till failure.
5. What is difference between fixed time maintenance and condition based maintenance? Give example of both.
6. What is predictive maintenance and what are its advantages and disadvantages?
7. What is proactive maintenance and why is it called so? Mention few techniques/ methods which are used in proactive maintenance.
8. What is condition based maintenance? Explain its objectives and methodologies.
9. What is design-out maintenance and when is it used? How is it different than design for maintainability improvement?
10. Explain briefly, the routine maintenance and opportunistic maintenance.

CHAPTER

6 CONDITION MONITORING (CM)

Condition Monitoring (CM) is determining the health and condition of equipments, machines and systems etc by observing, checking, measuring and monitoring certain parameters and signals etc. In broader sense, it is said as Equipment Health Monitoring (EHM). The concept of EHM is a simple one—Monitor the steady state characteristics of the **equipment** and learn those characteristics. If these conditions change in a negative way then generate an alarm, investigate the problem and make a correction before the fault becomes so serious that a plant is shut down, production is lost and costs spiral.

Condition monitoring (CM) is not a life-extending activity. Life-extending activities are things such as lubrication, alignment, balancing and operating procedures. CM only provides information on failures before there is a breakdown.

6.1 CONDITION SIGNALS AND MONITORING

While in use, most equipments/ machines emanates or indicates some signals; both "**Primary Signals**" and "**Secondary Signals**". Primary signals are generally those signals or parameters which are required to assess the performance of the equipments and which are designed to be emanated, such as oscillations in vibratory chutes/ screens etc. Monitoring of primary signals are termed as "Performance Monitoring" or "Performance Trend Monitoring". All other signals, which appear as loss output, like vibration, sound, thermal, chemical or physical changes etc, are termed as secondary signals. Secondary signals are, normally, not designed for. Monitoring primary signals alone does not help in efficient assessment of health and condition of equipments/ machines. As secondary signals are generally result or form of loss output, monitoring of these signals becomes inevitable and, often more necessary, for equipment health monitoring and technical diagnostics.

Every defect or malfunction in an equipment or system indicates significant frequencies or signals, which can be easily identified and measured or compared with suitable instruments/ devices. As a very rough estimate, about 70 % of damages/ defects can be characterized by such signals. Such signals can broadly be classified as "**Dynamic Signals**" or "**Tribological Signals**". Dynamic Signals constitute vibration, sound, leakages, heat etc and Tribological Signals include friction and wear characteristics, lubricant contamination and chemical changes etc. Most of these signals can be used for condition monitoring. Some defects can easily be detected by observing signals and comparing with general severity chart etc, but there are many malfunctions or defects which could be detected only by an indirect diagnosis through correlation techniques etc.

Condition Based Maintenance mainly depends on efficient condition monitoring. Pursuant to previous chapter, it is again emphasized that selection of condition monitoring parameters, monitoring points and sampling points, monitoring frequencies and techniques and analysis of monitored parameters/ signals are to be done timely and efficiently to enable maintenance personnel take timely actions.

Condition monitoring can be done "off-line" or "on-line". In offline system, monitoring equipments are used in parallel to the equipment/ machine to be monitored. Various monitoring points are provided for attaching such equipments as and when needed. In online system, monitoring equipments/ systems are built-in or installed in series (such as filter chocking signals) with the running equipments. Offline monitoring systems can be periodic or continuous. In periodic systems, monitoring instruments/ gadgets can be connected or made "on" at the time of monitoring (taking readings etc) and then removed or switched "off". Offline continuous instruments, generally, monitors (records readings etc) as long as equipments operate. However, few such monitoring systems (reservoir level/ temperature monitoring systems) operate/ monitor even when the main equipment is switched off. Online monitoring systems are generally continuous (with provision to bypass). All these are shown in the form of a block diagram below as Fig.6.1 Most CM instruments are connected with microprocessors or computers etc.

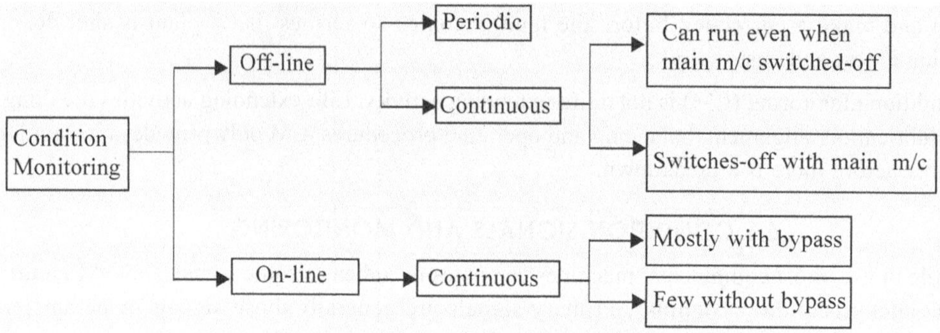

Fig. 6.1 Block Diagrams of Monitoring Systems

6.2 CONDITION MONITORING TECHNIQUES

Based on emanated secondary signals (dynamic and/ or tribological) various condition monitoring techniques have been developed and are in use. A number of strategies and tactics exist for collecting condition data and interpreting it for taking corrective action. Some involve simply analyzing feedback signals inside a control loop and any fault signals that might be generated by the controllers. Others take the more proactive approach of adding sensors to measure and track changes in a descriptive parameter. Often the traditional process parameters like temperature, pressure and flow are not enough, because the data is more static. Rather than being a string of repetitive events, these data are averages by nature and, so, contain no historical details. Consequently, they are useful only for making go-no go decisions. For example, an increasing temperature on a bearing does not point to the cause; it only suggests that corrective action is necessary. Although vibration analysis is the workhorse of monitoring the condition of rotating mechanisms, other techniques (lubricant monitoring etc) can either supplement it or sometimes even replace it entirely. Today, in industries, we use a combination of different techniques, often one complementing and supplementing others.

The evolution of information technology has done much for automating condition monitoring and making it practical for a wider range of manufacturers/ users. Advances in communications have made automating condition monitoring more practical. Today many consultants/ vendors are offering composite conditioning monitoring programmes in the name of **Integrated Condition Monitoring** (ICM) or **Machinery Condition Monitoring** (MCM) etc., all claiming to reduce unforeseen machinery breakdown, and therefore increase equipment's availability for increased

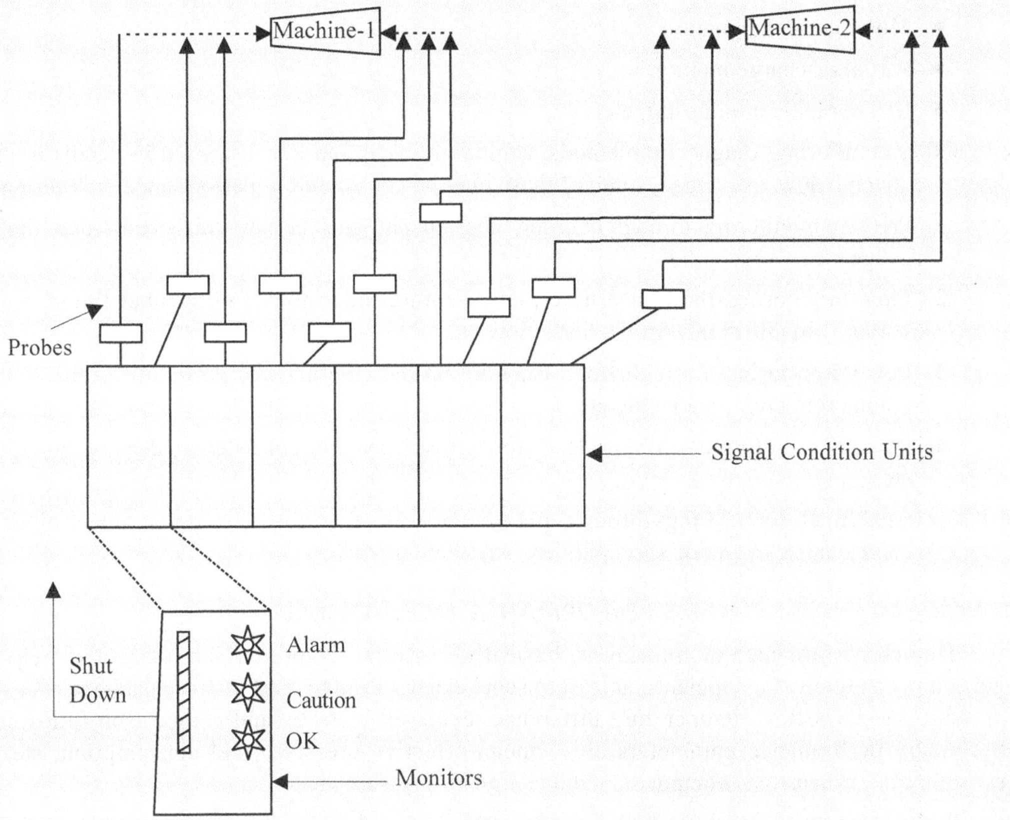

Fig. 6.2 Rough Schematic Diagram of a continuous off-line system

revenue earning; few of them also claiming to develop tailor-made monitoring instruments/ gadgets for their users. Few recently developed condition monitoring tools use multivariate statistical process control techniques, such as Principal Component Analysis (PCA) and Partial Least Squares (PLS), instead of more traditional methods like Statistical Process Control etc., to accurately detect and isolate several fault conditions.

However, we would consider traditional techniques, most of which can broadly be classified in following groups/ types—

- Performance Monitoring-
- Visual, optical, tactile, and aural Monitoring
- Temperature Monitoring
- Vibration Monitoring
- Lubricant monitoring
- Leakage monitoring
- Crack monitoring
- Thickness monitoring
- Corrosion monitoring
- Noise/ sound monitoring, Acoustic Emission; Ultrasound/ ultrasonic monitoring

- Smell/ odour monitoring;
- Motor current signature analysis;
- Partial discharge monitoring,
- Radiographic monitoring; etc

Other monitoring/ diagnostic methods, commonly used, but not discussed in details in this book are given below. Of-course, some of these may be considered in performance monitoring.

1. Pressure monitoring–through bourdon tubes, diaphragms, bellows, piezoresistance, strain-gauge etc,
2. Flow monitoring–through orifice plate, venture tube, nozzle pitot tube, flume, weir, vortex, variable area, coriolis mass flow etc,
3. Level monitoring–through float, capacity probe, conductance probe, ultrasonic echo, optical refraction, nucleonic etc
4. Physical monitoring– through load, force, displacement, consistency, density, humidity etc,
5. Analytical monitoring–through pH meter, selective ion electrodes, electrolytical conductance, magnetic susceptibility, chromatography etc.

6.3 PERFORMANCE MONITORING

Though performance monitoring is, basically, essential for production, these also help in indicating condition of equipments, at least to some extent. Primary performance parameters are— pressure vis-à-vis flow, temperature difference, cycle efficiencies and comparison ratios etc. Secondary performance parameters are— mean effective pressure, fuel consumption, steam consumption, exhaust gas composition and charge air pressure etc.

Performance monitoring strategies, most commonly used, are—

- Variation in production rate,
- Variation in product quality,
- nput/ output ration, also amount of waste generation,
- Power in and power out,
- Efficiency; etc.

In addition to measuring the values in terms of quality and quantity, it also includes simultaneously comparing two out-put parameters with a set of standard operating conditions.

6.4 VISUAL, TACTICAL AND AURAL MONITORING

These are the most commonly used systems. Such monitoring can be done using—

(A) Direct Methods

- Human Eyes, ears and other senses,
- Optical Probes
- Microscope, Borescopes, Stroboscopes, fiberscope, dye-penetrants, stress wave sensors, etc,

(B) Indirect Methods

- Optical Probes with Television etc.
- Photographs and videos,

- Radiographs,
- X rays, thermographs etc.

Followings are some of the characteristics of such visual monitoring—

1. They are extremely sensitive and may vary from person to person,
2. Those are, often, erratic,
3. Require long experience to judge,
4. Inexpensive (unless the equipment is remote),
5. Difficult to pass-on skill,
6. Distractions are a factor,
7. Basically a breakdown strategy.

6.4.1 Automatic Visual Inspection (AVI)

Automatic inspection techniques, normally in quality control and to some extent in condition monitoring, is done to justify the followings—

1. Releasing workers from tedious or repetitive tasks,
2. Enabling precise and continuous inspection in inaccessible environments or under hazardous conditions, such as measuring the thickness of hot steel,
3. Improving inspection system reliability, and
4. Iimproving inspection system predictability.

Automated optical inspection (AOI) may be another name or similar technique as AVI, as both use set of cameras and rapid image analysis techniques.

6.5 TEMPERATURE MONITORING

Most equipments operate on ambient temperature. Often defects occurring because of wear, friction, slippage, impact or other tribological reasons in part of energy getting wasted and converted into heat energy, which heats the connected components The difference between the ambient temperature and that of equipment operating under load is the temperature rise, and this temperature rise is monitored to detect any fault. The techniques used in such monitoring may be one or more of the followings—

1. Temperature crayons and tapes,
2. Thermometers and optical pyrometers,
3. Softening cones/ wax/ paints,
4. Thermocouples and Fusible Plugs,
5. Infrared meters,
6. Thermography (Infrared Radiation Scanner),
7. Bimetallic Strips,
8. Thermistors,
9. Vapour Pressure in Bulbs,
10. Mercury in Glass; etc.

Temperature monitoring by feel of hand or by simple measuring items/ instruments, like thermometers, temperature crayons and tapes etc is an age-old practice of finding out defects or defective components. Optical pyrometers and thermocouples are used for higher temperatures and places which can not be approached easily. A thermocouple consists of two dissimilar metals, joined together at one end, which produce a small voltage at a given temperature. When the thermocouple is heated, the voltage produced is approximately linear with temperature. Thermocouples can be handled manually for each measurements (like temtips for bath temperature during steel making) or can be permanently installed for continuous measurement and recording (e.g. reheating and annealing furnaces in steel plants etc). Fusible plugs, normally used in boilers etc, as safety measures, can also indicate defects or malfunctioning. Infrared Radiation (IR) meters or Infrared Radiation Scanner (Thermography) are non-contact temperature detectors, which work on the basis of intensity of the infrared energy emitted by the objects. Other techniques, mentioned above, are also used in industries as per need. The resistance method is useful for determining the temperature rise of electrical motors that do not have embedded detectors—e.g., thermocouples or resistance temperature detectors (RTDs).

Most temperature techniques are relatively inexpensive, require low skill level to use and can also be online and continuous. Various types of conventional, digital, contact or non-contact and intermittent or continuous temperature measuring/ recording instruments/ devices are available in market. Though such monitoring has wide application, they are regularly used for lubricant condition, electrical insulation, operating condition etc. Thermography is getting increased acceptance in maintenance and, as such, it would be discussed in more details later in this chapter. However, temperature monitoring is not much used for incipient damages.

6.5.1 Thermal Measurement Summary

The following table outlines the more common types of measurement with comments on applications and a brief technical description of the method.

Thermal Measurement Summary Table

Method	Description	Applications
Point/ Zone temperature	Usually a thermocouple or RTD. Often 'imbedded', can be provided as an encapsulated sensor for permanent fit.	Can be used on all accessible surfaces. Walk around versions take time to settle and are less robust than the pyrometer.
Area Pyrometer	Measures the emitted IR radiation from a surface. Often with a Laser sight or area indicator.	Very good for walk around temperature checks on machines and panels.
Temperature Paint / Stickers	Chemical indicators calibrated to change colour at a specific temperature.	Great for inspection rounds.
Thermography	Hand held still or video camera sensitive to emitted IR.	The ultimate, high-resolution thermal picture. Camera extremely costly; Camera use and interpretation does require 'good' training.

6.5.2 Thermography

Infrared, infrared thermography, Infrared Radiation Scanner, thermal imaging, infrared radiometry, infrared imaging, and IR condition monitoring, are all terms used for this fascinating, growing field. No matter what we call it, these techniques have achieved a status, probably, no other diagnostic instrument or inspection tool has attained - a diversity of applications from medical thermography to electrical inspections, thermal environmental studies to building roof thermography, IR refractory evaluations to inspecting mechanical systems, infrared condition monitoring to breast thermography, non-destructive testing to research and development, from security and surveillance to building diagnostics and aerial infrared thermography (AIT) is able to find latent leaks of hot water and other imperfections (malfunctions) etc.

For maintenance work in industries, thermography mostly means Infrared Radiation Scanner. Infrared Thermography (IR) operates by measuring the infrared energy of any object's surface and producing a detailed visual image showing its temperature profile. Since IR operates without physical contact, it offers a safe, on-line, and cost-effective means to evaluate the electrical and other operation of any property. While many borderline problem areas typically remain undetected during an off-line physical inspection, they can be easily pinpointed using IR. These work on the principle of thermal energy emitted/ radiated from objects. All objects, having a temperature above absolute zero, emit energy of radiation. Electromagnetic radiations, with wavelength from 0.75 to 1000 microns, are called infrared radiations. This emitted energy is used for infrared measurement, as it tells the temperature and condition of the object. As a perfectly emitting surface (Black Body) is very rare, most surfaces reflect/ emit varying amount of infrared energy and are called "Grey Bodies". Emissivity of such material is the ratio of actual emitted energy to theoretical black body energy at a specified wavelength. Emissivity of all materials (grey bodies) are less than one. In thermography or thermometry, an instrument using collecting optics, the radiation detector and some form of indicator, collects and detects signals (combination of reflected energy and emitted energy) from the object, amplifies, scans and indicates in different forms, including imaging. Followings are few factors affecting thermography and, as such, we should take care about instrument's location and adjustment—

- Possible obstruction between sensor and target (which is to be monitored) in the form of any object, particles, steam and gas molecules, which may absorb, reflect or scatter the radiation before it reaches the sensor,
- Background radiations from objects, other than target,
- Low emissivity; etc.

Infrared Thermography is ideally suited for detecting faults in electrical panels, breakers, switchgear, splices, insulators, starters, contactors, wiring, distribution systems, disconnects, and transformers, etc. With a lesser degree of reliability, IR can locate water leakage in roofs, steam trap malfunctions, ineffective HVAC air distribution patterns, motor and bearing defects, building energy loss, pipe blockages and underground pipe leakage - providing a visual representation of the heat generated or lost in almost any physical application. For most infrared inspections, a client's primary interest is to identify the hot spot locations presenting the most serious and immediate threat. However, other electrical and mechanical problems may exist which may also present a potential threat to reliable building and plant operations. A typical IR inspection may identify panel doors damaged, mismatched or over rated fuses in place, indicator lights out, inoperative safety mechanisms, as well as various other forms of potential electrical or mechanical threats and liabilities. Unlabeled panels, while not in themselves an operating problem, would likely slow an emergency response to one.

Infrared (IR) **thermal imaging** is a method or technique which addresses these issues by providing comprehensive two-dimensional maps of thousands of temperatures in a matter of seconds. This is accomplished without the need to make contact with the components. Thermal imaging enhances a company's ability to predict equipment failure and plan corrective action before a costly shutdown, equipment damage, or personal injury occurs. Connected techniques like thermology, thermal wave imaging, thermovision and thermal infrared night vision use thermal imagers or Infrared imaging cameras. Figure 6.3 provides an example of the detail available with thermal imaging of an electrical device with temperature around 108^0 C. Fig. 6.4 shows what may happen if the initial demonstrated thermal considerations are ignored.

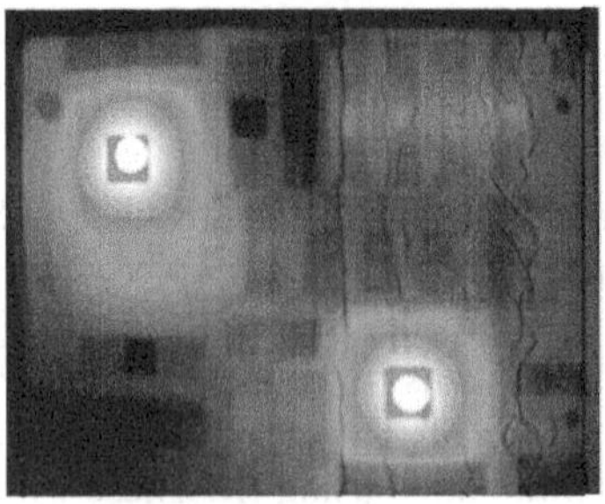

Fig. 6.3 Thermal Image-1

Fig. 6.4. Thermal Image-2

Table 6.1 gives few typical uses and indicates the condition to be detected for common uses in maintenance. Two colour Ratio Pyrometer, Infrared Cameras and many other sensors are commonly used. American Infrared (USA) (Info@AmericanInfrared.com), Avio (Nippon Avionics Co.,Ltd.) Japan (product@avio.co.jp), CEDIP Infrared Systems, France (http://www.cedip-infrared.com/contact.htm), XenIcs, NV, Belgium (http://www.xenics.com/), Infra Tec, Germany (web@infratec.de), Mikron Infrared Inc, Thermal Imaging Division, MI, USA (jon@mikroninfrared.com) and IRISYS Ltd, UK (sales@irisys.co.uk) etc are few of the suppliers for such equipments.

Table 6.1 Few Typical uses of thermography

Application	**Condition Detected**
Electrical Motors, generators etc	Overheating of windings and bearings, blockages in cooling passages, friction, damping, material deformations, brush contact problems, rotors
Electrical panels, breakers, switchgear, splices, insulators, starters, contactors, wiring, distribution systems, disconnects, and transformers etc.	Heating, wear and melting of contact points etc.
Drives/Conveyors, Pillow Blocks, Couplings, Gears, Power Transmission Belts, Pulleys, Shafts.	Overheated bearings or rollers, misalignment of shaft, pulley or coupling, lubrication failure
Pumps/Compressors/Fans/Blowers	uneven pressure. Overheated bearings, high compressor discharges temperature, high oil temperature, and broken or defective valve
Internal Combustion Engines	Valve or injector malfunction, blocked radiatortubes and oil coolers. Thermal distribution, high radiator inlet or outlet temperature.

Heavy Duty Equipment - Tires, Bearings, Brakes, Hydraulics, Kilns, Ball Mills, Paper Machines	Overheating brakes, tires, bearings, pulleys, gears, gear or pulley misalignment, and blockages in hydraulics.
Mechanical Drive Turbines and Small Turbine-Generator Units, Gas Turbine, Exhaust Ducts	High lube oil temperature, high bearing temperatures, faulty stop/control valve operation, uneven metal temperature, leaking shaft seals, gas turbine firing conditions, including deterioration in firing chambers, cross firing tubes.
Ovens, Furnaces, Kilns, Pipes	Location and severity of damaged insulation, location of steam leaks in buried steam lines.
Valves: Shutoff Valves, Relief Valves, Steam Traps	Leakage, blockage.

Advantages of Thermography- Followings are few of the advantages of thermography—

1. Quick problem detection without interrupting service,
2. Prevention of premature failure and extension of equipment life,
3. Identification of potentially dangerous or hazardous equipment,
4. Wide temperature range, from (-) 17^0 C to 900^0 C; extends upto 1600^0C, with filters,
5. Can monitor target in motion and also fragile target,,
6. Temperature profile can be recorded and displayed easily,
7. Can monitor targets electrically charged (high voltage equipments),
8. Can also monitor small and remote items; etc.

6.6 LEAKAGE MONITORING

Leakage is unintended discharge or oozing of fluids and gasses (both, external and/ or internal) from mechanical components, occurring because of increase in clearance or wear-out or erosion of parts like valve seats, poppets and spools etc or damage of seals and packings or because of incorrect selection/ fitting of seals and components or because of other reasons. Whatever cares may be taken, some leakages do take place some time, which is an indication of generation of some defects/ deteriorations in some components. Occasionally thinning-out of oils and lubricants (decrease in viscosity or dilution by less viscous fluids) induces or increases leakage, which again is an indication of some defects or deteriorations.

Leakages are both, external or internal. External leakages are relatively easy to detect as some fluid or gasses flow out of leaking components. Internal leakages are little difficult to detect as the leaked fluid/ gas goes back to tank/ reservoir directly or through existing pipelines and is normally not seen flowing outside. Their detection is done through correlation with malfunctioning or reduced output/ performance of components/ equipments or through contamination of the original fluid etc. In a large system/ plant or with fluids of evaporating or little invisible type, even the external leakages have to be roughly detected through correlation methods, before localizing the leakage correctly. Monitoring of internal leakages are done through regular checking (and analysis as per need) of the level and condition of the fluids in reservoirs and sumps or by installing leak test sensors. Monitoring of external fluid leakages are relatively more easy. In some equipments, leaked oil collection systems and their discharge passage is provided. In few big equipment systems, where regular physical inspection is not easily possible, or in systems using dangerous/ hazardous fluids (acids etc), suitable sensors are provided at strategic locations which react with the leaked fluids and give immediate alarm and/ or visual display/ indication of the leakage for taking immediate corrective actions.

For detecting fluid leakages of underground pipelines, one crude method is that the area around the leakage gets wet and the fluid starts coming/ oozing out from any soft location nearby. For very long lines, few sensors can be provided at some distances for timely indication of leakages. In many cases, pressure drops also indicate leakages. Few usual leakage detection/ monitoring methods are—

- Interstitial monitoring- A detector installed between the walls should activate an alarm in the event of liquid penetration due to leakage from one of the walls,
- Level Monitoring- Level gauges and alarms in tanks, sumps etc,
- Vapor Monitoring- Vapor sensors installed around underground tanks etc,
- Liquid Monitoring- such as cable sensors, which set off an alarm due to changes in conductivity or resistance, when contacted by leaked fluids.

Monitoring of air or gaseous leakages (mostly external) is not that easy, as that can not be easily seen. Following methods are generally adopted for their detection—

1. Sound; hissing or whistling sound coming out from leaked point, specially in high pressure system,
2. Smell; for coke-oven gas or other such gasses,
3. Using small flame or thin paper or fibre strips, which gets sucked in case of leakages in vacuum or low pressure systems,
4. Bubble coming out if the leaking components are located inside water or other fluids or if such fluids are poured over the leaking components,
5. Sensors installed at strategic points which react with the leaked gasses and give audio and/ or visual alarms/ indications; etc.

For further monitoring of these leakages, suitable devices can be installed (permanently or temporarily) to determine the rate and trend of leakages. As one example, for monitoring the leak rate in vacuum degassing systems in steel plants, suitable orifice plates are installed, known amount of air is injected in evacuated condition and the rate or decrease in vacuum is determined and compared with the rate without air injection (normal running condition).

Monitoring of gas leakages in underground pipelines also requires making of small holes (by inserting small pipes) over the pipelines and installing suitable probes/ sensors which would react with the leaked gasses to give signals. Such probes can be used for underground fluid lines also. Leak-gun 110 of M/s AI Industrial (M/s Nikat Associates), Chester, England, is one example which is portable and suitable for underground pipelines and pressurized cables. This uses miniature ionizing detector, which is mounted in-front of a hand unit. In case of underground pressurized cables lines, SF-6 is used as tracer gas and also serves as high speed insulation. Bar holes (seepage holes) are made in the ground above the cable line, by using one inch diameter steel rod. Such seepage holes are located at the same distance apart as the depth of the cable. In case of leakage, SF-6 gas is diffused in the soil The detector of the leakgun-110 is very sensitive to SF-6. By inserting the gun in these seepage holes, the presence and concentration of SF-6 gas can be found on the LCD (liquid crystal display) of the gun and also audible alarm is sounded. The hole showing the highest concentration reading marks the point of excavation for rectification.

Many leak monitoring instruments/ devices are available in the market for different application. The NAXYS (Naxys AS, Hegrenesveen, 5042 Bergen, Norway) Sub-sea Leakage Monitor is an ultra sensitive instrument that detects leakages of both liquids and gases to the sub-sea environment. Its operating principle allows both detection and localization of leakages at a range of up to 100 meters.

For a very large fluid distribution network, the technique of leakage monitoring requires the installation of flow-meters at strategic points throughout the distribution system, each meter recording the flows into a discrete area with a defined and permanent boundary. Such an area is called a Meter Area. The total network is divided into manageable Meter Areas. The difference in recorded flow, compared to scheduled flow, indicates the areas requiring attention for leakage control.

Leak detection and prevention is not only a desirable practice but in many cases it is a requirement by law. Few rules regulate refrigerant and other hazardous emissions resulting from the normal operation of equipment as well as those practices associated with the maintenance, servicing, repair, or disposal of the equipment.

Sec. 6.8 (Lubricant Monitoring) covers leakage monitoring also to some extent.

6.7 VIBRATION MONITORING

Vibration is essentially the heartbeat of all mechanical equipment. Capturing this vibration in a number of different forms allows analysts to diagnose equipment ailments, such as worn bearings or imbalance etc, by measuring the amplitude and frequency of the vibration. The purpose of vibration monitoring is to establish the running condition of machinery in a fashion which is objective and scientific. Vibration monitoring, normally, means **vibration measurement/ collection and analysis** programme and it monitors the response of the system or equipment/ machine to internal and external forces being applied. The response is measured by a general purpose transducer at the pump and motor bearings and other rotating/ oscillating components and is passed to an analyzer for interpretation. The results are downloaded to the asset management software/ technique for analysis and recommendations. It aims for early identification and correction of potential machinery problems.

Vibration monitoring is a vast subject. Newer software and hardware are being developed which are faster, easier to use, more effective and less expensive. Though out-sourcing to specialist is talk of today, big industries may like to keep the programme in the hand of internal personnel, whereas the smaller industries may like to out-source. The aim is to collect data automatically and transmit it to specialists very easily. It involves understanding of the followings—

- Vibration fundamentals and Vibration Signatures,
- Vibration Collection (Data Acquisition) and Analysis
- Vibration Transducers
- Causes of machine vibration; machinery vibration trouble shooting,
- Vibration signature measurement and display,
- Vibration signal frequency analysis (FFT- Fast Fourier Transform)
- Machine vibration standards and acceptable limits (Condition monitoring),
- Fault diagnosis based on forcing functions,
- Various online/ offline and automatic/ semi-automation monitoring techniques and instruments; Etc.

Few of the above will be discussed herewith.

6.7.1 Vibration Fundamental and Signature

Vibration is the most prominent secondary signal in most machines and is roughly defined as motion of rotating/ oscillating machine, back and forth from its position of rest or neutral and is considered in displacement mode and frequency mode and also in acceleration mode. Details about vibration fundamental can be found in author's earlier edition of this book or in many other books.

Machine vibration is very complex, consisting of many frequencies etc due to vibrations of a number of components. The overall level of the vibration is only an indication of the operating condition of the machine. It is difficult to locate any specific fault in this as the amplitude is not occurring at just one frequency. For fault location/ identification, it would be essential to first determine the individual amplitudes and frequencies. A plot of amplitudes and frequencies of all the individual vibrating components of the machine is known as vibration signature. These signatures are the characteristics of a machine. A change in such signature is an indication of impending failure and its location. Such signatures are, occasionally, called vibration spectrum.

6.7.2 Vibration Collection and Analysis

Correct and representative vibration collection and recording (Data Acquisition) is of prime importance. The recording of vibration data can be performed with equipment as basic as a metallic rod and a pencil and paper, advances in technology enable computer assisted data collection which provides much more detail in a fraction of the time. There are many vendors who can sell instrumentation, or you may choose to out-source. Such recording can continuous or periodic. Portable data collection instruments have long been used to collect vibration, temperature, electric current signatures, and other condition monitoring data for assessing the mechanical and electrical health of plant operating assets. While a large variety of commercially available data collectors exist, intrinsically safe designs have meant users either accept a larger, heavier device, or sacrifice the ability to collect waveform data – generally considered "must-have" for meaningful predictive maintenance. DI-225D Vibration Data Collector, DI-1100 Single Channel Vibration Data Collector (of M/s Diagnostic Solutions Ltd, Unit 1a, Rossett Business Village, Rossett, Nr Chester LL12 0AY, UK) are examples. However, most of the recent instruments do measuring/ recording and analysis etc.

Vibration analysis is a very effective method for detecting machine faults and diagnosing the nature and severity of the fault. However, there are hundreds of special terms, techniques and concepts to understand. Using a vibration analyzer, and some skill, personnel can pinpoint the causes of rough machinery condition. Such problems as rotor imbalance and misalignment make up a great proportion of mechanical deficiencies and can be identified and rectified. Other problems such as bearing wear are not only detected, but also qualified as to the severity of wear. In many instances, a historical case study can forecast the remaining life of bearings so that scheduled repairs can be prioritized. On theoretical aspects, common vibration analysis methods are—

- Spectral Analysis, including RPM Spectral Maps (Water falls) or Cascade Diagrams and Campbell Diagrams etc,
- Statistical Analysis and Kurtosis Method,
- Envelope Analysis,
- Spike Energy Method; etc.

Theoretical aspects or these methods can be seen in author's previous edition of this book or in some other books. Following analysis techniques are used more commonly in practice—

1. Natural Frequency and critical damping,
2. Harmonic forces and the concept of resonance,
3. Transient and impulse forces,

4. Vibration Isolation and vibration absorption,
5. Motor Current Analysis,
6. Real Time Analysis for vibration; Etc.

We would not go into the theory of these techniques, as these techniques are used and give desired results in the various vibration transducers and other instruments in simple understandable forms. However, the vibration analysis is broadly carried out in following three levels—

- Overall vibration level measurement to detect that the problem exists,
- Spectral or frequency analysis to locate where the problem is in the machine,
- Special techniques, at more detailed level, which can indicate what the problem is.

6.7.3 Vibration Measuring Systems

These fall into the following categories:

(A) Single Value Methods

1. Hand Held meters
2. SPM Units
3. Acoustic Emission Units
4. Vibration Pens
5. 4-20mA sensors

(B) Time Frequency Method

1. Hand Held Data-collectors
2. On line systems

(A) Single Value Measurements– The variety of single value measurements is endless, from the ISO filtered levels mentioned in ISO 10816 [1] to the 'magic numbers' offered by some of the technology vendors; promising to solve all of your plant condition problems within one simple solution. ISO 10816-3 is summarized below–

Single Value Method Summary		
Method	**Description**	**Applications**
ISO Filtered Velocity	2Hz – 1kHz filtered Velocity	Works as a general condition indicator.
SPM	Carpet and Peak related to the demodulation of a sensor resonance around 30kHz.	One of the better single value bearing indicator methods. Some problems on larger bearings and gear units.
Acoustic Emission	Distress and dB, demodulates a 100kHz carrier which is sensitive to stress waves.	Better general indicator than velocity, without the ISO comfort zone.
Vibration Meters / pens	Combine velocity, bearing and acceleration techniques (sometimes include thermal)	Look for ISO Velocity, envelope and high frequency acceleration for best performance.
4-20mA sensors	Filtered data converted to DCS/PLC compatible signal.	ISO velocity version available,envelope version still awaited. Can be used to 'home in' on specific problems by special order.

(B) Time Frequency Measurements— This type of measurement involves the detection and display of specific components of a time history sensor output. The use of specific frequency components lends itself to the detection of faults down to a single mechanical component (e.g. bearing, gear and impeller etc).

Various methods, techniques and signal conditioning systems, to detect specific components condition, may measure the followings—

1. Displacement (Low Frequency)
2. Velocity (Mid Frequency)
3. Acceleration (High Frequency)

There are two main types of envelope measurement, namely:—

(i) Band Pass Filtered Enveloping–Allows the time history associated with the excitation of a specific frequency band pass through to the demodulator circuit. This method avoids the possibility of phase cancellation and is better at isolating the transient activity

(ii) High Pass Filtered Enveloping– Sets a High Pass filter and allows all measured time data above the HP filter pass to the demodulator, this method is regarded as less effective by vibration practitioners.

6.7.4 Vibration Transducers

Transducer, basically, is a device that accepts an input of energy in one form and produces an output of energy in some other form, with a known, fixed relationship between the input and output. The output may be either in the same form as input or in other form (more in electrical signals). Vibration transducers convert vibration input of the machine to electrical signal outputs in various forms giving required information. Accelerometer, Piezoresistive accelerometer, Velocity transducers, FFT Analyzers, Low-resonance geophones, PZT (lead zirconate titanate) actuators, Vibration shaker, Vibration holographs (The measurement of laser holography to measure vibration) etc are some of the types of vibration transducers/ sensors. Few of these will be discussed very briefly herewith. Different types of transducers respond to different parameters of the vibration source, as shown below.

- Proximity probes are sensitive to displacement,
- Velocity probes are sensitive to velocity,
- Acceleration probes are sensitive to acceleration.

Accelerometer– A transducer whose output (electrical/mechanical) is directly proportional to acceleration forces. The output is usually produced by force applied to a piezoelectric crystal which generates a current proportional to the applied force. This current is then amplified and displayed as a time waveform or processed by a Fourier transform to produce a frequency display. Single integration of the acceleration signal will produce a velocity display and double integration of the acceleration signal will produce a displacement display. Accelerometers, generally, consist of a piezoelectric crystal and mass normally enclosed in a protective metal case. Few accelerometers feature variable capacitance or work as Combination Capacitive/Piezoelectric Sensors. These can be attached to the vibrating surfaces by bolts (for permanent applications), adhesive (adhesive reduces the operational frequency response range) and magnet (for temporary measurements with a portable data collector or analyzer). ST703-C (capacitance type) and ST703-P (piezoelectric type) sensors of M/s Sequence Technologies (www.sequenceti.com); 1100 series,5300, 6000 series etc of M/s Vibra-Metrics (sales@vibrametrics.com) and different models from M/s SENSONICS Ltd, UK (sales@sensonics.co.uk) and M/s.Pruftechnik, www.pruftechnik.com) are few examples. Fig. 6.7 shows one from M/s Sensonics.

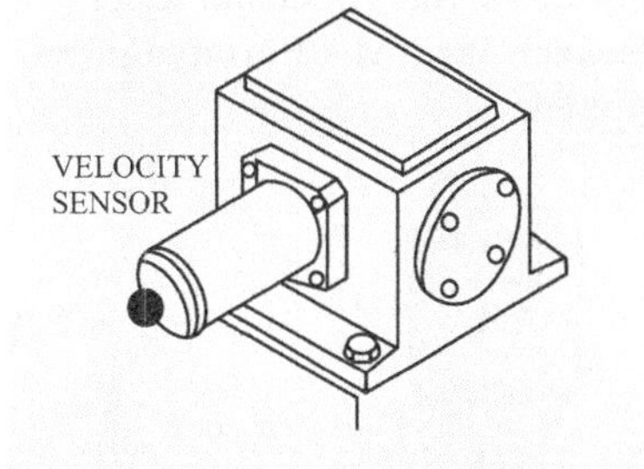

Fig. 6.5. Velocity Transducer

Velocity Transducer- It is an electrical/mechanical transducer whose output is directly proportional to the velocity of the measured unit. The velocity pickup is a very popular transducer or sensor for monitoring the vibration of rotating machinery. This type of vibration transducer installs easily on machines, and generally costs less than other sensors. For these two reasons, this type of transducer is ideal for general purpose machine applications. Velocity pickups have been used as vibration transducers on rotating machines for a very long time, and they are still utilized for a variety of applications today. Velocity pickups are available in many different physical configurations and output sensitivities. Generally, these consist of a magnet suspended on a coil, surrounded by a conductive coil. Movement of the transducer induces movement in the suspended magnet. This movement inside the conductive coil generates an electrical current proportional to the velocity of the movement. A time waveform or a Fourier transform of the current will result in a velocity measurement. The signal can also be integrated to produce a displacement measurement. Occasionally these are also called Velocity Transmitter and Velocity monitor and are compatible with piezo-voltage output accelerometer.

PZT (lead zirconate titanate) Actuators- PZT is typical piezoelectric material. Piezoelectric materials change dimension when an electric field is applied. This change is quite small, ~0.1%, but large forces are generated, ~ 100-1000 N. Since the dimensional change is moderately proportional to the applied field, piezoelectric materials or piezoactuators, are ideal for positioning applications where nanometer sensitivity is required.

Though PZT materials are used in many vibration transducers, PZT sensors are often used for vibration suppression/ cancellation/ damping in many engineering components. Piezoelectric actuators can be separated in two large families: the first one includes actuators that operate at low frequencies (off-resonance) and are typically used for positioning and vibration control applications; the second family regroups actuators that operate at resonance, and are used in a multitude of ultrasonic applications.

FFT (Fast Fourier Transform) Analyzer- The FFT analyzer is a device that uses the FFT algorithm to calculate a spectrum from a time domain signal, and is the most common type of spectrum analyzer available today. An FFT analyzer stores an input signal waveform as data by digitally (discretely) sampling it, determines the Fourier coefficients in a short time using FFT, and displays the results of this analysis. Since FFT basically involves resolving a signal into simple frequencies by expressing the value of frequency components (spectrum), the FFT analyzer is also called a frequency analyzer or spectrum analyzer. The FFT analyzer is a very useful device, and is available in a great variety of models with varying complexity. It is the heart of an cccy machinery predictive maintenance program. Fig. 6.9 is a conceptual diagram showing the relationship between the results of the analysis of the composite vibration waveforms from this rotary machine obtained using an FFT analyzer and the positions of the sources of vibration. Fig. 6.6 shows one data collector, analyzer and balancing unit (Vibscanner from M/s.Pruftechnik, www.pruftechnik.com).

Vibration Shakers- These generate vibration for structural research and modal testing. The vibration generator or "shaker" produces dynamic force operating on the mass reaction principle. Basically, a heavy mechanical component of the shaker is caused to oscillate by the drive signal from the power amplifier. A reaction force is generated, which excites the structure under test. Normally two types of shakers are manufactured– the low frequency electromagnetic shakers and the high frequency piezoelectric shakers. These are often used with Accelerometer and other transducers. M/s Bruel and Kjaer (M/s Josts Engineering, www.josts.com, in India) and **Wilcoxon Research Inc.** (M/s **Pyrodynamics,** www.pyrodynamics-india.com, in India) etc are few of the suppliers

Fig. 6.6. Vibscanner

Fig. 6.7. Accelerometer

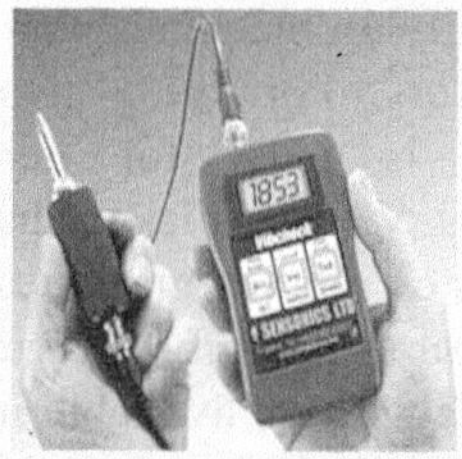

Fig. 6.8. Handheld Vibration Meter

Vibration holographs- It is one of the non-contacting vibration monitoring, where measurement of laser holography is used to measure vibration. Vibration measurement can be totally non-contacting if an optical excitation source is used. Laser light striking a solid heats it, producing a strain that is modulated in the same manner as the laser light is modulated, i.e. chopped. In addition to being non-contacting, optical excitation offers significant vibration potential for selectively exciting certain vibrational modes byaltering the region of excitation to enhance the response of one vibrational mode over another. Since the optical interference and the photorefractive effect occur throughout the photorefractive material, the single point detection method can be generalized to form an image of the vibration over the surface of the plate. The output beam intensity from the detection process can be measured directly by a suitable camera. However, such instruments are still not widely used for maintenance in industries.

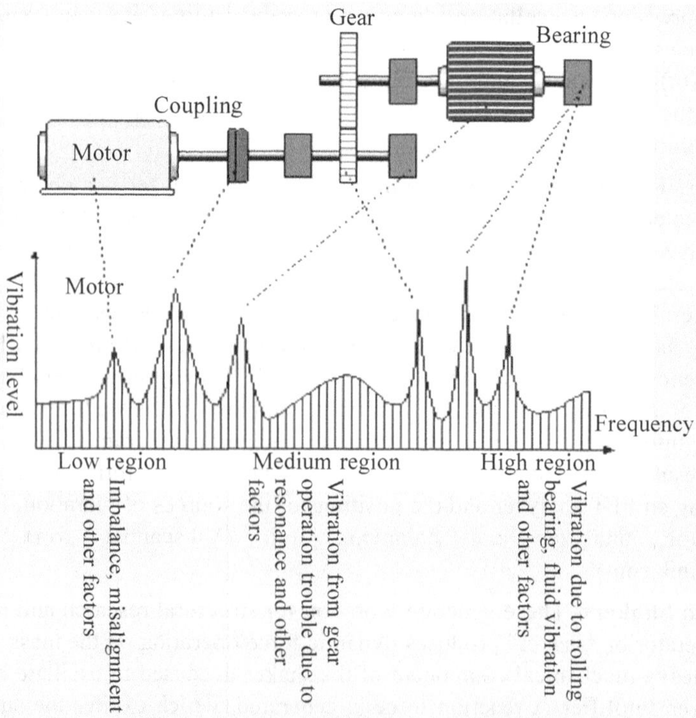

Fig. 6.9. FFT Analysis and Vibration Waveform

6.7.5 Machinery Vibration Trouble Shooting

As indicated earlier, all vibrations are not bad. Machines produce some oscillatory motion as part of their operation and these are not to be worried about. Such vibrations may be called "Benign" or normal vibrations. The presence of this benign vibration at "**normal levels**" provides a comfortable feeling that the machines are OK. Few of such vibrations are—

- Pure tones and hums of most electrical motors,
- Blade passing frequency,
- Broadband turbulence from fluid handling machines, like fans and pumps,
- Gear-mesh frequencies; etc.

However, these benign vibrations become cause of worry if these exceed the normal level and the increase is not corresponding to load change. In addition to these, there are many other serious vibrations. As such the vibrations to be worried about are the followings and "Vibration Troubleshooting" is needed for those—

1. Benign vibrations above the normal level, not corresponding to load change,
2. 1x RPM amplitudes above the balance limits, as roughly shown in Table 6.4,
3. Shock pulses,
4. Large shaking motions,
5. Abnormal noise; etc.

Fig. 6.10 shows a flowchart for vibration trouble shooting the diagnostic approach above the dotted line and remedial measures below the dotted line. Many other usual remedial methods may also be needed, such as bearing replacements, lubricant and lubricant schedule correction, structural stiffening to raise natural frequencies etc.

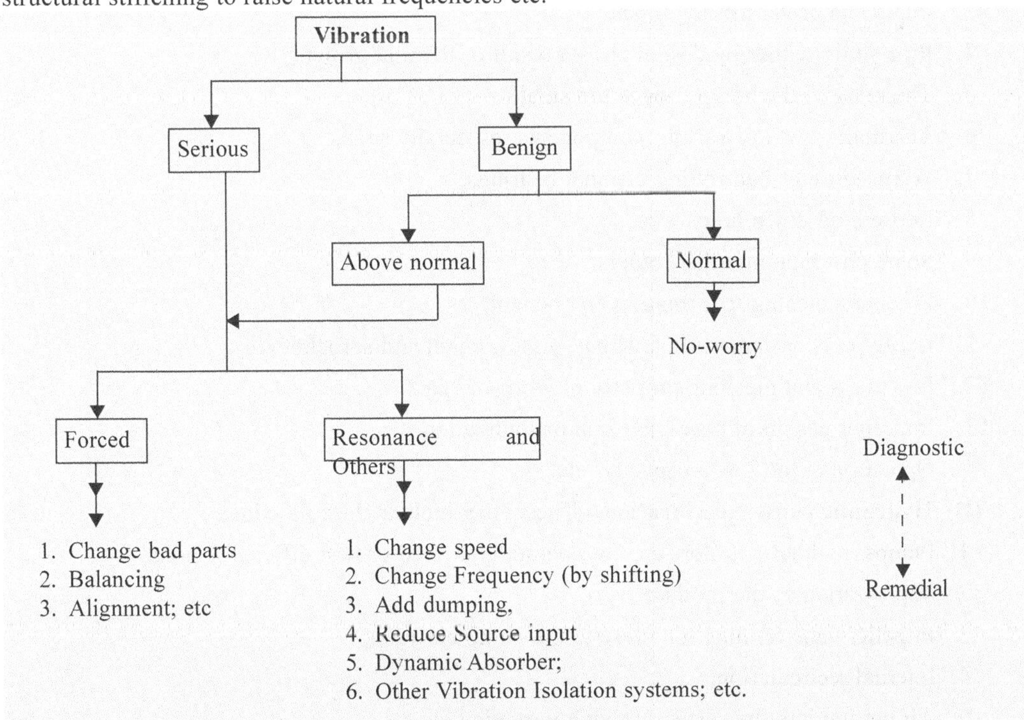

Fig. 6.10. Vibration Trouble Shooting Flowchart

Vibration isolation systems serve the purpose of limiting the transmission of vibration to surrounding structures or preventing ambient vibration from reaching machines performing precision work., especially for machines which are sensitive to shock or vibration from other machines or processes. One way to do it is using a large reaction mass foundation mounted on isolation padding was designed to reduce vibration transmissibility. The second form of isolation is employed when it is desired to limit the amount of vibration a machine transmits to the surrounding structure

Vibration troubleshooting is routinely applied to rotating and reciprocating equipment to assess general mechanical health. Bearing, motor, gearbox, driveline and foundation problems can be easily identified prior to becoming an operational problem. It is the most effective tool/ technique for identifying a problem's root cause. After the cause has been found, this testing is further used to gain a thorough understanding of why the problem exists. Corrective action can then be engineered to eliminate the problem at hand without creating a new one. Vibration troubleshooting, when coupled with strain gage testing, is an integral part of any failure investigation involving fatigue damage.

Table 6.2 shows the vibration frequencies vis-à-vis likely causes, arrived by vibration experts after through examinations and the table works as Vibration guideline chart.

Causes of Vibration– One important step of trouble shooting is to understand the causes of vibration. Most of the causes can be as followings—

(A) Mechanical Causes—These may include the followings-

1. Unbalanced rotating components. Damaged impellers and non concentric shaft sleeves are common,
2. A bent or warped shaft,
3. Pump and driver misalignment,
4. Pipe strain; either by design or as a result of thermal growth,
5. The mass of the pump base is too small,
6. Thermal growth of various components, especially shafts,
7. Worn, loose or bad rolling element bearings,
8. Loose hold down bolts,
9. Some part rubbing other/ others,
10. Product attaching to a rotating component,
11. Loose parts or damaged parts, e.g. pulley, chain and sprocket etc,
12. Bad gears, not meshing properly or worn-out etc.
13. Improper design of base-plate and/or foundation,
14. Operation at too low a capacity; etc.

(B) Hydraulic causes of vibration—These may include the followings—

1. Pumps/ hydraulic motors etc not operating at rated or best efficiency point,
2. Vaporization of the product,
3. Impeller vane running too close to the pump cutwater,
4. Internal recirculation,
5. Air getting into the system through vortexing etc,
6. Turbulence in the system (non laminar flow),

7. Water hammer;
8. Cavitation, clogged suction strainers or presence of air in hydraulic fluid; etc.
9. Parallel operation of poorly matched pumps; etc.

(C) Other causes of vibration– These may include the followings—

1. Harmonic vibration from nearby equipment,
2. Operating the pump at a critical speed: more prominent in variable speed and pulley pumps,
3. Seal "slip-stick" at the seal faces
4. A pump discharges recirculation line aimed at the seal face,
5. Electrical troubles,
6. Dynamic loading of mechanical components; etc.

Many of these cause can be found out from the Table 6.2

Table 6.2. Vibration Guideline Chart

Frequency (in RPM)	Most likely Cause	Other Probable Causes and Remarks
1× RPM (Running Speed)	Imbalance (Unbalance)	(1) Misalignment, (2) Bent Shaft (3) Eccentric journals, gears, pulleys (4) Resonance, (5) Bad belts (6) Reciprocating forces, (7) Electrical problems
2×RPM (First Harmonic)	Mechanical Looseness	(1) Misalignment, high in axial direction (2) Reciprocating forces, (3) Resonance (4) Loose bearings and parts, (5) Bad belts
3 × RPM (2nd Harmonic)	Misalignment (shaft to shaft)	(1) Excessive axial clearance (2) Internal misalignment (machine parts) (3) Electrical, (4) Air gap
4 × RPM	Coupling Problems	(1) Coupling Problems (2) Bearing looseness, (3) Resonance, possibly in slower machines, (4) Machine part problem
5 × RPM	Machine part problem	(1) Vane pass in pump (2) Excessive clearance
6× RPM	Machine part problem	(1) Vane passing (2) Rotor slippage on shafts
Sub-rotative	Oil Whirl	(1) Lubrication, (2) Oil whip (3) FTF fundamental train frequency
Half running speed (1/2 x RPM	Rub	(1) In a sleeve bearing- rub, (2) Drive belt, (3) Resonance drive belt (4) Defective machine parts)
		(1) Rotor bar problem, (2) Eccentric rotor

Line frequency (50 Hz or 60 Hz)	Electrical problem	(3) Unequal air gap, (4) Unbalanced phases, (5) Insulation breakdown; etc.
2× Line frequency (100 or 120 Hz)	Rotor problem	(1) Rotor resonance, (2) Torque pulses (Relatively rare phenomenon)
Upper frequencies (Harmonic)	Bad Gears	(1) Gear wear, Side bands of defective shaft speed, (2) Aerodynamic force (Fan speed x # blades (3) Hydraulic force (No. of impeller vanes x RPM, (4) Cavitation, (5) Machine part
Upper frequencies (Non- harmonic)	Bad bearings and defective lubrication	(1) Bearing inner-race, outer-race and cage Problems, (2) Bearing ball spin, (2) Cavitation, recirculation and turbulence (3) Rubbing

Table 6.3 gives such guideline chart (Vibration Identification), in slightly different form.

Table 6.3 Vibration Identification Chart

Cause	Amplitude	Frequency	Phase	Remark
Unbalance	Proportional tounbalance, largest in radial direction	1× RPM	Single reference mark	Most common causes of vibration
Misalign-ment in coupling or bearing	Large in axial direction, 50% or more of radial vibration	1 × RPM usual, 2 × RPM or 3 × RPM sometimes	Single, double or triple	Best found by appearance of large axial vibration. Use dial indicators or other Methods for positive diagnosis. Balance rotor in case of no coupling misalign-ment and sleeve bearings.
Bad anti-Friction bearing	Unsteady, use velocity measurement if possible	Very high, Several times RPM	Erratic	Bearing responsible, most likely the one nearest point of largest high frequency vibration.
Eccentric journal	Usually not Large	1× RPM	Single mark	If on gears, largest vibration in line with gear centres. If on motor or generator, vibration disappears when power is switched off. If on pumps or blowers, attempt to balance.
Bad gears or gear noise	Low, use velocity measure if	Very high, gear teeth times RPM	Erratic	Gear wear, broken teeth, eccentricity, bent shaft, dust, powder of materials, foreign piece, backlash.

	possible			
Mechanical Looseness		2 × RPM	Two reference mark, slight erratic	Usually accompanied by unbalance and/ or misalignment.
Bad drive Belts	Erratic or pulsating	1, 2, 3 or 4 times RPM	1 or 2, depending on frequen-cy, Usually unsteady	Strobe light is best tool to freeze faulty belt.
Electrical Problems	Disappears when power switched-off	1× RPM; 1 or 2× synchronous frequency	Single or rotating double mark	If vibration amplitude drops off instantly when power is switched off, look for electrical problems
Aerodyna-mic, hydraulic forces	1× RPM or num-ber of blades of fan.or impeller × RPM			Rare as a cause of trouble, except in case of resonance.
Reciprocat-Ing forces		1. 2 or higher times RPM		Inherent in reciprocating machines; can only be reduced by isolation or design changes.

For rectification of the problems/ causes, as detected in trouble shooting, very many tools, gadgets, techniques and software are available in market. **Alignment** is, probably, most important job in trouble shooting and for that many laser shaft/ coupling alignment systems, Belt and sheave alignment system, Turbine and Linebore Laser Alignment Systems and different dial indicator kits (both for shaft and face) etc are available. For proper alignment, following points must be considered—

1. Offset and Angularity,
2. Alignment Tolerance,
3. Past Alignment records,
4. Pre-alignment Procedures,
5. Base and Foot Problems,
6. Generic Precision Alignment Procedure,
7. Rim and Face Alignment Methods,
8. Reverse Dial Alignment Methods,
9. Laser System Alignment Methods,
10. Problems with Dynamic Movement; etc.

Similarly for balancing, vibration damping, absorption and isolation etc, many tools and techniques are available.

6.7.6 Machinery Vibration Standard, Severity Chart and Acceptable Limits

Quite a few vibration standards are available. VDI 2056/ 1964 (German) or its equivalent BS 4675, ISO 2372 or ISO / IS 3954 etc are used for overall level measuring equipments, normally for vibration signals within the frequency range of 10Hz to 1000 Hz. Indian standard IS 4729 is applicable to rotating electrical machines with power ranging from 0.15 KW to 1000 KW or more. Many of these standards give normal vibration levels and damage factors for various machine tool spindles etc. These standards guide us to set vibration alarms. Such alarm levels can be based on followings—

1. Fixed alarms based on above mentioned standards,
2. Percentage change alarm,
3. Trend based alarms,
4. Narrow band alarm,
5. Statistical, by point, bearing, machine, direction or machine family, etc.

From standards and experiments, few acceptable vibration limits have been derived for various machines/ equipments. Table 6.4 shows acceptable vibration limits for few equipments (taken from

Table 6.4. Acceptable Vibration Limit refer

Vibration Limits Equipments Electrical motors	Balance condition Displacement (mils peak to peak at 1x RPM	Overall Velocity (Inch per sec. Peak 10~1000Hz)	Overall acceleration Peak (0 ~ 5000 Hz)
(1000 ~ 2000 RPM)	2.0	0.2	0.5
Above 2000 RPM	1.0	0.2	1.0
Generator	2.0	0.2	0.5
Centrifugal Fan			
< 600 RPM	4.0	0.3	0.5
600 ~ 1000 RPM	3.0	0.3	1.0
1000 ~ 2000 RPM	2.0	0.3	1.5
> 2000 RPM	1.0	0.3	2.0
Vane-axial fans	1.0	0.2	0.5
Blowers	1.0	0.3	0.5
Pumps	2.0	0.2	0.5
Centrifugal	1.0	0.2	3.0
Compressors		0.2	3.0
Cooling Tower	3.0	0.4	2.0
Gear Boxes			
Reciprocating	5.0	1.0	10.0
Engines			
Turbines	1.0	0.2	0.5
Twin Screw	1.0	1.0	15.0
Compressors			

Vibration Severity Chart- For conventional vibration overall measurements, there exist a number of general machinery vibration severity charts developed through the years. However, it is almost impossible to develop a universal overall level Spike Energy severity chart for general machinery applications. This is due to the fact that too many variables involved, such as different machine types, operating conditions, accelerometers, mounting methods, and ambient conditions. On the other hand, it is possible to develop an overall Spike Energy severity chart based upon empirical data for certain type of machines. General machinery vibration severity chart is used by vibration analyst for setting alarm level at different points of machine.

Table 6.5 shows one such chart; IRD General Machinery Severity Chart. ISO-2372 gives vibration severity chart in different form (Good, Useable, Still acceptable and Unacceptable) for small machines (Class-I), medium machines (Class-II), large machines (Class-III) and turbo-machines (Class-IV).

Table 6.5. IRD General Machinery Severity Chart

Machinery Condition	Velocity (Inch/sec peak)	Velocity (VdB)
Very Rough	V > 0.628	VdB > 121
Rough	0.314 < V , 0,628	115 < VdB < 121
Slightly Rough	0.157 < V < 0.314	109 < VdB < 115
Fair	0.0785 < V < 0.157	103 < VdB < 109
Good	0.0392 < V < 0.0785	97 < VdB < 103
Very Good	0.0196 < V 0.0392	91 < VdB < 97
Smooth	0.0098 < V < 0.0196	85 < VdB < 91
Very Smooth	0.0049 < V < 0.0098	79 < VdB < 85
Extremely Smooth	V < 0.0049	VdB < 79

6.8 LUBRICANT MONITORING/ OIL ANALYSIS

Oils and lubricants are as important as blood for human being. As such, oil and lubricant analysis is widely used in condition monitoring of lubricated machineries and hydraulic systems. Most of such tests/ analysis can be grouped in following three categories—

(A) **Fluid Properties Analysis-** Information on a lubricant's physical and chemical properties such as viscosity, acid number, demulsibility, additives,

(B) **Fluid Contamination Analysis-** The presence of contaminants (uninvited guests) such as dirt, glycol, soot, fuel, water etc in the oil.

(C) **Fluid Wear Debris Analysis-** Concentration and characterization of wear metals and other contaminants, suspended in used oil, mainly from the machine components, through which the oil interfaces and generates some wear metals and wear particles and not about oil itself. This is also called Machine Health analysis.

(D) Fluid Chemical Properties (TBN, TAN, additives, contamination, % water)

Category (A) and (B) are more used for assessing the condition of lubricating and hydraulic oils and their ability to perform their intended functions. Viscosity monitoring is an important part of oil/ lubricant monitoring as decrease in viscosity may indicate fuel dilution or contamination by water or other fluids and increase in viscosity may indicate thermal breakdown and oxidation of oil or using wrong oil. But category (C) is used for equipment health monitoring. As this category, mainly, works on contaminants (wear debris or wear particles) generated due to interaction between the various components/ parts of the machine and carried away by the lubricant to sumps etc, these are also known as **"Contaminant Monitoring"** or **"Wear Debris Analysis"** or **"Wear-particle Analysis".** However, contamination monitoring is more synonymous for "Environmental Monitoring"; as such for general maintenance purposes, the term "Lubricant Monitoring" is used, which includes all these.

6.8.1 Components and Techniques of Lubricant Monitoring

We mostly consider following components/ techniques in lubricant monitoring, also referred as **Wear Debris Analysis** —

1. Sources of wear debris,
2. The distinction between amount, size, shape and chemical breakdown,
3. Numerical Methods Monitoring—Spectrametric/ Spectroscopic oil analysis, light emitted / absorbed by excited atoms, Particle counting (light / laser blockage) etc,
4. Direct Reader Ferrography- (light blockage or ferrous only); Particle Quantifier (magnetic field changes or ferrous only),
5. Visual Method Monitoring— Analytical Ferrography (magnetic separation or ferrous only0, Rotary Particle Depositor (magnetic separation or ferrous only)
6. Filter Debris Analysis (filter separation); and Filtergrams (filter separation and clarification),
7. Scanning Electron Microscope
8. Magnetic plugs - Ferrous only wear particles
9. Used Engine oil Test Kit and other oil test kits,
10. Lab tests - oil quality, wear and contamination of particles and fluids; etc.

    Few of these are discussed briefly herewith.
11. Lubricant Contamination Control.

6.8.2 Types and Sources of Contaminants

Solid particle contaminants vary in hardness, friability and ductility, depending on the composition of the particle. Common contaminants like rust and black iron oxides have a Mohs hardness rating of five to six (on a scale of 1 to 10, with 10 being the hardest). Environmental dust, mostly silica, has a hardness rating of two to eight. Quarry dust has a rating of five to nine. From the manufacturing process, tool steel has a hardness rating of six to seven. Silicon carbide and aluminum oxide have a hardness rating of nine. Diamond tops the scale with a hardness rating of ten. The size, hardness and friability or ductility of the particle influence the amount of damage that the particle can cause. Though solid contaminants do more damage, many fluid contaminants (water, acid etc) and gaseous contaminants (air, smoke and other gasses etc) are also harmful by their dust concentration or reacting with components.

In another way, contaminants (mainly generated) can be classified as followings—

- **Rubbing Wear particles**– normal wear, small flakes of metal etc,
- **Fatigue chunks**– sub-surface stressing in gears or shafts, large chunks of metal (up to 5000 μm in size) etc
- **Sliding wear particles**– scuffing wear, metal-to-metal contact with no lubricant film, gear teeth pistons, hydraulics, loose bearings, parallel striations on surface etc,
- **Cutting wear**– abrasive contact between two surfaces, 2 body wear (metal/metal) softer metal is cut, 3 body wear (metal/contaminant/metal), curls of metal like lathe swarf etc,
- **Laminar wear**– sub-surface stressing in rolling element bearings, reworked wear particles, flat, thin flakes of metal,
- **Oxides and rusts etc.**

Sources of contaminants, commonly encountered in industries, can be grouped into following categories—

1. **Built-in Contamination**—These include built-in manufacturing debris, permanent debris and other damage, which occur during run-in etc,
2. **Internally Generated Contamination**— These particles are usually wear debris from gears, bearings, splines or other components resulting from micropitting, macropitting, adhesion, abrasion or fretting corrosion wear modes,
3. **Ingressed Contamination**—These may include moisture, acids or other fluids and different kind of fumes and gasses from atmosphere or nearby industrial environment etc. Though breathers are used to vent internal pressure when air enters through seals or when air within the gearbox expands and contracts during normal heating and cooling, ingress is often possible to faulty filters of breathers or from faulty seals and packings etc.
4. **Contamination Added During Maintenance**—Various contaminants are, inadvertently, added during maintenance because of improper maintenance cleaning, protecting and unclean environment etc. Breathers, filters, drain ports, fill ports, sample ports, dip sticks, magnetic plugs and inspection ports should be designed and located for easy maintenance and minimization of contamination
5. Contaminants due to **accidents and malicious acts,**
6. **Biological Contamination**—Bacterias, Microbes, Algae, Fungi and other Micro-organism in water, hydraulic and cutting fluids and other industrial fluids; during handling, production, processing, maintaining and serving operations.

Evaluating these debris (mainly wear particles) in hydraulic, lubrication and other fluid systems can provide valuable information about abnormal contaminant ingression, accelerated component wear, impending component failure and/or fluid break down etc..

6.8.3 Magnetic Plugs

These are simple magnetic probes/ plugs positioned in the lubrication systems for maximum catch efficiency of the wear debris and other metallic particles from the oil. These are generally fitted (screwed or bolted) at the bottom portion of sump, reservoir, bigger pumps, gearboxes and centrifugal compressor castings etc. so that these can be removed periodically, along with collected metallic debris, and replaced by spare plug or same plug after carefully removing all debris and cleaning. Some metallic plugs are non-detachable type and, in that case, a suitable inspection window is provided near by to remove the collected debris periodically. The debris, so collected, is quantified and examined visually or by other means discussed later in this chapter. Magnetic plugs also helps in cleaning the oil. In earth-moving equipments and few other equipments, magnetic plugs also work as drain plug. If necessary, we also should consider neutralization of the magnetic attraction. Occasionally, we consider fitting of a plastic or metallic sheath or use of wax to remove the debris.

6.8.4 Filter Debris Analysis (filter separation) and Filtergrams

Filters are commonly used in lubricating and hydraulic systems to keep the fluid clean. Filter captures the particles generated since the last filter change and leading up to the failure event or next filter change. The debris collected in the filters is often examined for wear prediction etc.

Few recent filter debris analysis methods have been designed as **diagnostic filter system**. In such systems, the debris is removed from the filter element onto an analysis membrane by back flushing with solvent or solvent-air mixture, by ultrasonic removal in a solvent, or by flushing debris off the filter element with a solvent or by other means. The debris collected on the analysis membrane is evaluated for amount, size distribution and morphology. The chemical elemental composition is determined by spectroscopy or by infrared (IR) spectroscopy or ferrography etc. Such diagnostic filter system is comprised of a diagnostic filter element having a pull-out diagnostic layer and a custom-designed contaminant analyzer, that allows for on-site analysis of debris captured on the diagnostic layer. The diagnostic layer can be easily removed on-site for convenient visual inspection of the captured debris. The porosity of the diagnostic layer can be adjusted, within the constraints of the regular filtration medium pack, so that the captured debris is appropriate for diagnostics.

The advantages of employing filter debris analysis are— the full-flow nature of the filter element allows all the fluid in the system to pass through the filter element; the coherent surface for capturing fluid system debris efficiently, results in a high concentration of debris compared to debris in the fluid or in magnetic plugs/chip detectors; and the filter element retains debris of all types (metallic and nonmetallic).

Filtergrams are the images of debris segregation and concentration. These are prepared by filtering a sample of oil through a 3 to 8 mm cellulose nitrate membrane filter. A drop of clarifying solution is added and the slide is baked in an oven. This sets the cellulose nitrate upon the slide and optically clarifies it. Filtergrams are, often similar to Ferrograms, when used for ferrous materials.

6.8.5 Ferrography

It is an analytical method of assessing machine health by quantifying and examining wear particles suspended in the lubricant or hydraulic fluid. It is one of the modern condition monitoring technique based on wear debris analysis. It provides a comprehensive non-intrusive evaluation of the health of lubricated components while the equipment is in running state. The wear debris (ferrous as well as non-ferrous), present in the lubricant sample are precipitated on a glass substrate which can be viewed through a powerful bichromatic microscope. Under a specified condition of magnification and lighting filters used for transmitted and reflected light the shape, size, concentration and colour of wear debris are observed. The four major findings from ferrography are— **wear mode, rate, severity** and **location**. The identified wear modes include abrasion, impact, fatigue, erosion, corrosion, scuffing, severe sliding, etc. The concentration of wear debris indicates rate of wear. Size of debris indicates the severity of wear. The colour of the particles identifies the type of material which pinpoints the affected component(s).

Glass substrate, or ferrogram analysis, is one common method of particle identification. A combination of incline, sample preparation and a magnetic field ensure that all particles present in the lubricant sample are deposited on the substrate for examination. Particles ranging in size from less than 1 micron to greater than 2,000 microns are released on the substrate. The particles are released in a general order of size, with the largest ferrous particles being collected at the entry end of the substrate. Non-ferrous wear particles are released in a random manner throughout the length of the substrate, often appearing between the strings of ferrous particles. Contaminants, such as sand and dirt, fibers and friction polymers also are distributed in an irregular fashion throughout the length of the substrate.

Identifying these particles and the wear mechanisms that generated them can effectively demonstrate the equipment's operating history and current state of performance, as well as generate alarms to future wear conditions. Normal rubbing wear produces platelet particles typically ranging in size up to a major dimension of 15 microns. Bearing platelet wear has a flaked appearance and is easily misidentified as normal rubbing wear. Bearing wear typically occurs in larger formations than normal rubbing wear. The morphology of these particles may be of greater significance than their size. Gear wear particles are typically very large compared with other particles. Their composition may often be of greater significance than their size. The progression from high-carbon alloy steels to low-carbon alloy steels indicates wear severity. Severe sliding wear is identified by parallel striations on the particle surface and sharp fractured edges. These become more prominent with wear severity. Excessive load or speed are common root causes of these particles.

Ferrography is generally, of two types— Direct-reading and Analytical.

Direct Read Ferrography (DRF)- It is a quantitative analysis of ferrous particles present in used oil. It provides an index of large ferrous particles (DL as density of particles larger than 5 microns) to small ferrous particles (DS as density of particles smaller than 5 microns). Because it can identify the amount of ferrous wear occurring within a unit, it is most helpful in illustrating developing trends. In DRF, a sample is to be suitably diluted. We can arrive following equations—

Wear particle concentration (WPC) = DL + DS,

Percentage of large particles (PLP) = ((DL-DS) / WPC

The **WPC** is indicative of the rate of wear and is compared to industry averages for the specific type of equipment being tested. The PLP is often used to indicate the onset of abnormal wear conditions.

Analytical Ferrography (AF)- It can predict potential equipment failures or be an effective tool in determining the root cause of failure. AF is a qualitative, rather than quantitative analysis that provides digital imagery of the actual particles present. It utilizes microscopic analysis to identify the composition of the material present. This technology will differentiate the type of material contained within the sample and determine the wearing component from which it was generated. Powerful magnets trap the ferrous particles, which are then placed on slides for microscopic analysis. Particles are analyzed based on metallic or non-metallic, alloy via heat treatment, shape, size, color, and if possible, source

RPD Rotary Ferrograph- It combines magnetic and centrifugal separation. It is faster than both the AF and DRF techniques and presents a sample ready for microscopy in few minutes. Additionally, unlike DRF technology, the RPD Rotary Ferrograph does not require the sample to be diluted and does not suffer from interference due to carbonaceous material in the sample. The RPD Rotary Ferrograph offers a rapid and simple method of debris separation. A measured volume of sample is applied, by pipette, to a glass substrate located on a rotating magnet assembly. Particles of debris are deposited radially as three concentric rings by the combined effects of rotational magnetic and gravitational forces.

6.8.6 Spectroscopic Oil Analysis Program (SOAP)

It is a procedures for extracting fluid samples from operating systems and analyzing them spectroscopically to determines the concentration of key elements represented in the entrained fluid contaminant. Molecular spectroscopy uses principles of chemical physics to determine molecule concentrations. This has some distinct benefits for tracking additives. First, because molecular concentrations are being measured directly, additive decomposition, like the hydrolysis of ZDDP cited above, become immediately apparent. Secondly, many molecular spectroscopic tools are well-suited for determining concentrations of organic molecules. Molecular spectroscopy may cover a multitude of analytical techniques, such as Fourier transform infrared (FTIR) and gas chromatography etc.

Spectroscopy is the most widely applied technique for debris monitoring. It provides a quantitative, multi-elemental analysis of wear debris in lubricating oil. The elemental concentration of as many as 20 elements are reported in parts per million (ppm). Wear metals such as iron, aluminum, chromium, copper, tin, lead, silver, titanium and nickel are detectable, as well as lubricant additives such as calcium, barium, zinc, phosphorus, magnesium, boron and molybdenum. Certain contaminants such as silicon, sodium and potassium are also routinely detected.

Spectrometric (or Spectrographic) Oil Analysis (Process) are also similar techniques and are also called SOA (SOAP). Table 6.6 shows typical machinery problem, indicated by SOAP.

Table 6.6. Typical Machinery Problems Indicated by SOAP

Elements identified by SOAP	Typical Problems
Aluminum	• Ingress of Aluminum dirt • Wear of oil wetted Aluminum components, e.g. Aluminum bearings, light alloy castings, pistons etc
Iron	• General wear of gears, rolling bearings, cross-slides, housings etc • Wear of engine cylinder bore, if the debris is taken from engine sump,
Copper	• Wear of brass/ bronze bushes and slides; bearing cages of rolling bearings and clutch lining etc.
Chromium	• Wear of chrome plated rings, cylinders, pistons etc,
Lead	• Wear of lead-based plain bearings (white metals) etc,
Silicon	• Ingress of siliceous dirt (air intake in system), • Residual Casting sands, etc.

Fig. 6.11 compares the three commonly used wear debris analysis techniques (magnetic plugs, ferrography and SOAP, with respect to their scope and limitation.

6.8.7 Particle Quantifier (PQ)

It exposes a lubricant to a magnetic field. The presence of any ferrous metals causes a distortion in the field, which is represented as the PQ Index, an arbitrary unit of measurement that correlates well with DR ferrography. Although PQ does not provide a ratio of small to large ferrous particles, if the PQ Index is smaller than ferrous ppm, it's unlikely there are many particles larger than 5 microns present. If the PQ Index increases dramatically while the Fe ppm remains consistent or goes down, larger ferrous particles are being generated and further testing or diagnostics are recommended.

6.8.8 Used Oil Test Kits

Many portable used oil test kits are commercially available and used in industries, but their main job is assessing the oils, generally on site, for finding their deterioration vis-à-vis usability, with the aim to extend oil drain intervals, i.e. to extend the use of oil. Finding out the machine component deterioration may be their secondary job. These kits contain visgauge and some reagents, in a suitable cabinet, though which maintenance staff can do viscosity test, additive/ acidity test, reverse alkalinity test, detergency/ dispersancy test and water content test etc on spot and know the essential properties of that oils. Such kits are very useful for big engines, gearboxes, generators etc where lubricating oils can be changed on condition-based in place of fixed-time basis, in order to save costly oils. Of-course few Used Oil Test Kit packages also detect premature wear and their trend.

Limiting Factors	**SOAP**	**Ferrography**	**Magnetic Plug**
Efficiency of each technique depends on particle size	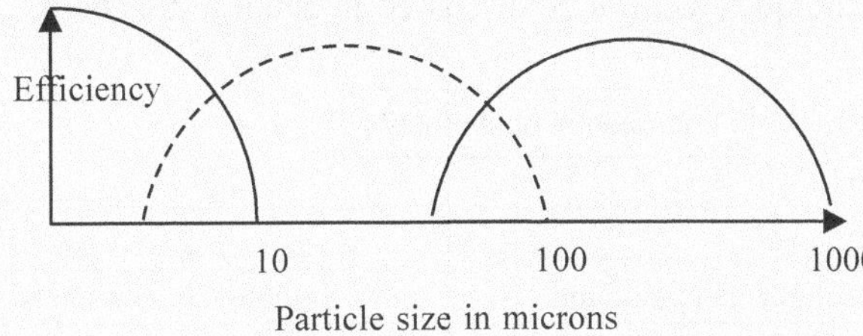		
Type of Debris			
• Ferrous	Suitable	Suitable	Suitable
• Non-ferrous	Suitable	Suitable for few	Not-suitable
• Non-metallic	Suitable	Suitable for few	Not-suitable
Warning time for gradual failure	Very Good	Very Good	Good
Type of surveillance	Periodic (based on samples)	Continuous Possible by online DR Ferrograph	Continuous possible by online electronic chip detector
Possibility of Alarm	Not-possible	Possible	Possible
Use for Large volume Systems	Difficult for dilution of debris concentration	Difficult	Good

Fig 6.11. Comparision of three Wear Debris Analysis Techniques

6.8.9 Lubricant Contamination Control

Maintaining clean oil is one of the best investments maintenance departments can make. Equipment manufacturers increase the speed and reduce the size of their equipment, thereby decreasing component clearances. As such, extending the lifecycle of equipment and maximizing lubricant performance are on-going objectives in a contamination control programme.

Contamination control requirements depend on both the environment and the application. Hydraulic power system oils and lubricants and bearing lubrication have some of the most critical demands for contaminant-free fluids, Mechanical filtration is the primary technology in most contamination control programs and will continue to be so for the foreseeable future. However, the pore sizes of mechanical filters are quickly approaching their limits and it is evident that this technology does not solve all contamination control problems. One of the biggest limitations of mechanical filtration is its inability to remove soluble contaminants, as well as insoluble contaminants under 1 micron. As the demand for cleaner oils and lubricants increases, technologies beyond mechanical filtration are required. Technologies that can remove fluid degradation by-products, both for conventional (Group 1) and high performance oils (Group-2), will be setting the trends for contamination control in the future. Fig. 6.12 shows the range of contamination control techniques and highlights the trendsetting technologies for today's most demanding contamination control requirements. The technologies, shown in bold in the Fig. 6.12 are recent and trendsetting ones. The actual machine/ system will decide the extent of control needed.

Though the main aims of contamination control are Category (A) and Category (B) of Sec. 6.8, it would also reduce the generation of wear debris and equipment deterioration. As such, these control techniques will not be discussed in detail in this book.

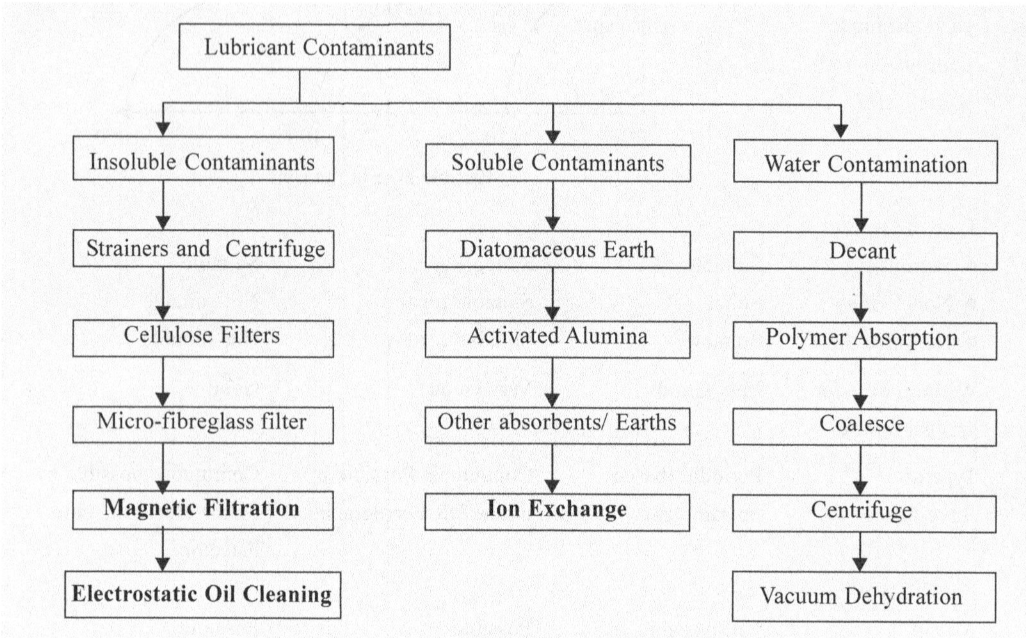

Fig. 6.12. Lubricant Contamination Control Technologies

6.9 THICKNESS MONITORING

The wall thickness monitoring system is often designed to operate in-line as a production control tool. Even for products which move at high speed, can not be touched, or has pass line variation (the product position can not be held constant), in-line monitoring of thickness is possible with two laser triangulation sensors, mounted on opposite sides of the product, with the difference in sensor readings determining thickness variations.

However, thickness monitoring is very effective and useful technique for assessing the thickness (and thus condition) of the pipelines, pressure vessels, tanks, bottles, cylinders, boat-hulls, radar domes, aircraft wings and body panels etc. Most thickness monitoring equipments work on ultrasonic system. A sound pulse, generated by a probe, travels through the material, bounces-off the back surface of the material and returns to the probe. The instrument accurately measures the length of time taken between transmitting and receiving back the pulse, automatically multiplies that by the sound velocity of the material under test and converts the result into an illuminated digital readout of the thickness of the material. The operator only has to select the sound velocity in the instrument, corresponding to the material under test, as the sound velocity changes from one grade of material to another. Fig 6.13 gives a simple sketch of this system.

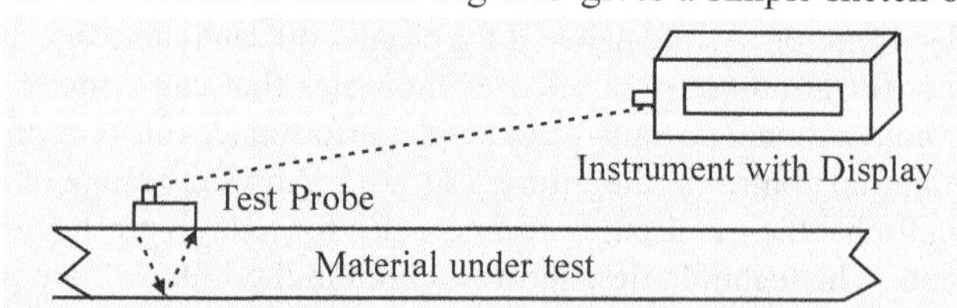

Fig. 6.13. Ultrasonic Thickness Monitoring

Quality assurance also uses thickness monitoring for detecting defects etc. Ofcourse few thickness monitoring systems work on other principles such as radiographic or eddy current etc. Large numbers of models and suppliers for such equipments are available in the market.

6.10 CRACK MONITORING

Crack monitoring is more used for quality assurance and metallographic analysis to assess the quality of metals and quality of procedures during making, shaping and treating of metals in industries. However, crack detection/ monitoring is also used in CBM and condition monitoring of many metallic components.

Crack monitoring programmes measures not total crack depth and width but change in crack width. This change in crack width is called crack displacement. The crack displace-ment measured by the sensors may be driven by any combination of the factors listed below—

1. Differential thermal expansion,
2. Structural and machine overloading,
3. Chemical changes in various components of machine,
4. Shrinkage and twisting of different components temperature and humidity changes etc,
5. Fatigue and aging of components, etc.

Various techniques used for crack monitoring are— Dye-penetrant Test, Magnetic flux (Magna-flux), Electric resistance, Eddy current, Ultrasonic and Radiographic tests etc. Based on these, Strain gauges, Extrinsic Fabry-Perot Interferometry (EFPI), Distributed Strain Sensing (DSS) techniques, IRIS (Internal Rotary Inspection System) tube testing, RFT (Remote Field Testing) tube testing and many other ultrasonic and eddy current models etc have also been developed.

6.11 CORROSION MONITORING

The principles of corrosion monitoring equipment is based on corrosion or chemical wear of the material. The use of such techniques for condition monitoring of machines/ components (for helping in condition based maintenance or other maintenance jobs) is very limited and selective. Again, some of these techniques may not give accurate machine deterioration rate.

However, few common corrosion monitoring techniques are enumerated below—

- Weight Loss Method,
- Electrical Resistance Method,
- Linear Polarization Resistance (LPR) Method,
- Galvanic or Zero Resistance Method,
- Hydrogen Monitoring Method; etc.

6.12 NOISE/ SOUND (ACOUSTIC) MONITORING

Acoustic is a general word for noise and sound. Noise and sound are basically the same except that noise is considered as harsh, unpleasant and undesirable sound.

In its simplest form, noise/ sound is the vibration of molecules through a medium like air, fluid or metal/ solid. Vibrations can be broken down to energy levels at discrete frequencies and when the human ear detects these levels and frequencies, it translates them into intensities and tones which are then transmitted to the brain. The human ear can detect frequencies between 20 Hz and 20 kHz. This range is referred to as the audible, or sonic, range. Frequencies above this range are referred to as ultrasonic or ultrasound.

Noise monitoring is very important for controlling noise pollution and environmental protection as noise affect human-being both ways, physically and psychologically and prolonged exposure to high noise level can lead to permanent hearing loss. As such, many statues and laws have been promulgated and many equipments/ instruments are available for noise monitoring. Occasionally acoustic enclosures are provided an economical controlled acoustic environment.

Noise may be seismic, animal/ mammal made, man-made or unidentified. Noise in industries may be considered as manmade. Ambient noise from factory machinery can be very difficult to eliminate at source. Such noise is the result of transient vibrations of the structures and components of the machines and such vibrations are induced by rapid energy release, rapid pressure and temperature rise, cavitation and air ingress, leakages and other malfunctions. As such, noise monitoring can also be used, to some extent, to monitor the health and condition of machines. For identifying the noise sources, following techniques may be used—

- Subjective assessment,
- Acoustic ducts (such as horn etc),
- Surface intensity approach (using accelerometer on vibrating surface and a microphone)
- Acoustic intensity approach and sound-pressure monitoring (using microphone devices),
- Lead wrapping technique (using sound absorbing porous material blanket, pasted on lead sheet and placing on machine vibrating surface so that the porous material is sandwiched between the lead sheet and machine surface),
- A-weighting (curves relating to the measurement of sound pressure level, as opposed to actual sound pressure),
- Impulsive noise monitoring,
- Infrasonic noise monitoring and microbarograph; etc.

After identifying the noise source and identifying the intensity, necessary repair action can be planned. As an example, a bad engine exhaust system may cause noise upto 90 dB. However, for CBM and CM, vibration monitoring and Lubricant monitoring and ultrasound monitoring have overshadowed noise monitoring.

6.12.1 Ultrasound/ Ultrasonic Monitoring

Ultrasound or ultrasonic has already been defined as high-frequency sound waves, which are above the range of human perception. Usually they start at 20 kilohertz (kHz) and go up to the megahertz range. Airborne or structure-borne devices cover frequencies from 20kHz to 300kHz.

Airborne and structure-borne ultrasound instruments provide information through multiple paths; qualitatively with their ability to hear ultrasounds through a noise-isolating headphone, and quantitatively via incremental readings on a meter/display panel. Digital instruments provide on-board data storage for data logging and for viewing baseline data. Some newer versions of the instruments also include on-board sound recording for spectral analysis.

Such instruments allow inspectors to confirm a diagnosis on the spot because they clearly discriminate among various equipment sounds. An electronic process called "heterodyning" accurately converts ultrasounds sensed by the instrument into the audible range where users can hear and recognize them through headphones. This process enables users to record sound events through various recording devices.

Since the magnitude of ultrasound wavelengths are smaller than those in the audible range, they have characteristics that are conducive to condition analysis/ monitoring. One advantage is that the amplitude of a generated ultrasound falls off exponentially from the source, making the emission localized and easily isolated for detection and analysis. Most ultrasound systems have an output that can be connected to a vibration data collector.

Good analysis can be performed by using a system of enveloping which is somewhat similar to heterodyning. Enveloping is a demodulation process that is used predominantly to measure the incipient defects that are present when a rolling element bearing starts to fail.

(A) Use of Ultrasound Noise Monitoring– Such monitoring is very effective for industrial applications because a large number of systems, equipment, components emit more ultrasonic noise when problems develop or failures likely to occur. A few of such applications are given herewith—

1. Leak detection in pressurized or vacuum systems– When a gas (or air), escapes from high pressure to low pressure through an orifice or vice-versa, the flow becomes turbulent, generating ultrasonic noise. The noise is characteristically identified as a 'rushing' or 'roaring' sound with few discernible frequencies.

2. Steam-trap inspection and analysis– All steam systems, no matter how well insulated, will lose some steam to condensation. When enough condensate collects, it will begin to ripple into waves due to the steam passing rapidly above it, much like a lake ripples when the wind blows over it. As the rippling continues the size of the 'waves' will grow until the steam will actually lift the condensate, forming it into a type of plug, and push it through the pipe; and thus cause water-hammer.

Such steam-trap discharge cycle can be monitored by listening to the outlet line and corrective action can be taken before reaching the stage of condensate lifting or water-hammering.

3. Mechanical inspection and analysis– One of the main sources of ultrasonic noise in equipment is friction between two objects moving relative to each other. This is most commonly seen in bearings. Whether due to improper lubrication or improper lubricant or mechanical defects, metal-to-metal contact will generate significant amounts of noise. Traditional vibration analysis has shown that most mechanical defects (pits, cracks, spalls, etc.) generate noise in the sonic range.

But by that time enough damage has been caused to the components and the component normally need to be changed at suitable time. However, ultrasonic guns which have frequency adjustment and are sensitive enough can detect such defects much earlier, before the defects in components have reached alarming (sonic) range and corrective active may be taken timely and component change may not be needed. This is the **difference between Acoustic vibration monitoring and normal vibration monitoring.** Gears and other such components can also be monitored suitably.

4. Lubrication improvement– Though this may be a type of mechanical inspection, it is mentioned separately to emphasize the point. Over-lubrication or under-lubrication, both are bad. An under-lubricated bearing will allow metal-to-metal contact and wear, while an over-lubricated bearing will heat up and possibly pop the bearing seals. Under-lubricated bearings will sound like general noise (little impacting) and will generate noise around 25 kHz, though heavy loads on bearings may tend to lower this somewhat. To effectively lubricate a bearing, a baseline reading at 25kHz should be established. Then as further checking indicates a dB rise at 25kHz, monitor the bearing while lubricating. As lubrication is added, the noise levels should decrease. When the dB levels return close to the baseline, cease lubricating. Do not over-lubricate a bearing. It may take some time before lubrication spreads itself throughout the bearings inner surfaces, so proceed slowly.

5. Valve inspection and analysis– Valves are inspected (somewhat similar to steam traps) by checking sound intensity levels both upstream and downstream of the valve. If the sound level downstream of the valve is greater than upstream of the valve, then the valve is, depending on the type, probably partially open or maybe completely open. To be more sure, use a sonic detector downstream of the valve (or maybe on the valve stem) because laminar flow will generate sonic noise. If listening with a sonic sensor and an ultrasonic sensor yields no sound downstream of a valve, then the valve may be closed.

6. Crack detection and monitoring– Checking for cracks in metals by ultrasonic is extensively used in industries. The sound energy is introduced and propagates through the materials in the form of waves. When there is a discontinuity (such as a crack) in the wave path, part of the energy will be reflected back from the flaw surface. The reflected wave signal is transformed into an electrical signal by the transducer and is displayed on a screen.

(Refer to sec. 6.10 also).

7. Thickness measuring/ monitoring– Refer to Sec. 6.9.

8. Electrical inspection– One application for ultrasonic monitoring is the detection of arcing (electricity traveling through the air), corona (ionization of air around electrical conductors), or electrical insulation breakdown. Searching for electrical problems is similar to searching for gas leaks, in that, with the airborne sensor in, pan the gun around through electrical equipment and listen for electrical fault sounds like popping, buzzing, or crackling. Then move toward where the sound is the loudest.

Beside industrial application, ultrasound is used at many other places, e.g. medical diagnosis physiotherapy etc.

(B) Pitfalls–Though ultrasonic is very useful, it call for few cautions. It is generally difficult to trend the outputs of ultrasonic detectors, the units are not scalable and will vary depending on how far the sensor is positioned from the bearing (or other component) being examined. Again, the ultrasound detector is directional and care has to be taken to ensure that the source of ultrasound is effectively monitored.

6.13 SMELL/ ODOUR MONITORING

Of the five senses, the sense of smell is the most complex and unique in structure and organization. Currently, the human nose is the best instrument for measuring odour. Measurements may be conducted in the laboratory using an olfactometer (panel of certified evaluators) or in the field with trained odour sniffers.

Efficient odour monitoring systems have, probably, not yet been developed and used for condition monitoring, though smelling through nose is used, since ages, to determine the leakage and presence of few gasses and smelling substances. However, few gas detection systems (online or offline) and instruments are available in the market which, mostly, work on chemical, electro-chemical and infrared actions/ reactions. Few examples are given below—

1. SO_2 Analyzers (based on heated UV instruments) to measure SO_2 in flue gas in sulphuric acid plant and paper/ pulp mills,
2. CO, CO_2, NH_3 and CH_4 detectors/ analyzers based on infrared technique, using gas filled detectors,
3. Hydrocarbon Analyzers for monitoring stack gasses of boilers/ furnaces. Many of these work on flame ionization detectors,
4. NO_x Analyzers based on chemiluminescent techniques, etc.

6.14 MOTOR CURRENT SIGNATURE ANALYSIS (MCSA)

It is a system used for analyzing or trending dynamic, energized systems. It is a condition monitoring technique that is now widely used to diagnose problems such as broken rotor bars, abnormal level of air-gap eccentricity, shorted turns in low voltage stator windings and few mechanical problems.

MCSA is the online analysis of the current to detect fault in a three phase induction motor drive while it is still operating. A basic MCSA instrument will consist —

1. A current transformer (CT) to sense the signal,
2. A resistive shunt across the output of the CT (CTs are available with internal shunt),
3. An MCSA instrument(or spectrum analyzer) to produce the current spectrum signature.

MCSA uses the electric motor as a transducer, allowing the user to evaluate the electrical and mechanical condition from the Motor Control Center (MCC) or disconnect. For accurate analysis, MCSA systems rely upon FFT analysis, much the same as vibration analysis. Most MCSA systems also rely upon analysis of demodulated voltage and/or current, which involves the removal of the fundamental frequency (Line Frequency or LF).

6.14.1 Electrical analysis and monitoring

MCSA does the following electrical analysis—

(A) Power quality monitoring—It includes–

1. Voltage quality, over or under voltage; Voltage unbalance, which causes unbalanced current in the motor resulting in overheating of the winding.

2. Harmonic (voltage harmonic and current harmonic) distortion is another area of concern and is normally caused by electronic switching systems which cause standing, negative and positive rotating fields within the motor.

3. Power factor is represented, in an inductive circuit, as how the peak current lags behind the peak voltage. The result is additional current requirements for the same load as current lags further behind voltage

(B) Rotor Analysis— Broken rotor bars, static eccentricity and dynamic eccentricity are three basic types of rotor issues that MCSA can evaluate.

(C) Stator Analysis— Stator winding problems are found by first identifying stator slot passing frequencies.

6.14.2 Mechanical analysis

MCSA does the following mechanical analysis—

(A) Bearings– In order for bearing problems to appear in MCSA, the condition will be severe. Bearing problems found through this method should be considered urgent and addressed as soon as possible. It requires some calculations involving bearing dimensions and line and running frequencies etc.

(B) Mechanical Imbalance– Mechanical imbalance is found by determining the rotor bar and center frequencies, as in static and dynamic eccentricity,

6.14.3 Driven Equipments

Driven equipment frequencies can also be detected and monitored. Belt and Sheave drives, gear-mesh drives, direct drives, fans and impellers can be monitored, considering the various frequencies and vital dimensions.

Identification of following defects are tackled by the analysis and monitoring covered earlier in this section—

1. Broken or cracked rotor bars,
2. High resistance joints in squirrel-cage windings,
3. Casting porosities in aluminum die-cast rotors,
4. Poor brazed joints in copper fabricated rotors,
5. Rotor winding problems in slip-ring induction motors,
6. Static and dynamic air-gap irregularities,
7. Unbalanced magnetic forces,
8. Mechanical unbalance, bent shaft or bearing irregularities; etc.

MCSA technology can be used in conjunction with other technologies, such as **motor circuit analysis (MCA),** in order to provide a complete overview of the motor circuit.

6.15 PARTIAL DISCHARGE MONITORING

This is another new electrical condition monitoring system. Partial discharge (PD) phenomena is evidence of a degrading insulation capacity. As partial discharge often precedes an insulation breakdown of high voltage equipment, its detection and monitoring is of vital importance.

High-precision modular acquisition analysis toolkits are available for detecting, recording and analyzing partial discharge events in many applications. It can be used for Transformer testing, large motor and generator testing, Cable testing respectively on-site assembling, Gas insulated switchgears (GIS) testing etc.

6.16 RADIOGRAPHIC MONITORING

Radiation can pass through solid objects, but is attenuated or reduced in strength, denser or thicker the material, the greater the attenuation. If X ray or gamma ray passes through a metal and then through a crack or hole in the metal, the hole or crack will attenuate less radiation. By knowing the material density and thickness and the equivalent energy of radiation, it is possible to calculate how much exposure is needed to produce a radiographic image of the part of the structure.

Though radiographic analysis and monitoring is of great interest in first-aid-medical, veterinary-medical applications, physiotherapy and other medical diagnosis and treatments, it has enough use in industries also. X-Ray equipments are in use in industrial radiography for quite some time. Many of the monitoring equipments, mentioned earlier in this chapter, use industrial radiography. **Industrial radiography** means work involving the examination of the structure of materials by non-destructive methods, utilizing ionising radiation, including all procedures which are significant from a radiation safety point of view, such as the activating of X-ray units, the monitoring of radiation levels, etc

Radiographic wall-thickness measurements of insulated piping in the chemical and petrochemical industry are typically carried out by exposing film-cassettes with Ir 192 (Iridium) as x-ray source. The radiographic image of the tube is then evaluated off-line after the film is developed. Few mobile planar array-sensor (Radiation Image Detector) are capable of generating radiographic on-line images. These images can be shown on-line on a PC-screen and can be evaluated directly on-site during the examination. Erosion in pipelines can also be monitored.

Neutron Radiography works on the basis of neutron attenuation by the hydrogen in the corrosion product itself and is, occasionally, preferred over other methods (e.g. ultrasonic or eddy current methods etc) for its potential for early detection of corrosion, before substantial attack and thinning of the material has occurred.

Radiographic Non Destructive Testing (RNDT) gadgets are in use for many other industrial monitoring. Of-course many of those are used for quality control. However, great safety precaution is to be taken while using or disposing radiographic instruments.

6.17 BENEFITS OF CM

The main aim of condition monitoring is to timely predict the nature and extent of in-coming defects in any equipment. Today condition monitoring is used in almost all machines, equipments, systems and plants in all kind of industrial applications to monitor their health and behaviour pattern. Many of the applications are already mentioned earlier in this chapter. We can further illustrate the benefits as below–

1. Predicts (and therefore assist in avoiding) unplanned equipment failures.

2. Helps to take a holistic view of equipment condition. For example, if Vibration Analysis indicated that a bearing failure on a particular pump was imminent, wouldn't it be preferable to be able to assess the condition of all the other components of the pump (impeller, seals, back-plate etc.) in order to determine whether any of these items should be replaced or refurbished at the same time as the bearings? Traditionally, this assessment is left until the pump has already been shutdown,

3. Condition monitoring reduces the cost of condition monitoring. This sounds ridiculous, but the fact is that modern good low cost monitors either do away completely few skilled workers to monitor the condition of equipments or need fewer and lesser skilled worker to do the same level of checking.

4. It will improve the consistency and accuracy of failure diagnosis, as there will be less chances of human error due to repeated nature of job and fatigue.

5. Helps in improving equipment and component reliability and to optimize the equipment performance. For example, if vibration analysis indicates abnormal vibration from the bearing, the source may be either from the bearing (worn-out balls or races etc) or from external source (misalignment, imbalance, pump cavitation etc) and rectifying those will improve MTBF and thus reliability.

6.18 OTHER USES OF CONDITION MONITORING

In addition to benefits mentioned above and the use of condition monitoring for various fault detection and monitoring at various stages, condition monitoring has few other applications in industries. A few of those are given below—

6.18.1 Mechanical Measurements

CM techniques are also used for mechanical measurements, (loads, strains etc) by using Strain Gauge and Load Cells, Accelerometers and Displacement Sensors.

Load cells are normally a structure with a known elasticity which is then strain gauged. The load is then proportional to the strain gauge output. **Strain Gauges** are of two types—

1. Thin film strain gauges which are normally made of stainless steel or aluminum and which work by the principal of metals changing their resistivity with strain,
2. Vibrating wire strain gauges.

Few suppliers for such strain gauges are Strainsert Company (http://www.strainsert.com/) and Kistler Instrument Corporation (www.kistler.com).

6.18.2 Condition Monitoring using Laser Alignment of Shafts in Rotating Equipment

Near to perfect alignment of the shafts of rotating equipment is very important on high speed rotating equipment where misalignment can cause expensive equipment failure. Laser shaft alignment provides the most accurate method of this type of condition monitoring.

6.18.3 Wireless Condition Monitoring Application

Remote condition monitoring can benefit just about every industry where electromechanical machines are vital to production. Several applications where wireless condition monitoring is particularly effective include monitoring of pumps in wastewater treatment plants, drives used on oil/gas drilling rigs, drives on assembly lines in automotive plants and overhead cranes in hot metal mills. Another application is power plant cooling fan monitoring, where cooling fans are mounted in very harsh areas where hot steam is always present.

6.18.4 Thermal Monitoring

Maintaining boiler flue gasses, coke-oven gasses and high pressure and high temperature water and steam, heat exchangers, cooling towers etc is possible only by remote condition monitoring.

QUESTIONS

1. What is condition monitoring and equipment health monitoring? What types of condition monitoring are normally used in industries?
2. Differentiate between primary signals and secondary signals emanating from machines. Which type of signals are more used in vibration monitoring and why?
3. What is leakage monitoring? Explain some of the leakage mediums used for condition monitoring.
4. What is lubricant monitoring? Why and where this analysis is used more effectively?
5. What is wear Debris analysis? What are the three wear debris analysis techniques commonly used and compare their performance and use.
6. What is thermal monitoring and what different thermal monitoring are used in industries? Explain principles and uses of thermography.
7. What is electric current monitoring in relation to motor defect determination?
8. What is vibration analysis and what are its principles/ methods? How is it used for equipment health monitoring?
9. What are the causes of machine vibration and what vibration measurement and analysis tools and methods commonly used for rotating machines?
10. How do you monitor the condition of cross country pipe lines?
11. Give few examples of use of ultrasound for checking the condition of equipments.

CHAPTER

7 MAINTENANCE PLANNING AND SCHEDULING

Having known the fault lists, result of condition monitoring and having decided the type/ strategy of maintenanee, the next step is to plan and schedule the maintenance jobs so that the jobs can be executed properly and desired results can be obtained in time. Planning and scheduling functions are the key deliverables of the planning role. This is where the most gains in execution have the potential to be made and acted upon. In some larger organizations these are split, allowing more adequate resources for each role, but often planning and scheduling is done by same person or group. Often problem arises in that case, as the maintenance planner does not do much planning and jumps to scheduling. Very often, the planner bogs down to work as expedite material, work on tools, plan emergency work or work as relief supervisor and gives very little thought to work-order planning. The two, planning and scheduling have to have completely different approach. The role of the planner needs to cover the full range of the work order system, from input into coding, prioritization and a degree of autonomy in execution.

7.1 PLANNING

Planning of jobs or Job-planning can occur at any stage during the life of a works order or maintenance job. An indicator (electronic or otherwise) in the work order systems needs to be able to identify the work-order by status of planning. In this manner works orders requiring parts, procedures, documents, skills or equipment can easily be focused upon. A work order cannot be considered planned until all of these have been considered. However, detailed manpower planning/ deployment is not done at this stage.

Planning of maintenance jobs basically deals with answering two questions, "**what**" and "**How**" of the job; i.e. "what jobs/ activities are to be done" and "how those jobs and activities are to be done". While answering these two questions, many supplementary questions are to be answered, e.g. "Where the job is to be done" and "Why the job is to be done" etc, but all these will be helping in developing "what" and "how" of the job. As such, it is very essential that engineering knowledge must be applied extensively to maintenance jobs for development of appropriate job plans using most suited techniques, tools, materials and special facilities etc.

As the job planning forms the basic foundation, over which the efficiency and cost of further actions (e.g. scheduling, execution and control etc) depends, persons responsible for job planning should have adequate qualities/ capabilities, such as, knowledge about jobs and available techniques facilities and resources, analytical ability, conceptual logical ability and judgmental courage etc.

7.1.1 Steps of Job Planning

The main steps to be followed for proper job planning are, generally, as follows—

1. ***Knowledge Base-*** It includes knowledge about equipment, job, available techniques, materials and facilities etc. The planner should have through understanding of workorder requirement and should know and refer to—

- Correct equipment number and code,
- Work description request or equipment symptom and failure analysis reports etc,
- Refer to CMMS for details and previous history of the job, if any,
- Drawings, instruction manuals and Maintenance Manuals, including assembling, Dismantling and packing instructions etc,
- Job Manuals,
- Experience of same or similar machines,
- Departmental maintenance workshop and that of other departments,
- Plant's Captive or centralized workshops or repair cells,
- Nearby ancillary industries, etc.

2. ***Job Investigation at Site-*** It gives a clear perception of the total job and also helps in ascertaining the followings—

- Physical access and space limitation- This may call for jobs like removing covers, guards, stoppers or cutting a portion machine housing etc for better approach,
- Assessing if available lifting and handling tackles/ facilities are enough or special cranes/ facilities are to be brought; also helps in deciding scaffolding etc.
- Facilities for disposal of water, oil, gases and hazardous materials which may leak or come out during dismantling,
- Space for keeping the dismantled parts and safety enclosures for machine under repair.
- Proximity of other jobs, going on at potentially the same time,
- Safety issues, e.g. requirement of permits, clearances, putting danger tags, any hot or cold or wet situation there, etc,
- Identify any reference point or benchmark or match-mark, for ease of assembly,
- Take necessary photographs before dismantling, for checking later-on.

3. ***Identify and document the work-*** Knowing earlier two steps and knowing the needs of preventive, predictive and other maintenance jobs, a complete job/ work list is made. Pending and leftover jobs from previous schedules may also be included. The planner often consults concerned operating and maintenance personnel and PdM and CBM inspectors also for any special problems and cross-checking. A trained planner may review the inspection results for common problems (e.g. misalignment, imbalance etc) and also not so common problems (e.g. resonance etc).

4. ***Development of Repair Plan-*** Preparation of step-by-step procedures which would accomplish the work with the most economical use of time, manpower and material. It may include making of sketches, line diagrams and rough networks etc. For first time jobs, provision of still and video photography can also be planned before starting the job and during critical dismantling. Weight of each item to be determined/ estimated before hand and planning should be done to avoid double or multiple handling of the same item. The total job should be broken in small measurable activities at this stage.

5. *Preparation of Tools, Tackles and Facilities List-* In smaller plans, like Fig.7.1, this can be done in the plan sheet itself, but for bigger jobs, a separate list may be attached indicating the needs of special tools, tackles and facilities needed, with sketches etc, such as special porter bar and other lifting tackles, arranging special long boom crane and transfer cars, special torque wrenches and big spanners etc. Advance action has to be taken for most of these. Surveying equipments may also have to be arranged to mark the level of machine foundations (if earlier markings are not seen), before dismantling.

6. *Estimation of Time Required to do the Job-* For smaller jobs, as shown in Fig 7.1, it is easier, but for bigger jobs it may need knowledge of "work simplification", "work measurement" and Critical path analysis etc.

7. *Work-order Feedback forms/ plan-* Two types of feedback forms are used; one for supervisor of planner to approve the plan and other for scheduler to report after completion.

Fig. 7.1 shows a sample plan for a small job of changing V belt of blower- X. In the figure SH-1 means Store House No. 1. Cost column can be filled at initial planning stage or, better, at review stage, after knowing exactly the number of parts changed.

Planning Sheet

<table>
<tr><td colspan="3">Equipment Name- Blower X
Equipment Code—</td><td>W.O.Number
—</td><td>Planner
Mr, —</td><td>Date
—</td><td colspan="2">Seq.
—</td></tr>
<tr><td colspan="3">Job Title- Change V belt (cracked and worn)</td><td></td><td></td><td></td><td colspan="2"></td></tr>
<tr><td>Job Scope</td><td></td><td></td><td colspan="5">Materials, Tools and Equipments</td></tr>
<tr><td>Job Steps</td><td>Crew</td><td>Est. Hours</td><td>Description</td><td>Qty</td><td colspan="2">Stock as on——</td><td>Cost</td></tr>
<tr><td>1. Look out, disconnect and Tag Motor starter,
2. Remove belt guard, loosen and Jack motor, Remove bolts,
3. Check both sheaves for side-Wall wear and radial run-out (should not be more the +1mm for big and +0.25 for small ends), Replace worn-out sheaves if needed,
4. Install new bolts, align sheaves, tighten motor bolts.
5. Tension belts, using force deflection method, recheck alignment,
6. Reinstall guards, remove locks, clean up and release for operation</td><td>2 men</td><td>0.25
0.75
1.00
1.0
0.5
0.75</td><td>Sheave gauge (Tool kit)
Mag. base Dial Indicator
• Big Sheave (Part No.)
• Bushing (Part No.)
• Small Sheave(Part No)
• Bushing (Part No.)
Bolts—— (Part No)
Belt alignment tool
Belt tension gauge</td><td>1 set
1 set
1 set
4 sets</td><td colspan="2">4, SH-1
4, SH-1
3, SH-1
3, SH-1
8, SH-2</td><td></td></tr>
<tr><td>Drawing No–</td><td colspan="3">Total Time———</td><td colspan="4">Total Cost———</td></tr>
</table>

Fig. 7.1. Changing V belt of Blower-X

Fig. 7.2 roughly shows a Planning Process, indicating connections with different agencies.

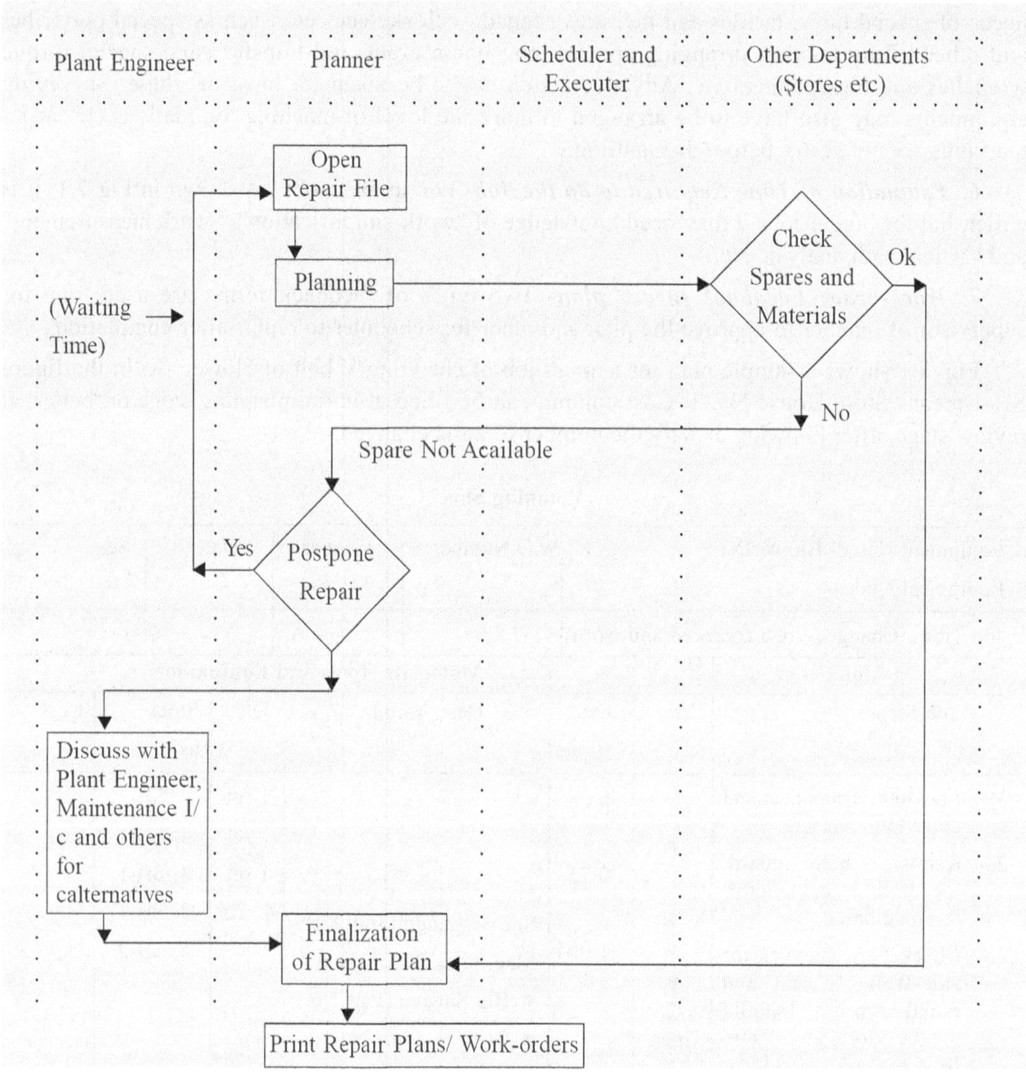

Fig. 7.2 Planning Process

7.1.2 Planning Techniques

There may be numerous methods/ techniques, a planner may use to plan and putdown that on paper or computer etc. Few of those are mentioned below–

1. Work breakdown structure– This involves breaking the total job down into lower and lower levels of tasks that have to be carried out. A good test is, can you easily estimate how long the task will take and who will be required. There is never a wrong or a right answers to how you do a Work Breakdown Structure. It is for the planner or planning manager to come up with what they consider the best solution. The lowest levels of the tasks should be capable of being clearly defined. The pre-requisite for a planner is–

- should be able to estimate resources (time, people material etc),
- should be able to assign work and

- should think about reporting cycle.

2. Work Measurement—It is the careful analysis of a task, its size, the method used in its performance, and its efficiency. The objective is to determine the **workload** (the amount of work assigned to or expected from a worker in a specified time period) in an operation, the time that is required, and the number of workers needed to perform the work efficiently.

The purpose of measurement is to collect real data about actual events.

Components– A work measurement system has three components– 'preferred methods', 'time values', and 'reporting'. Preferred methods are not always the most efficient or fastest way to do a task. This should enhance safety, quality, and productivity.

The time (Time value), that a job should take, is determined not on the basis of speeding up the motions a worker normally makes but on the normal pace of the average worker, taking into consideration allowances for rest periods, coffee breaks, and fatigue.

A reporting system is important to the success of any work measurement method. Supervisors and managers must have access to labor-management information that is both timely and complete. Timely information can be used to manage and shift labor hours to areas where they are needed and to correct problems or at least prevent them from becoming a crisis.

Work Measurement Methods— All work measurement systems are based on the same, simple three-stage procedure— 'analysis', 'data collection and measurement', and 'synthesis'. They differ in the nature and degree of analysis, the nature and level of data collection and measurement, and the nature of the synthesis process.

There are four work measurement methods— historical data, work sampling, time study and predetermined motion/time systems. Each of the four methods have their strengths and weaknesses.

Work Simplification technique can also be used, but that is used more for mass production operations.

3. Work Packages– It involves details of what work is required and when it is to be delivered, negotiation to agree for the work with the line manager and for necessary clearances, arranging the broken down works in some logical sequence, determining what works can be done in parallel and putting them on paper or computer etc. A sample shown below–

Task/ work/ activity	Start time	Duration	Agency	Note

3. Gantt Chart— Though these are basically planning technique, these are extensively
4. PERT and CPM— used in scheduling and monitoring, as such are discussed later.

7.1.3 Planner's Tool Kit

For field inspection and notes, a planner should have a note book for writing (and or a digital voice recorder), a small digital camera 9for taking pre-dismantling photos), a strong torch light, a marker and a measuring tape etc. In the office, he/she should have computer, printer and other essential things.

7.2 JOB MANUALS

Job Manuals are almost permanent about methodologies, tools, tackles and facilities for all maintenance jobs, which may have to be done in future. This involves saving the different plans and modifying and expanding those plans for other similar machines. Generally following steps are involved in preparing Job Manuals—

1. Make a list of all major and medium jobs of the plant (preferably equipment. Shop wise) and codify (numerical codes) for proper identification,
2. For each coded job, a separate job manual is to be made in the form of job plan, indicating job steps, tools, tackles, spares, consumables and facilities needed and also indicating safety and statutory clearance etc.
3. Each job manual, thus prepared, should be cross checked and cleared by Maintenance I/c,
4. Different Job manuals should be bunched and sent to respective potential users,
5. Necessary updation of job manuals, as needed, in consultation with respective users,

These job manuals are quite useful for maintenance organization and plant. Few main uses are–

- For actual planning and scheduling of any job, the job manuals provide ready information for use as it is or for further micro-planning.
- The executing agencies refer to the job manual before starting the job, to ensure that they follow correct procedures and to take necessary tool and spares etc to avoid frequent visits to stores and thus save time,
- Material Department may also use job manual for better material procurement strategy.

7.3 SCHEDULING

Scheduling is the function of coordinating all of the logistical issues around the issues regarding the execution phase of the work. This can also uncover some areas of planning deficiency, which needs to be addressed. Scheduling is best performed in a capacity-scheduling manner. Most modern systems have the capacity to output data to spreadsheets or in similar forms. Only planned job can be scheduled properly. A schedule without a planned job is like "wish list".

Scheduling of maintenance jobs basically deals with answering two questions—"**Who**" and "**When**" of the job, i.e, "who would do the job" and "when the job would be started and done".

Effective scheduling essentially needs realistic thinking, based on substantial data and records. However, majority of scheduling work needs to occur in areas such as –

- Overhead labour hours (such as safety and toolbox meetings, break times and training times etc) are to be gathered, along with holidays and off-days and scheduled as standing works orders for future analysis of these,
- Hours for PM completion to be deduced form data in the CMMS. This focuses on ensuring the equipment is maintained to its best levels,
- Addition of corrective and approved improvement actions as dictated by the prioritization system and operations plan. These are to be planned works orders only. A guide could be: "Age of works orders against priority (As a measure of the priority systems effectiveness)",
- The combination of corrective, preventative and improvement work needs to total the levels set for planned / scheduled work. However, a workable level may be 70%- 80% in the initial stages; etc.

7.3.1 Pre-requisites for schedulers

A scheduler should also have knowledge about job, techniques, facilities, analytical ability and judgmental courage. The scheduler must obtain knowledge/ information about following facts/ aspects, before starting his job—

1. Manpower availability by trade, location, shift, crew arrangement and permissible overtime limit etc,
2. Man-hour back log on current or unfinished jobs,
3. Availability of the equipment or area where the work has to be performed,
4. Availability of proper tools, tackles, spares, consumables, structural and other required materials,
5. Availability of special equipments, jigs/ fixtures, special lifting and handling facilities and cranes etc. This should also include labour and time saving devices like pneumatic hammers and excavators etc.
6. Availability of external manpower and their capabilities; these may be from other shops/ departments of the plant or from contractors (local, nearby, ancillary etc).
7. Starting date of the job; also often completion time of total job is predetermined and, in that case, resources are to be arranged accordingly.
8. Past schedules and charts (updated) if the same job has been done earlier; etc.

7.3.2 Scheduling Principles

Following principles greatly contribute to the success of scheduling, especially in a bigger organization–

1. Job plan should provide number of workmen required with the lowest required skill/ craft level and the job duration,

2. Entire plant must respect the importance of schedules and job priorities,

3. Special or heavy demand jobs can not be scheduled unless backlog is addressed by providing additional resources or by relaxing/ reassigning priorities,

4. Lobs will not be scheduled unless all the planned materials (parts, tools, special facilities etc) are available in time and in required quantity.

5. Each available maintenance trade must be scheduled for a full day of productive work for every day of availability.

6. Scheduler should develop a one week schedule for each crew based on craft hour available forecast that shows highest skill level available and job priorities. Consideration is also given for multiple jobs on the same equipment or system and of proactive versus reactive work available.

7. The one week schedule assigns work for every available work hour. The schedule allows for emergencies and high priority, reactive jobs by scheduling a reasonable amount of work hours on easily interruptible tasks. Preference is given for completing higher priority work by even underutilizing available skill level over completing lower priority work.

8. The crew supervisor develops a daily schedule, one day in advance, using current job progress, the one week schedule and new high priority reactive job as a guide. The crew supervisor handles the current day's work and problem, even to the extent of rescheduling the entire crew for emergency works.

9. **Wrench Time** (It is a measure of crafts personnel at work, using tools, in front of jobs. Wrench time does not include obtaining parts, tools or instructions, or the travel associated with those tasks or traveling to or from jobs etc) is the primary measure of workmen's efficiency and of planning and scheduling efficiencies. Again, schedule compliance is the measure of adherence of one week schedule and its effectiveness. Wrench time can't improve if the crew is not given enough work.

10. **Advance Scheduling**- It is actually more of an allocation of work and may not be a detailed schedule of exact personnel and time assignment. Advance scheduling provides enough work for the entire crew and sets goal for maximum utilization of available craft hours.

Scheduling differs from 'Job Loading'. **Job loading** is the jobs listed in sequence of agreed upon importance during the weekly or other coordination meetings. This sequence may not be chronological within the week or that period.

7.4 SCHEDULE TYPES AND TECHNIQUES

Different types of schedules are made suiting the respective job plans and different techniques are used for making and following those schedules. The first step of all scheduling is to break the job into small measurable elements, called activities and to arrange them in logical sequences considering the preceding, concurrent and succeeding activities so that a succeeding activity should follow preceding activities and concurrent activities can start together,

Arranging these activities in different fashion makes different types of schedules. The schedules may be Weekly General Schedule, Daily Schedule, Gantt Chart, Bar Chart, PERT network and CPM or CPA schedules etc.

7.4.1 Weekly General Schedule

This is made to provide weeks worth of work for each employee in an area. The shutdown (S/D) date column indicates the best day in the week to perform some job. As such, operation should also get a copy of this schedule to help arrange shutdown. However, maintenance supervisor has to be attentive for individual's deployment. Table 7.1 shows a portion of one such schedule

Table 7.1 (Sample Weekly Schedule)

Schedule of work, beginning Sept. 7, 2004				
Area-D, Available Hours- 180 Hrs + 52 PM Hrs.= 232 Hrs.				
W.O.No.	**Equip. No.**	**Description**	**Est. Hrs.**	**S/D date**
P 14351	26C420	PM, Compressor, Quarterly	4	Sep. 9
P14318	Gen.	Greasing, Route D-4	11	
P14321	Gen.	Oil, Route D-5	16	
P14322	MOB23	PM, Pickup Truck Safety Check	2	
P14324	Gen.	PdM, Vibration Monitoring, Week 27	7	
P14326	37T910	PM, E-13 D Transmitter Calibration	4	Sep.11
03426	47R200	Repair Minor Leak at Reactor Seal	14	
03398	37R907	Replace Rupture Disc with low pressure	7	Sep.11
03369	05T905	Replace Frayed Hose	6	
03414	26B501	Increase fan speed, Change sheave and motor	11	Sep.9
03415	10B501	Replace fan bearings, per vibration analysis	9	Sep.8
—	—	—— ——	—	—

7.4.2 Daily Schedule

Daily schedule is developed to free maintenance supervisor from some of the coordination problems. Effort is made by the scheduler to provide a day's work for each maintenance employee of the area. Using the CMMS, the scheduler lists the number of times each job has been placed on schedule. It is normally the responsibility of team leader or maintenance supervisor to assign the right person or persons to each job. The job that is not completed in a day, may be placed on schedule of next day or left for another suitable day of week. Table 7,2 shows one such sample schedule. The table shows original estimated time just for reference. It can be deleted or, additionally, one more column can be introduced showing least time taken in past for the same job.

Table 7.2 Sample Daily Schedule

Daily Schedule										
Area- D	Day- Tuesday	Date- 7, Sept. 2003				(Unit- Hrs.)				
WO No.	Equip. No.	Description	Mr A	Mr. B	Mr C	Mr D	Mr E	Mr F	Org Est.	Actual Hrs.
03415	10B601	Replace Fan Bearings				4.5		4.5	7.0	
03403	10M210	Replace worn mixer blades			5.5		5.5		8.0	
P1432	Plant	Oiling, Route 06			2.5		2.5		7.0	
03465	37T973	Relocate Level Controls				3.5		3.5	8.0	
P1431	Plant	Grease, Route 04		x					2.0	
03464	37P101	Replace Guards and Nipples	6.0						8.0	
03481	35P201	Emergency- Replace pump seals	10							
03480	37M120	Emergency, Replace Valve	2.0							
		Straight Time Hrs.	8.0	8.0	8.0	8.0	8.0	8.0		
		Overtime Hours	0	2.0	0	0	0	0		

7.4.3 Gantt Chart and Bar Chart

Gantt charts are, normally, a project planning tool that can be used to represent the timing of tasks required to complete a project. Because Gantt charts are simple to understand and easy to construct, they are used by most project managers for all but the most complex project and are also used by maintenance engineers for scheduling bigger maintenance jobs. In a Gantt chart, each task/ activity takes up one row. Dates run along the top in increments of days, weeks or months, depending on the total length of the work. The expected time for each task is represented by a horizontal bar whose left end marks the expected beginning of the task and whose right end marks the expected completion date. Tasks may run sequentially, in parallel or overlapping. As the work progresses, the chart is updated by filling in the bars to a length proportional to the fraction of work that has been accomplished on the task. This way, one can get a quick reading of project progress by drawing a vertical line through the chart at the current date. Completed tasks lie to the left of the line and are completely filled in. Current tasks cross the line and are behind schedule if their filled-in section is to the left of the line and ahead of schedule if the filled-in section stops to the right of the line. Gantt charts made with Microsoft Excel are easy to update and maintain.

Bar chart can be defined as a graphical comparison of several quantities in which the lengths of the horizontal or vertical bars represent the relative magnitude of the values. It is the type of chart most commonly used by technical analysts. The horizontal scale on the bottom of the chart indicates the passage of time. The time scale is usually shift-wise, daily, weekly or monthly. Bar chart is concerned with two major tasks. Firstly, it visualizes the time flow of the individual processes for all project workspace members. The second task of the bar chart is to provide the project manager or maintenance personnel with functionalities for individual processes.

Bar Charts are of different types for different applications. In maintenance scenario, Bar Chart is a Gantt Chart on slightly lower scale for scheduling maintenance jobs and employing the same principles as Gantt Chart. Maintenance Bar Charts are of two types—

1. Overall Chart—This broadly shows all the jobs /activities of a repair plan in one chart. However for a big repair plan, this may become confusing.
2. Individual Charts—This, again, can be of two types—

- Individual section/ area/ execution agency wise; such as separate charts for shop maintenance crews, heavy maintenance crews, contractors and captive shops etc.
- Individual crew or workmen wise chart within each individual section agency.

Individual workman chart is also called **Manpower Chart.**

Fig.7.3 shows a portion of Bar Chart for 900 Mill shutdown. In the chart, A,B C are the three working shifts of a day, HM means heavy maintenance, "Opr" means operation and "Contr" means contractor. The duration of each activity is shown by the lower portion of each bar. The portion above each bar in the same row is used to fill up during review/ follow-up, indicating the actual time taken.

Sl No.	Activity (Job)	Time Agency	1-1-94			2-1-94			3-1-94			4-1-94			
			A	B	C	A	B	C	A	B	C	A	B	C	A
1	Remove Rolls	Opr													
2	Remove top and bottom spindles and remove top carrier	H M													
3	Remove bottom spindle brg. and bottom spindle lift cylinder	H M													
4	Remove tilter mechanism and entry right side guard	H M													
5	Remove oil of Lube system 1	Contr.													
—	——— ———	———													

Fig 7.3 Bar Chart for 900 Mill Shutdown Repair

Limitation of Gantt Chart- Though Gantt charts are very useful for maintenance activities, it has some limitations or disadvantages, such as—

1. They can become quite unwieldy for major repairs or projects with more than about 30 activities. Larger Gantt charts may not be suitable for most computer displays.

2. Gantt charts communicate relatively little information per unit area of display. That is, major repairs or projects are often considerably more complex than can be communicated effectively with a Gantt chart.

3. Gantt charts do not represent the size of the total job or the relative size of work elements, therefore the magnitude of a behind-schedule condition is easily miscommunicated.

7.4.4 PERT, CPM and CPA Networks

Though Bar Charts /Gantt Charts have lot of merits, one big demerit is that they do not clearly define interdependencies of activities, i.e. which activity must precede and which activity must succeed concerned activities and how much delay in one activity actually affects other activities. To overcome this demerit, many **Network Techniques** have come up, like **PERT (Programme Evaluation and Review Technique), CPM (Critical Path Method) or CPA (Critical Path Analysis)** etc. Though most of these were developed for monitoring big project works, they are equally suitable for maintenance jobs.

All these Network Techniques have common methodologies and approach. The networks are made of "Activities" and "Events". Activity tells the actual job to be done and is in the form of straight line. Event tells the actual beginning or ending of an activity (job) and is designated by a circle, or sometimes a square. Events do not consume any time but activities do consume time. All events are numbered and an activity is indicated by two connecting events. For plotting a network, following guidelines are followed—

1. Total Repair/ Shutdown job is broken into small measurable activities. Occasionally a Work Breakdown Structure (WBS) is used for that. WBS is a hierarchic decomposition or breakdown of a major activity into successive levels, in which each level is a finer breakdown of the preceding one.
2. All activities and events are arranged in logical sequence and plotted on paper so that succeeding activity and event must follow the preceding activity and event, i.e. succeeding activity must be on the right side of preceding activity,
3. The network is not drawn on time scale, but it has time-sense; i.e. each event, starting earlier than few events, must be on left (west) side of those events. Similarly, an event, finishing later than few events, must be on right (east) side of those events.
4. From one event many activities can start (including concurrent activities) and many activities can end at one event,
5. There can be only one activity between two events. If there is no activity between two events, they may be connected by dummy activity (dotted line) to show the time-sense.
6. The activity line may or may not have arrow head. In case arrow head is indicated, it must be from left to right (West to East),
7. Based on the activity time, the "Earliest Start Time, T_{ES}" and "Latest Finish Time, T_{LF}" is calculated and indicated on network (at least for smaller networks). Similarly "Latest Start Time, T_{LS}" and "Earliest Finish Time, T_{EF}" are also calculated. Based on these, the "**Slack Time**" for each activity (amount of time that the activity can be delayed past its earliest start time or earliest finish time, without delaying the job/ project) and "**Float**" for each event (the amount of time an activity can be delayed without affecting subsequent activities) are calculated. Float is, also called **node** by some users. Event number, earliest finish time (T_{EF}) and latest finish time (T_{LF}) are either written inside the event (if the event size is big to accommodate, as shown in Fig 7.5) or on the top of the event.
8. Each network will have at least one critical path, having zero slack and zero float. These critical paths are specifically marked on the network.
9. The total duration of Repair job is, thus, obtained by adding all the activity times on the critical path.

PERT vis-à-vis CPM- CPM and CPA are same with little different emphasis and both would be taken as CPM. PERT is used for bigger works, involving some uncertainties and first time jobs. It is a network model that allows for randomness in activity time. As such PERT uses three time estimates; "Optimistic", "Most-likely" and "Pessimistic" and based on these, Expected Activity Time is calculated—

Exp. Activity Time = [Optimistic Time + 4(Most Likely Time) + Pessimistic Time] / 6

CPM charts are similar to PERT charts and are sometimes known as PERT/CPM. CPM is a deterministic method that uses a fixed time estimate for each activity. While CPM is easy to understand and uses, it does not consider time variations that can have a great impact on completion time of a complex project or Repair Plan. In most maintenance application we use PERT/CPM model networks.

CPM may be considered as a smaller PERT and is more often used in maintenance.

Different types of PERT/CPM networks are being used in maintenance today; three of which are shown here as Fig 7.4, Fig.7.5 and Fig.7.6. Fig. 7.4 and Fig.7.5 are the same network except that in Fig 7.5 activity lines are suitably folded to make the network easier to be made on smaller paper. Again in Fig 7.4 and Fig. 7.5, events have been shown as squares for ease in making, but in actual use these should better be circles. In Fig7.6, the same activities have been arranged in squared network and $T_{EF,}$ T_{LF} etc can be mentioned within squares . All these networks need separate "Activity Lists" which consists of activity number, agency, time (duration), activity description and may or may not include earliest start time and latest finish time. Table 7.3 shows one activity list.

Table 7.3. Activity List for 900 Mill Shutdown

Sl. No.	Activity number	Agency	Time (Hrs.)	Activity Description	Earliest Start time	Latest Finish time
1	0-1	HM	8	Remove rolls		
2	1-2	HM	8	Remove Top and Bottom Spindles and Top Carrier		
3	2-3	HM	8	Remove Bottom Spindle Brg. and Bottom Spindle Lift Cylinder		
4	3-4	HM	8			
5	0-9	Contr.	32	Remove Oil of Lube Oil System No. 1		
6	0-7	FM		Remove Front Working Table Covers		
—	—	—		—		

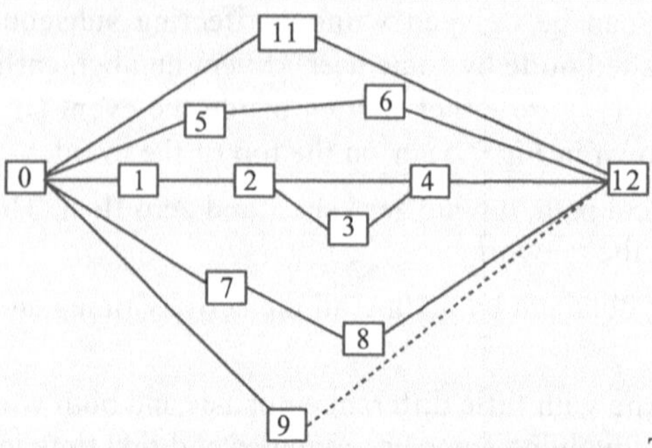

- Slack of activity 2-4 = T_{LS} -T_{ES}
  = T_{LF} -T_{EF}
- Free Float of activity 2-4
  = (T_{ES} of event 4 - T_{ES} of event2- Activity time)
- Total Float of Activity 2-4
  = (T_{LF} of event 4-T_{ES} of event 2– Activity time)

Total Float of any event can also be said as T_{LF} - T_{LS}

Fig. 7.4. PERT/CPM Network, Conventional-1

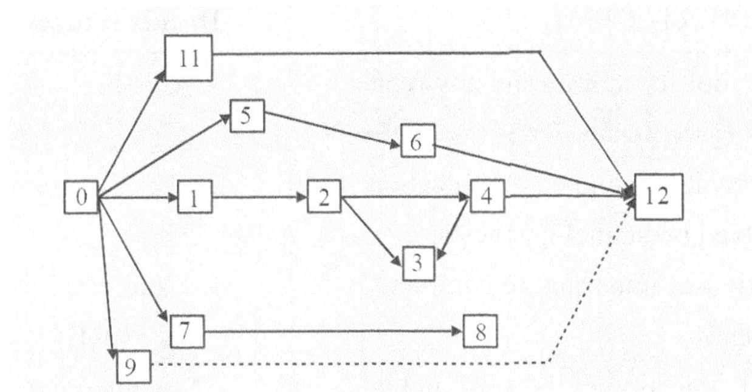

Fig. 7.5. PERT/CPM Network, Conventional-2

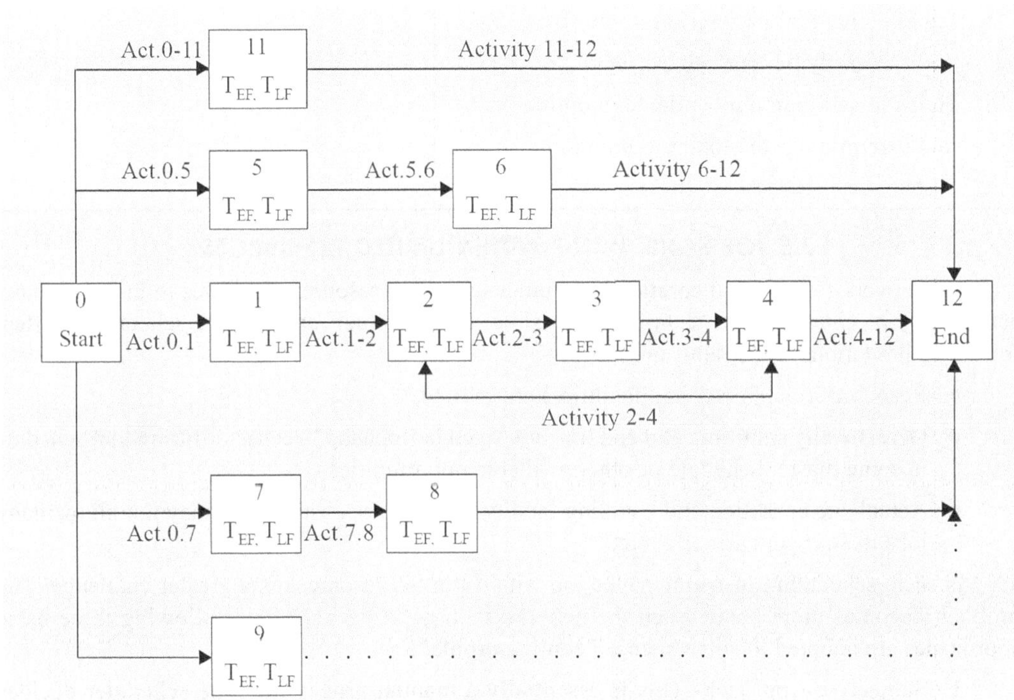

Fig. 7.6. PERT/CPM Squared Network

As for historical development, PERT was first used by the U.S. Department of the Navy in planning and carrying out the development of the Polaris submarine-launched missile programme in 1950 and CPM was developed by M/s Du Pont and Remington Rand around the same time. However, these are now extensively used in project and maintenance jobs.

There are some limitations in network techniques also. Though it has time-sense, it does not give any time comparison between the duration of individual activities, as the activity lines length depends on ease of making network. As such, in actual practice, a combination of Bar Chart and PERT/CPM network are used; Bar chart for execution and PERT/CPM chart for making bar chart and monitoring.

Advantages and Disadvantages of PERT and CPM— These are shown in the following chart–

Advantages (PERT/ CPM)	Disadvantages
1. Very effective tool for monitoring any repair plan or project,	1. Costly
2. Helps in understanding and communication amongst involved personnel/ agencies,	2. Time consuming,
3. Highlights activities important to complete the total job on time,	3. Poor estimates in activity times in PERT,
4. Shows interrelationship in workflow and is useful in determining labour and resources needs in advance,	4. Inclination to use pessimistic time estimates for activity time (in PERT)
5. An effective tool in controlling the cost and can easily be computerized,	
6. Helps in avoiding duplication and omissions and determining the total job duration systematically.	

7.5 JOB SCHEDULING WITHIN LIMITED RESOURCES

The network technique theoretically assumes unlimited resources and tries to make as many activities/ jobs concurrent as technically possible. But in actual maintenance scheduling, often, following limitations/ constraints are faced—

- Resources (manpower and facilities) are limited,
- Occasionally start time and end time are fixed beforehand because of fitting that schedule in some master schedule or other available opportunities,
- Actual site condition and handling facility constraints/ obstructions may not allow many jobs to start concurrently; etc.

As such, scheduling of maintenance job within limited resources is a greater challenge. The problem becomes more acute when the network is large. As such, often, following three basic approaches are adopted to solve such scheduling problems—

1. Subjective Approach- This is essentially a manual process and the scheduler decides subjectively which activity is to be deferred or to be made from concurrent to succeeding,
2. Mathematical Programming Approach- In this approach, considering all the constraints and limitations, a systematic search is made for optimal solution amongst various possible alternatives. The optimal solution is the schedule of activities which minimizes the completion time, while observing the activity precedence and resource limitation. This approach may use inter-programming and dynamic programming etc.
3. Heuristic Approach- This approach is also told as "Rule of Thumb Approach" and is, probably, the most suited approach for maintenance scheduling within limited resources. It involves devising simple procedures of activity scheduling that do not affect much the critical path activities and do not deviate much from the optimal ones. The heuristics, which may be used, include—

- Minimum job/activity slack time method,

- Minimum late finish time method,
- Greatest resource utilization method,
- Most job utilization method, etc.

The thumb rule, for priority allocation while scheduling, may include 'allocate resources to activity having least slack', 'if a tie, allocate resources to the activity that is in progress', 'if tie again, allocate resources to the activity requiring larges resource' and 'allocate resources to the activity that precedes minimum resource time requirement' etc.

7.6 SHORT TERM PLANNING

Like in other walk of life, in maintenance also we often use short term and long term plans. Short term plans are generally short term schedule and its duration may be hours, days, weeks or months and these plans can be adjusted or changed frequently. Short term plans should cover 10% to 15% of the schedule of total job and it should demonstrate continuity. These jobs should normally be derived from master schedule of total job. Short term planning of maintenance jobs generally includes small preventive maintenance, diagnostic and predictive maintenance, small defect rectification, adjustments and small corrective maintenance etc, which are planned and scheduled on day-to-day basis or at shorter notice. The persons/ group, engaged for short term planning, also ensures timely resource availability, coordination amongst connected agencies as well as proper documentation. Some of the short term jobs are given herewith—

1. Lubrication plans and schedules,
2. Small defect rectification,
3. Vibration monitoring of few critical equipments,
4. Lubricant monitoring, thermography and other monitoring as per need,
5. Small PM and PdM jobs, etc.

The planning and scheduling of such jobs are done, taking into consideration of plans of other departments, without necessitating the stoppage of running equipments or on scheduled off days or on days the equipment is not used. Occasionally, small shutdowns, for some urgent repairs, may also be included in short term planning.

7.7 LONG TERM PLANNING

Long term plans are generally for future and the duration may be months or years. Whereas short term planning is "tactical planning", long term planning is "strategic planning". Long term maintenance planning includes the jobs/ actions which are helpful in maintaining original level of performance in terms of output, efficiency and reliability. It also includes maintenance jobs which lead to improved availability and capacity of equipments (may be slightly better than rated/ designed figures), improved maintenance facilities and safety and environmental protection. This also calls for development of life cycle plans with anticipated projection of major replacement of components/ assemblies over the entire life cycle of the plant. These jobs call for more resources, coordination with many agencies and planning much in advance. Some of the long term maintenance plans, commonly used in plants and industries are given herewith—

7.7.1 Major Repairs and Capital Repairs

The term "Major Repair" is often confused with the term "Capital Repair". Major Repair includes jobs required to restore a seriously deteriorated or broken-down equipment to a condition of usability for its desired functions. In addition to maintaining the essential qualities of the equipment, these repairs may include minor modification/ redesign. **Major overhaul** may be another term to mean the same. Frequency of such repairs is normally once or twice a year, but may vary depending on the duty to be performed and reliability needed.

"Capital repairs", normally, means major dismantling and nearly rebuilding of the equipment /plant, needing lot of resources. The value of asset, normally, builds up original value or very near to that after capital repairs. At the time of capital repairs, jobs of major repair nature or otherwise may also be included. Frequency of such repairs is once in two years or even later.

7.7.2 Annual Overhauls and Statutory Overhauls

This apparently appears like major repairs/overhauls and may include job of similar nature but is generally taken on yearly basis (often in the form of "Action Plan' and are mainly taken for following reasons—

1. Annual maintenance cleaning, painting and corrosion prevention of equipments, machines, tanks, reservoirs, filters and heat exchangers etc,
2. Statutory overhauls and inspection and certification (by governmental or other authorized agencies) of equipments like boilers, pressure vessels, weighbridges etc.
3. Annual Operational Cleaning and painting jobs, like removal of mill-scales, dust and debris, cobbles and crop-pieces and reconditioning of deflectors, diverters etc.

Of-course, these repairs also includes jobs of major repairs nature and leftover jobs.

7.7.3 Renovation and Revamping

Renovation and revamping are, nearly, similar jobs. These differ vastly from capital repairs as renovation/ revamping jobs are taken rarely (after many years), after the equipment condition has deteriorated to such an extent that the amount of major/ capital repairs, PM, PdM and other maintenance repairs do not yield desired result and maintenance becomes uneconomical. These include almost total rebuilding of the equipment complex, using either similar type of technology with some redesign or improved technology and equipment. Up-gradation of production capacity and productivity may or may not be the main aim of these plans but up-gradation of equipment availability and reliability are definitely the aims for such plans.

7.7.4 Modernization

Modernization of the whole plant/ industry or a section/ shop of the plant is planned when the operation and profitability of the plant or section uneconomical or uncompetitive and no improvement is expected by plans mentioned earlier. Modernization jobs become essential when the existing plant/ technology has become outdated /obsolete, productivity is not competitive and other better/ cheaper alternatives has become available. However, plant engineering or project departments are more responsible for planning such jobs and maintenance department plays the supporting role.

7.7.5 Five Year Rolling Plan

In industries, generally, five year rolling plans are made for major overhauls and replacement of components. Such plans are essential as procurement of few spares/ components/ assemblies may need long lead-time and few may be expected to be obsolete. Every year this five year plan is reviewed and updated for next five year. These plans are mainly indicative and budgetary and from these, annual overhauls, major repairs and other long term plans are planned and scheduled in detail. The period of such rolling plans may vary from three years to ten years in few industries.

7.7.6 Turn-around Planning

This terminology is used for maintenance and repair/ reconditioning jobs of entire plant or few shops/ areas, which are nearly capital repair type jobs or jobs between capital repair and revamping natures. Turnaround jobs are planned much in advance as changing of long lead-time components are also involved. A committee, consisting of members of production, maintenance, project and technical services departments is formed and is headed by a "Turnaround coordinator". The committee oversees all activities from preparation of drawings to execution. The term turn-around was initially introduced for planning total plant operations for reversing the downward trend.

7.7.7 Corporate Maintenance Plans

This is also an indicative and reference plan, like 5- year or 10-year rolling plans, but is made to suit the corporate planning goals for total operation of its major plants/ industries, mainly including jobs like major repairs, replacements, modifications and revamping etc, which may be taken at different plants at different times. Many of such jobs need lot of money and need corporate approval.

7.8 MAINTENANCE TURNAROUND AND OUTAGE MANAGEMENT

These terms are recently used often in industries. **Maintenance turnaround** is as the planned outage of the shop/ plant to completely overhaul equipments and plant so that it will become almost new and will have a full operating life. Turnaround is phase during which actual maintenance is performed as per plan.

The term '**Outage Management**' has been taken from is taken from computer system to assist restoration of electrical power. For general maintenance system and for Turnaround, 'Outage Management' means effective management of total shutdown period.

With the introduction of advance technology, sophisticated and complex equipment, today any plant outage/ downtime (including turnaround) is very costly and, as such, need of turnaround has to be examined carefully and turnaround period has to be as small as possible.

Cost-effectiveness is the main criteria for turnaround. Following points to be considered—

1. Present condition and its impact on cost, may be due to more downtimes or quality or waste loss,
2. Possibility of cost-effective turnaround,
3. How much of the overall maintenance budget should be allocated to turnaround,
4. Amount of profit loss due to plant being off line during turnaround and below target production for few days after turnaround.

As such thorough advance preparation has to be done to keep the turnaround as short as possible. This is done in following phases–

A. Phase-1; Initiation – Period during which turnaround parameters are defined and core persons are allocated. This includes reviewing past records, monitoring present performance, setting objectives, delegating authority and allocating personnel. This phase can spread over a period of months. Often a turnaround in-charge is allocated with steering group, consisting maintenance, operation, engineering and project department representatives.

B. Phase-2; Preparation– This includes different failure analysis, planning, scheduling, material/spares, tools and special facilities procurement and reserving those, indigenous and outside manufacture of necessary components, necessary outsourcing, issue of work-orders, job cards, work permits, necessary safety and environment aspects etc. This period can spread over from 3 to 18 months, depending on the size and complexity of turnaround, during which a large quantity of data (both, technical and non-technical) are to be considered.

C. Phase-3; Execution– Period during which actual work is done as per schedule. This also includes various level monitoring, coordination and control and necessary rescheduling of resources and resolving arising issues etc. This period may vary from 2 weeks to 8 weeks, depending on the size of turnaround and, during that constant watch and control is done about work quality, safety, logistics and cost aspects.

D. Phase-4; Termination– This period starts immediately after turnaround jobs are completed and plant starts and the period may vary from 1 to 3 weeks. During this period, analysis of performance and effect of turnaround jobs are monitored, necessary learning and changes are established, disposal of special tools and facilities, disposal of wastes etc and preparation of final reports are done, with necessary noting/ remarks which may help next turnaround.

Necessary auditing of activities and cost are done at each phases. Almost all the activities/ works mentioned at the four phases above, are dealt with some details in different chapters of this book.

QUESTIONS

1. What do you mean by maintenance job planning? Discuss various steps of maintenance job planning.
2. What is a job manual and what are its uses?
3. What is difference between job planning and job scheduling? What different scheduling techniques/ methods are used in industries?
4. What are PERT and CPM and what is the difference between them?
5. In an activity 3-4, the earliest start time and latest start time is 6 and 11 hours respectively, the earliest finish time and latest finish time is 19 and 24 hours respectively, the three times, optimistic, most likely and pessimistic times are 5, 14 and 17 hours respectively. Answer the followings–
   (1) Show the activity 2-3 on paper with all the given details,
   (2) Find out the activity time,
   (3) Find out the slack and total float times
6. What is Gantt Chart and Bar Chart and how do they differ from PERT/ CPM.
7. Define slack time and Total float in PERT and CPM networks. How do you calculate Total float and free float?
8. What is long term plan? Discuss briefly few long term plans.
9. What is network technique in relation to maintenance job planning? What are the different networks commonly used?
10. What is Maintenance Turnaround?

CHAPTER

8 CONDIFICATION, CATALOGING AND SYSTEM APPROACH

management process, in addition to many others. The **Holistic Approach** to maintenance clearly identifies key maintenance and reliability activities, explains their interactions and how they can be integrated into the whole management process. It enables optimization of the whole maintenance process rather than focusing on individual elements or jobs. A good maintenance management process can be considered as having six phases, as illustrated below—

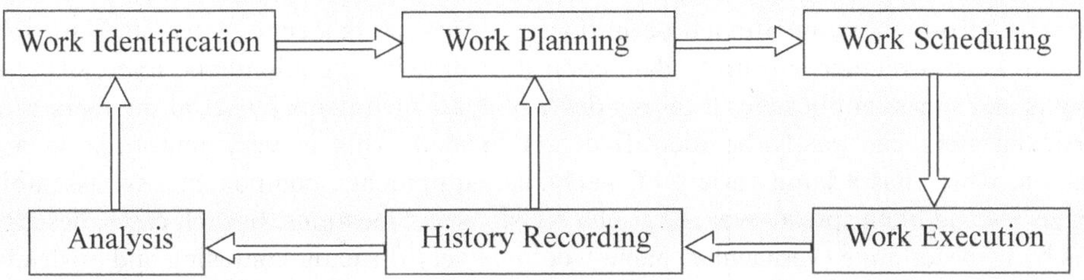

For proper identification and communication of all these six phases, a 'System Approach' to maintenance is developed which include proper cataloguing, codification and computerization of all the actions /activities, assets and materials related to maintenance of all the departments and work areas and integrating them into one system. The system approach provides an enterprise database that enables to capture and analyze data about current and historical maintenance work. It also helps keep track of the cost of maintaining any piece of equipment, work orders and labor time, and key performance indicators (KPIs) and benchmarks throughout the maintenance operation. CMMS has been discussed in next chapter.

8.1 DEFINITION

The system approach for maintenance starts with codification, classification and cataloguing. They are key value added and waste reduction functions of any activity or logistics in maintenance. Of these three key functions, the primary one is codification.

Codification– Though the term 'Codification' basically means the act or process of reducing to set laws and codes, it may mean slightly differently in different scenarios, such as the act or process of codifying (setting of written rules or principles or laws) for arranging in a systematic order (esp. in industrial or material scenario); act, process, or result of stating the rules and principles applicable in a given logical order to one or more broad areas of life in this form of a code (esp. for medical scenario); the process of collecting and restating the law of a jurisdiction in certain areas, usually by subject, forming a legal code (esp. in legal scenario); the process of standardizing and developing a norm for a language (linguistics scenario) etc. We would be taking the first meaning in this book.

Codification may be considered, in some extent, as classification or categorization, assigning codes (alpha-numeric, numeric or otherwise) for each categorized item or process or procedure for quicker identification in computerized or other documents. Codification manages information that is complex or could be ambiguous or misinterpreted, enabling its efficient, effective and timely use in business decisions. Information is streamlined through the consistent use of methodical and ordered principles and rules. Incorrect codification or no codification can significantly add to waste in terms of time, maintenance efforts, quality and money.

Cataloguing- It is a process of making a list or itemized display, as of titles, course offerings, or articles for exhibition or sale, usually including descriptive information or illustrations etc. Catalogues is a list or booklet or record of all the goods or processes or procedures. It is a complete enumeration of items arranged systematically with descriptive details. Catalogue is slightly different than Manual, which is, normally, a book which gives you practical instructions on how to do something or how to use or work with something, such as a machine.

8.2 CODIFICATION AND CATALOGUING

Extracting from above mentioned definition, codification is a time tested standardization technique to keep a regular and up-to-date control of drawings, equipments, materials, parts, components and subassemblies etc. It means describing all such items by set of numbers so that any particular item can easily be identified and located. This is very important in a big organization, which has a large variety of machines, equipments, components, subassemblies, parts/spares etc and often spread over at various divisions and locations. In such cases, describing the items by its descriptive (vernacular) name would be very difficult, confusing and misleading.

As an example, the same item may be called M-12 bolts at one place, may be called 12 mm Metric threaded bolt at another place and adjusting screw of carburetor at yet another place. It would be difficult for a normal man to understand that the items at all the three places are same, but computers will always that the three as three different items. For a very small organization, using manual records, such descriptive nomenclature may work to some extent, but for bigger organization, it is impossible.

As such, use of numerical identification (or alpha-numerical at some places), i.e. codification, is pre-requisite and absolutely essential for any industry of today.

Advantages of Codification- Following are some of the advantages of codification—

(*i*) One item has only one description and there is no chance of confusion or mistake,

(*ii*) The items can be described fully in much less space,

(*iii*) Searching and retrieval of any item (either manually from the list/ statement or from computer printouts or on computer screen) becomes much easier and quicker,

(*iv*) Each part code will give the part number and the part address/ location all the time,

(*v*) Flexibility of using spares from one division to another division of same industry and thus reduces chances of downtime as well as less inventory cost,

(*vi*) Codification is helpful for communization (standardization) of spares, variety reduction and import substitution etc,

(*vii*) Reduces the overall maintenance cost because of above mentioned advantages and otherwise; Etc.

Cares for Codification—-Following cares (or prerequisites) to be taken while developing or using codification—

- All concerned person of the industry should be made aware of the codes and codification system and there should be one system for all,
- For standard bought-out items/ components, confirming to national, international or reputed manufacturer's own codes and catalogues, a fresh code number, confirming to user industry's own existing centralized system may be given but some suitable characters/ digits should be introduced and a cross-linking has to be documented.

Such codification is not only used for maintenance but also by Materials Department, Finance Department, Project Department and others of that industry. For maintenance functions, following categories of codifications are used extensively—

- Drawing Codification,
- Equipment Codification,
- Parts/ spares Codification,
- Maintenance Jobs and Maintenance Defects Codification

8.2.1 Drawing Codification

Codification of drawing is linked up with material (equipments, subassemblies, spare and parts etc) catalogue/ codification. In a new plant/ industry, initial drawing cataloguing is done on the basis of supplier's part and drawing catalogues. But subsequently, with the start of plant's own codification of equipments and parts/ spares, the drawing codification may have to be adjusted/ modified suitably. However, ideally, the drawing codification is to be done first and equipment and part codification and cataloguing should follow.

An example (Fig. 8.1) of ten (10) digit codification of drawings (usually followed in some plants of SAIL, India) is illustrated herewith—

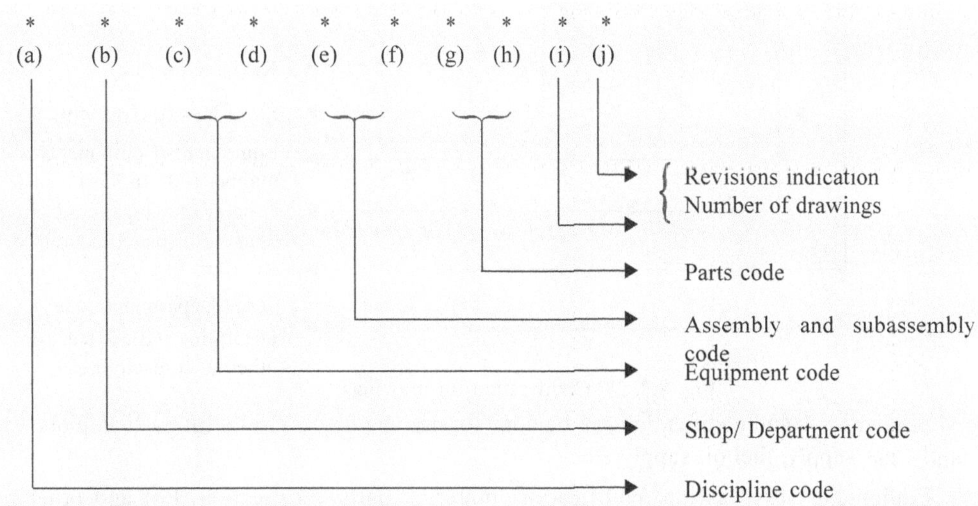

Fig. 8.1 Drawing codification

In the above figure (Fig.8.1), (a) to (j) are ten digits. The first digit, (a), indicates the discipline (say1 for operation items, 3 for mechanical items, 4 for electrical items etc). The 2nd digit, (b), indicates the shop number (say 0 for Steel Melting Shop, 1 for Blooming Mill, 2 for Sheet Mill etc), assuming maximum 10 shops/ departments in that industry. Digits (c) and (d), taken together, shows the equipment numbers in that shop, assuming maximum 99 (01 to 99) equipments in that shop/ department. Digits (e) and (f), taken together, shows the assemblies and subassemblies of any equipment, assuming that there can be maximum 99 assemblies/ subassemblies of ant equipment. Digits (g) and (h), taken together, indicates the part numbers of any assembly/ subassembly, again assuming a maximum of 99 parts of any assembly/ subassembly. For common items, which are used or located in many shops/ departments, the first two digits, (a) and (b) are taken together to describe the type of item (say 52 for EOT cranes, 63 for Lathe machines, 74 for Forklift trucks etc).

By adding last two digits, (i) and (j), the total 10 digits are converted to drawing codes. The digit (i) indicates the number of drawing for that part and digit (j) indicates revision number of that drawing (say 0 indicates initial (unrevised) drawing, 1 indicates the first revision, 2 the second revision etc)

Generally operational, mechanical and electrical items (including common items) are covered with this logic of codification. However, occasionally, for electrical circuit diagrams, slightly different codification is used, as shown in Fig.8.2, for following reasons–

(*a*) For shops alphabets (A to Z) are used in place of digits (0 to 9), as whole electrical power and circuit systems can be divided in many areas/ sections, which can not be covered by ten digits (0 to 9),

(*b*) One main drawing may have more than 99 sheets and, so, three digits are used,

(*c*) For last place at revisions, again alphabets (A ~ Z) are used, which permits more than 10 revisions in the long run.

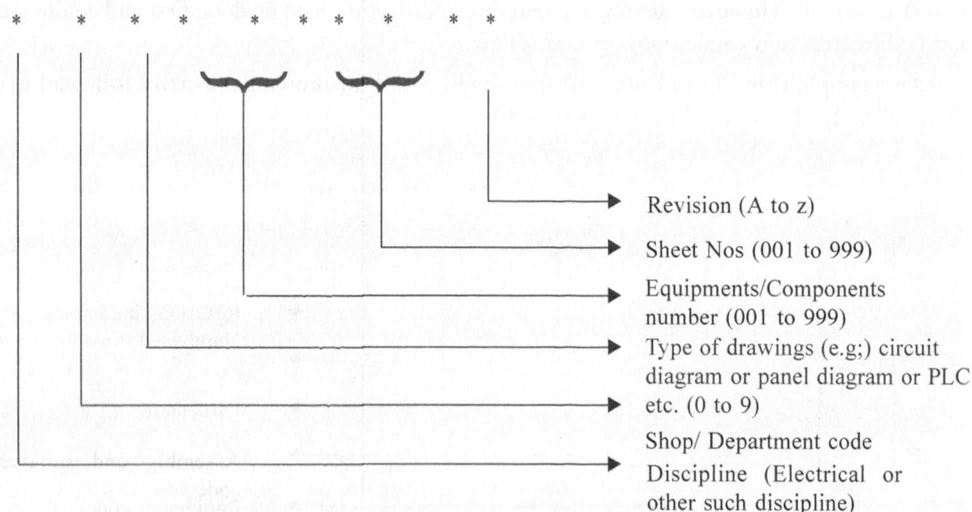

Fig. 8.2. 2nd Codification for drawings

Similar logic and codification can be done for drawings of central utilities, like, plant's water supply, gas supply, fuel oil supply etc.

Equipment classification/ codification, material/ parts codification, PM and other repair codification etc have been dealt in chapter 09.

8.3 HISTORY CARDS/ SHEETS

After codification/ cataloguing, next important thing is preparation of 'history cards/ sheets/ records' of each systems, equipments and major assemblies. This gives few essential information about acquisitions of asset/ material, which are permanent records and few essential information about maintenance and upkeep of the asset/ materials, which may have to be updated periodically/ regularly. These can be in the form of manual records like cards, sheets etc or in computers or even on company's websites.

Fig. 8.3 shows one typical 'History Sheet' from one of the steel plants of SAIL. In this, the top portion is the permanent records and the bottom two portions are to be updated regularly.

Such history cards are very essential and useful for maintenance. Few of the use are ——

(*i*) It keeps permanent records of the machine/ system and also of the major jobs done and constrained etc,

(*ii*) It provides ready guide and reference for planning repair jobs, shutdowns, tank cleaning, oil changing and other jobs,

(*iii*) It serves as a guide for spares planning and inventory control,

(*iv*) It monitors the consumption control of consumables like lubes and oils; Etc.

History Card/Sheet

1. Name of the Equipment—Lube Oil System of 3 Hi Mill
2. Code No.— — **********
3. Location—— Plate and Sheet Mill, ASP
4. Tank Capacity (including pipelines) —— 7,500 Liters
5. Type of Oil Used— —Servomesh SP-320
6. Pump Capacity and Pressure — —25 Liters/ min. at 2 Kg/ cm^2
7. Recommended Operating Temperature——40^0 C
8. Filtration/ Purification Details— — —********
9. End Use of the System— Lubrication of main reducers and pinion stand
10. System commissioned in 1968, by Mitshubishi, Japan

Oil Changed and Monitoring Details

Oil Topped-up		Oil Changed on	Oil Sent for Analysis		
Date	Amount		Oil sample taken on	Report received	Condition of oil
....	....		....	....	....
....	....		....	....	....

Major Jobs done

Sl. No.	Jobs Done on	Nature/ Details of Job	Spares Used	Constraints faced
....	....	....	....	....

Fig. 8.3. Sample History card for an Oil System

8.4 INSTRUCTION MANUALS

Instruction manuals or user's guide are commonly used by everyone for a variety of household items, games and other such items. Similarly, in industries, instruction manual is essential technical documents/ literature, which has to be supplied along with or at the time of receipt of any equipment/ system, be it small or big. In industries, many copies of instructional manuals are supplied just before the receipt of equipments/ systems (or along with equipment for smaller items) so that the same can be used right from the time of receipt (including de-packing, storing, installation etc). When newly developed equipment/ system is to be supplied, it is common practice to provide 'Provisional Manual' along with or at the time of receipt and 'Final manuals' (with required modification/ additions done at site during commissioning) are supplied after commissioning. The instruction manual may not be only in the form of a book, but may include many drawings, blown-up views, work cards, wall diagrams, video cassettes, micro-films, Power point or any combinations of these, occasionally both, as printout and electronically. These manuals must provide a permanent record of clear technical information, needed to understand, operate, maintain, dismantle, re-erection, along with any special precautions to be taken etc. For smaller equipments, instruction manual and operating manual may be clubbed together, but for bigger equipments, those are separate, may be in many volumes.

8.4.1 Information Required from Manual

The information needed from such manuals may be of following types—

(*i*) What is the equipment/system/plant and what is it for?

(*ii*) What does it consist of?

(*iii*) How to use it?

(*iv*) How does it work?

(*v*) How do we handle it? How to unpack, install and commission it?

(*vi*) How to maintain it?

(*vii*) What cares, precautions and safety measures are to be taken?

(*viii*) Details of improvements/ changes from specifications,

(*ix*) Who is the supplier, supplier's subcontractors and their credibility; Etc.

8.4.2 Sections of Instruction Manual

Based on the above mentioned requirements, the manuals are, generally, divided into following sections—

(*i*) **Technical data (Specification)**- This includes—

- Supply requirements, performance data (capacity, size, quantity and rate of output, delivery, volume, pressure, speed etc), dimensions, weight and other essential details about specification,
- Environmental factors, special safety and hazard warnings and precautions in connection with the use and handling of that equipment,
- List of data (reports, drawings, manuals/ leaflets of subcontractor's supply of bought- out items,

(*ii*) **Scope of Supply-** Items supplied by supplier/ manufacturer and items/ facilities needed from user/ purchaser.

(*iii*) **Operating Information**- This includes—

- Description of operating modes,
- Operating procedures (normal and emergency), with safety instructions. This also includes pre-operation and post-operation jobs/ cares,
- Operation's fault diagnosis and also necessary minor repairs, adjustments etc,
- Procedures for monitoring and reporting defects/faults/failures etc,

(*iv*) **Technical Descriptions-** This includes—

- Full technical details of equipments/ systems,
- Full technical descriptions of parts/components/sub-assemblies of equipments/ systems,
- Full technical description of supporting equipments/ instruments and facilities (power and drive sources, test gears/rigs, preparatory facilities, like roll changing rigs and installed condition monitoring equipments etc),

All the above mentioned details/ descriptions are supported with necessary drawings, blown-up views etc.

(*v*) **Handling Instructions-** This includes information about—

- Handling, transportation and unpacking, including removal of holding bolts/ fixtures,
- Erection, installation and connection details,
- Testing, commissioning and certifying (acceptance) procedures,
- Relocation, storage or disposal instructions, as applicable,

(*vi*) **Maintenance Instructions-** This includes—

- Various performance checklists and fault diagnosis instructions (daily, weekly, monthly and quarterly etc),
- Procedure about minor repair jobs, adjustments and calibration etc,
- Procedures about repairing and overhauling (dismantling, repairing and changing parts, reassembling) and testing etc. This also informs about need of any special tools/fixtures for dismantling/ assembling.

(*vii*) **Maintenance Schedules-** This includes-

- Periodical oiling and lubrication schedules,
- Preventive maintenance schedules and procedures,
- Calibration, overhaul and repair schedule,

(*viii*) **Spare Part Identification and Requirements-** For smaller equipments, spare parts are mentioned in the same manual, but for the bigger equipments, separate spares part catalogue/ lists are supplied. It generally includes–

- List of commissioning spares and essential spares for running for 2 ~ 3 years,
- Equipment identification drawings,
- Total spares part lists with manufacture's/ supplier's catalogue and drawing numbers; occasionally there are two lists—one for fast wear-out parts and another for rest of the spares,
- Illustrated list of specialized tools, rigs and fixtures etc.

8.4.3 Operating Manuals

Normally for smaller or medium equipments instruction manuals also contains necessary operating details and information, but for bigger and sophisticated equipments, a separate 'operating manual' is also supplied, along with instruction manual. Such operating manuals give more details about operating practices, procedures and parameters about the equipments and systems. In addition to those mentioned in point (3) of Sec. 8.3.2, operating manual emphasizes the following details—

(*i*) Procedure about initial starting and running of the equipment/ system,

(*ii*) Jobs and cares during regular running of the equipment/ system,

(*iii*) Procedure for shut-off of the system/equipment for temporary outage, including blanking etc of the furnaces/ stoves to preserve heat and for restarting after temporary/ short-term outage,

(*iv*) Procedure for shut-off of equipment/system for long period (longer outage) and for restarting after longer outage. Boilers, Steel Making Furnaces, Coke Ovens etc are some of the examples for this and previous point.

(*v*) Details about operating tools, tackles, jigs, guides, dies, rolls, cutters etc,

(*vi*) Regular safety measures, like blow-downs in boilers,

(*vii*) Jobs, cares, tests etc relating to quality outputs; Etc.

Frequent reference of instruction manuals and, to some extent, operating manuals is very essential for maintenance personnel. Instruction manuals help in planning and scheduling of various maintenance, repair, overhauling jobs, for erection and commissioning etc, for spares planning and control, for fault/ failure diagnosis and other related jobs.

For smaller items and household items, such manuals are often called as **Owner's Manual**.

8.4.4 Maintenance Manual

For medium or bigger equipments/ systems, often separate maintenance manuals are also provided by the suppliers/ manufacturers, which gives detailed information and instructions for repairs, dismantling, re-assembling, blown-up views of parts, spares and components, for fault diagnosis and for troubleshooting etc.

Basing instruction manual, operating manual, maintenance manual and on the experience of maintenance personnel and others, standard maintenance practice and departmental maintenance manuals (discussed in Sec.8.4 and Sec.8.5) are prepared by the department.

8.5 STANDARD MAINTENANCE PRACTICE (SMP) AND WORK INSTRUCTION

SMP can also be called as 'Standard Operating Practice (SOP) for Maintenance Personnel'. SMP is also called **Essential Maintenance Practice (EMP)** by some experts. However, EMP is used more in relation to health aspects. Whatever we may call these, these are prepared by user plant and department, based on available information as mentioned in previous section, for smooth, trouble-free and safe maintenance of equipments/ systems/ plant and for trouble-free operation and maintenance of central maintenance services and captive shops and, thus, ensure better equipment availability and reliability and also to ensure production of required quantity and quality.

In maintenance organization, both, SMP and SOP may exist. SMP for maintaining the equipments and facilities and SOP for operating maintenance facilities and captive shops (boilers, compressor house, fuel oil supply, fuel gas and other gasses storage and distribution, Central Machine and Repair shops etc). Each set of SMP tells its purpose, scope, reference, responsibilities, details of working procedures and safe working practices etc and these are for circulation to and use by concerned departmental head, area in-charges, section and crew in-charges, supervisors and others. Some SMPs may be specific to a particular equipment or a group of similar equipments and some SMPs may be of general nature.

Few SMPs of general nature may be following types—

(*i*) Lubrication procedure,

(*ii*) Equipment inspection procedure,

(*iii*) Breakdown maintenance procedure,

(*iv*) Planned repair and maintenance procedure,

(*v*) Procedure for taking shutdown on equipments and safety measures, before taking up maintenance; also handing over procedure after completion of repairs/ jobs.

(*vi*) Toxic waste and other waste disposal procedure,

(*vii*) Procedure for testing and calibration of condition monitoring gadgets,

(*viii*) Procedure for outsourcing in maintenance department; Etc.

Again, in maintenance organization, there are different '**Work Instructions**' to carry-out specific jobs. These are generally for individuals or smaller groups and are prepared and approved for maintenance services, facilities and captive shops, where the job is of specific and repetitive nature and where the operation and maintenance is done by the same department/ section. It is difficult to make 'work instructions, for breakdown maintenance or planned maintenance etc, which are covered by respective SMPs. Again, work instruction is different from job description as each work instruction tells its purpose, scope, reference, responsibilities, details of procedures and details of calculations etc. Few typical work instructions, commonly used in some plants/ industries are of following types—

- Refilling of LPG storage tanks,
- LPG distribution,
- Rich coke-oven gas analysis,
- Waste gas analysis
- Calibration of inside and outside micrometer; Etc.

8.6 DEPARTMENTAL MAINTENANCE MANUAL

Departmental maintenance manuals are detailed instructions of repairing, maintaining and overhauling etc of specific equipments. Some of the information about maintenance, upkeep, dismantling, erection and commissioning are given in the 'Instruction manual' and occasionally separate 'maintenance manual' supplied by different equipment supplier/ manufacturers. Departmental maintenance manual encompasses and modifies (if necessary) those and gives a comprehensive and detailed information, with necessary drawings, about minor and major repairs, overhauling and condition monitoring of equipments, both for bought-out items (e.g. pumps, compressors, electrical motors, switch gears etc) or specially manufactured items. Separate maintenance manuals are made for each such item. '**Workshop Manual**' is another name for such maintenance manuals. These manuals are, generally, set of information and instructions, with the help of which each such equipment/ component can be repaired, serviced and overhauled completely. For many smaller bought-out items, the suppliers 'maintenance manuals' are adopted as 'departmental maintenance manuals'.

8.7 DEPARTMENTAL MANUAL

Generally it is a permanent policy documents approved by competent authority, and it includes organization descriptions; delegations of authority; and policies, procedures, and standards for administrative, legislative, informational and other programme/ activities of the department. Departmental manual for maintenance organization is a document (normally separate for mechanical, electrical, civil and instrumentation maintenance/ departments), the purpose of which is delineate the organization structure of that department in the plant/ industry and to define the responsibilities and authorities of different sections of that department and of their supervisors and executives. Departmental manual also presents an overall view of the capabilities of that department. It normally covers the following areas and documents—

(*i*) Activities- Brief descriptions of different sections and areas of that department,

(*ii*) Organization Structure- Generally from the chief of the total maintenance organization (GM or DGM or AGM or I/c Maintenance etc), upto the level of section in-charge or crew in-charge,

(*iii*) Responsibilities and authorities of personnel of that department, upto section or crew in-charge (upto the lowest decision making level),

- List of maintenance procedures- It includes different SOPs for maintenance, SMPs, instruction manuals, maintenance manuals and other established procedures and systems for maintenance of all equipments and shops they are required to do, including the central services and captive shops etc,
- List of Work-instructions/ operating manuals/ SOPs for operating the equipments of central maintenance services and captive shops of the department,
- Highlighting emphasis on modern techniques used in the department, e.g. CMMS, TPM and other such techniques/ systems,
- List of Records/ documents to be maintained by the department-These include agreed and numbered records, e.g. Shift Log Reports, Daily Inspection Reports, Breakdown and Delay Reports, Equipment History Cards/ Records, Equipment Damage Report, various manuals etc. Of-course some sections/ areas of the department may use some other records/ documents also, which are not numbered and not part of departmental manual and such records may not be assessed/ audited by central agencies.

8.7.1 Maintenance Time Standard (MTS)

Maintenance Time Standard (i.e. standard time for various maintenance jobs) is an effective means for planning, scheduling, controlling and evaluation in order to cut down costs, not only in terms of money but also in terms of mental stress and strain and frustration, that a chaotic control causes.

However, occasionally it is questioned that is it feasible to set maintenance time standard and often following arguments are put forward for that–

(*i*) In maintenance the variety of jobs are to many and many of those are non-repetitive, unlike the case in production/ operation and, as such, setting time standard is very difficult,

(*ii*) The work-condition is not fixed and may vary from time to time. As an example, in one case one bolt may be opened in one minute, using simple tool, while in another case, the same or similar bolt may not open even by special more robust tool and in that process, the bolt head may get sheared and more elaborate process may have to be taken to extract the stuck broken bolt. This could happen either by excessive jamming that time or inferior bolt was used last time. As such, how uniform maintenance time standard can be established even for repetitive jobs.

(*iii*) Another problem may be the variability in the method used (even within SMP for that) by different workmen. In maintenance, it is more difficult to specify the job, motion by motion, as it may be possible for other repetitive jobs.

Yet it is possible to develop maintenance standard time with reasonable accuracy. The above mentioned apprehensive even if they occur, their percentage will be much less and that should not deter to fix MTSs. **Comparative Estimating** is a work measurement technique which can be used for setting time standards on low repetitive tasks, such as maintenance jobs. It has been in use for quite sometime and has probably been more commonly known as **Universal Maintenance Standards (UMS).** In view of developing technologies, such MTSs or UMSs should be revised every 15 to 20 years.

Apart from the modern techniques for setting time standard mentioned above (Comparative Estimating and UMS etc), followings are few of the age-old methods for setting-up (arriving at) maintenance standard time—

- **Breaking down the job into smaller measurable elements-** Even if most of the job may be of non-repetitive nature on overall basis, if the job is broken down in small measurable elements (like tightening M12 bolts, removing cover plates, welding 50mm long seam of 8mm fillet etc), time standard can be set for each element and logical sum of all such elements will give the standard time for that job. Necessary help of O.R. or Motion and Time Study can be taken.
- **Range of Time-** Instead of setting up exact time standard, a close range of time is set up as MTS for that job. As an example, a workmen may be allowed 5 ~ 6 hours for total job of changing one 4 inch valve in a pipeline system. The range of time is not the optimistic and pessimistic time of PERT, but a closure range. Though 95% of jobs may be completed during this range, some problems may be faced in major repair jobs, when times for all jobs are added to find the total time for the complete work.
- **Statistical Analysis of Past Records-** This would need segregation and analysis of past records for considerable period, during which each job was done at least for 3~4 times. However, MTSs, calculated in this way, will be less challenging, as all the past delays will be legalized and assumed to occur again.

In the methodology of setting up maintenance time standards, analysis of all (100%) jobs may not be possible as either that would be too much of a job or some job might not have occurred so far. Generally a number of key jobs (benchmark jobs) are analyzed and their time standard is setup. These standard times or key job times are then synthesized in some formulae to find out the standard times of other jobs. Alternatively, the work content of unmeasured jobs are compared with the work content of measured key jobs and then the MTS of unmeasured jobs are computed. While setting up such time standards, time likely to be consumed in non-productive jobs (e.g. obtaining instruction, travel time, time for collecting tools etc) are also to be considered. Wrench time (discussed in chapter 7) is also given due consideration.

Such maintenance time standards (MTS) are quite beneficial on following counts—

(*i*) Increased performance with decreased cost,

(*ii*) Reduction in delays through improved planning and scheduling,

(*iii*) Better supervision; etc.

8.8 MAINTENANCE-OPERATION LIAISON

Effective communication and coordination with operating personnel or the user if the equipments is very essential at all level of maintenance hierarchy. This is done by maintaining proper liaison with operating personnel at all levels. This will reduce confusion and delays considerably. This horizontal communication creates a feeling of personal touch and ensures better cooperation.

Few illustrative benefits are given herewith at three different levels—

(*i*) At lower level, i.e. direct contact between workmen/supervisor/ junior executive levels, which lead to—

- Issue of requisition of service (adjustment, repair, calibration etc) by operational supervisor timely, as well as informing personally,
- Better regular inspection of running equipments by maintenance personnel,
- Immediate feedback of any deviation in operating condition and onset of any defect, as observed by operators/ operating personnel,
- As, often, maintenance personnel for any particular equipment change but the operators for that equipment, normally, do not change, the new maintenance personnel can easily know the past history and behaviour of that equipment.
- Joint discussion with operation personnel on any improvement or betterment for the working of equipments/ systems; etc.

(*ii*) At planning level, where maintenance planning personnel liaise with operational planner, not only for initial programme, but also on a regular basis to discuss the followings and necessary changes in plans and schedules in time—

- Additional requests for services/ jobs by operating personnel,
- Work postponement, as a result of emergency demand/ job to maintenance personnel,
- Work delayed by changed operation's requirements or overshooting of time by operators,
- Work not able to start or held-up, because of non availability for some materials or facilities, so that operators could utilize that time; etc.

(*iii*) At senior level, i.e. direct contact between manager or in-charge of maintenance and operation of that shop/ area, which leads to—

- Advise on technical matters and improvement/ betterment of equipment's performance,
- Steps for improved safety measures for operating and maintenance personnel. Also for improved facilities for operating or maintaining the equipment,
- Minor adjustments between maintenance and operation budgets to accommodate unforeseen jobs; etc.

(*iv*) Other Benefits, because of effective communications between maintenance and operation and thus avoiding any strained relationship, may be followings—

- Reduction in number of calls to maintenance men, as operators develop a habit of fault diagnosis to some extent,
- Improved operation's appreciation for technical features of equipment and need for safe Operation of equipments,
- Improved energy conservation and economy,
- Cooperation/ help in fault finding and repair, which is one step in the direction of total productive maintenance (TPM),

- Identification of variations in input of materials or in environmental status, which may adversely affect the equipment or workmen,
- Improved handling and availability of maintenance and operation records,
- A higher standard of cleanliness for equipments, working areas and surroundings; Etc.

Maintenance department should have similar liaison and communication with other departments, such as, Technical and Engineering Services, Plant Development/ Project Department, Materials Department, Finance Department etc.

8.9 MAINTENANCE WORK ORDER AND WORK PERMIT

Though the term 'Work Order' is extensively used for capital shops utilization, for maintenance jobs, the term is used as 'Maintenance Work Order'. It is used to direct and authorize maintenance persons to perform a required task. It can also be said as a format used for detailed information about carrying out a job/ work. All maintenance jobs are initiated through a written work order (in the form of card or sheet or computer print-out), monitored till completion and finally documented into history system. Often the planning, scheduling and monitoring of all the jobs/works are achieved through a 'Work Order Control Center'. Such work orders can be issued by various wings of maintenance department (for planned preventive/ predictive maintenance, inspection and surveys and overhauling etc), by other departments needing maintenance services (breakdown or repair jobs) or by Work Order Control Center (Central Planning) etc, based on requisitions/ requests and inspection reports etc, but all work orders are numbered and coded by control center.

A maintenance work order generally gives following information/ requirements—

(*i*) Work order number and code,

(*ii*) Issuing department's /section's address and code,

(*iii*) Issue Date- Date the work order is issued,

(*iv*) Approved by— Sign of the competent officer which gives authority to concerned maintenance personnel for carrying out the job,

(*v*) Receiving date— date the concerned maintenance group/ section received the work order,

(*vi*) Priority— If the work order is to be taken on priority; in some plants priority type and code are also mentioned,

(*vii*) Location— location where the job has to be done,

(*viii*) Equipment— Equipment (with equipment code) on which job to be done,

(*ix*) Description— Brief description of job to be done, with job code if available,

(*x*) Materials (spares etc) needed or to be used—

(*xi*) Special facility or clearance needed-

(*xii*) Completion Date and Report—

The work order may also have scope of some other information, e.g. manpower used, cost of maintenance jobs done, other important jobs to be taken up (if any) and Work Permit issued etc.

'Work Permits' are also part of work order management. Work Permit is a document that certifies that the issuing authority has cleared the necessary equipment to allow safe maintenance work to begin. Such work permits can be issued by two departments—

- By concerned operation shop/ department to their maintenance counterpart, to begin their maintenance jobs. Before issue of such permits, the issuing shop ensures that equipment is shut-off and safe. Often, smaller jobs do not need separate work permit and the maintenance supervisor mentions that on the work order, after ascertaining the safety etc.

- Maintenance Department issues work permit to different executing agencies (mostly to outside parties for outsourced jobs), permitting them to start their jobs.

A work permit mentions 'work permit number', 'work order number' on basis of which work permit is issued, department/ section/ agency (with designation) from whom shut down is to be taken and many essential information as mentioned in the corresponding work order.

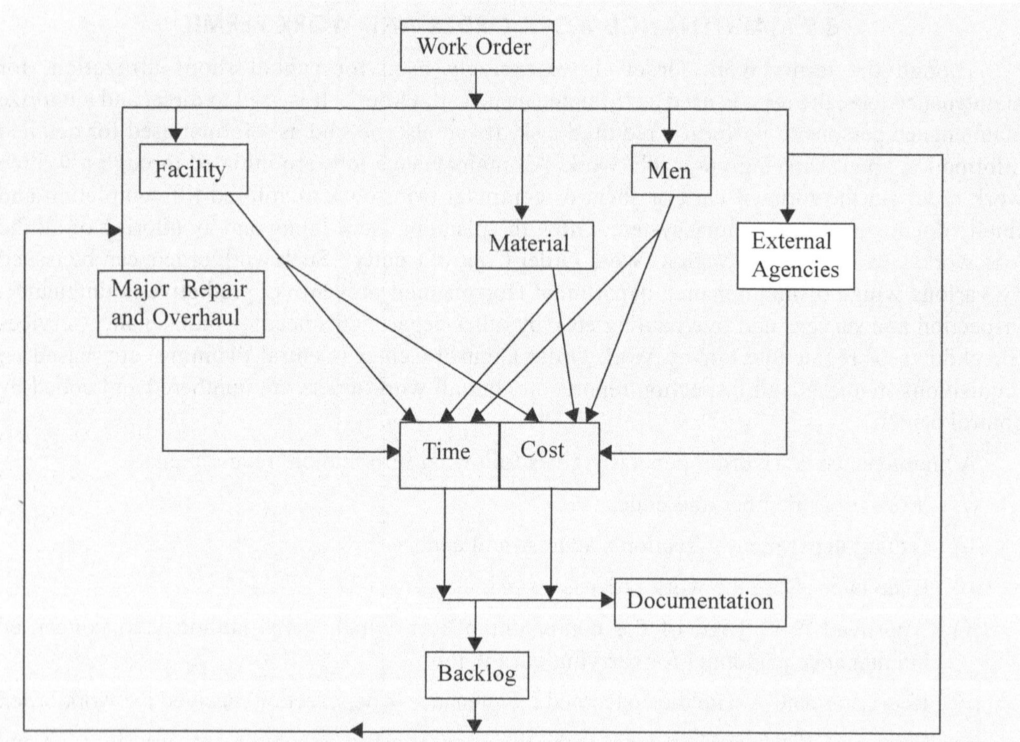

Fig. 8.4 Block Diagram of Work Order Control.

Fig. 8.4 shows a block diagram, depicting briefly a 'work order control system'. It also shows the 'Back Log' of work order. The 'work order back log' reflects how best the work order scheduling has been done and how best the resources have been utilized. The problems may be quite a few. The main problem is to keep the work order back log within limit which, in-turn, reduces the work orders offloaded to external agencies/ contractors. The work order back log should also be discussed periodically with concerned agencies if there is any ambiguity in the work order or if the need of job still exists and if rescheduling or additional resources are needed.

Fig 8.5 shows a 'Work Order Tree' for a very big/major work order, which indicated the areas and resources to be planned, arranged and scheduled for execution of the work order.

Based on work orders, detailed job schedules and job cards are made. Fig. 8.6 shows another view of work order and work permit control, indicating how and when work stops and again restarts.

8.10 JOB CARD AND JOB CARD PROCEDURE

'Job planning', 'job scheduling' and 'job manual' have already been discussed in chapter 7 of this book. Job card differs from job manual to the extent that job manual is a little general

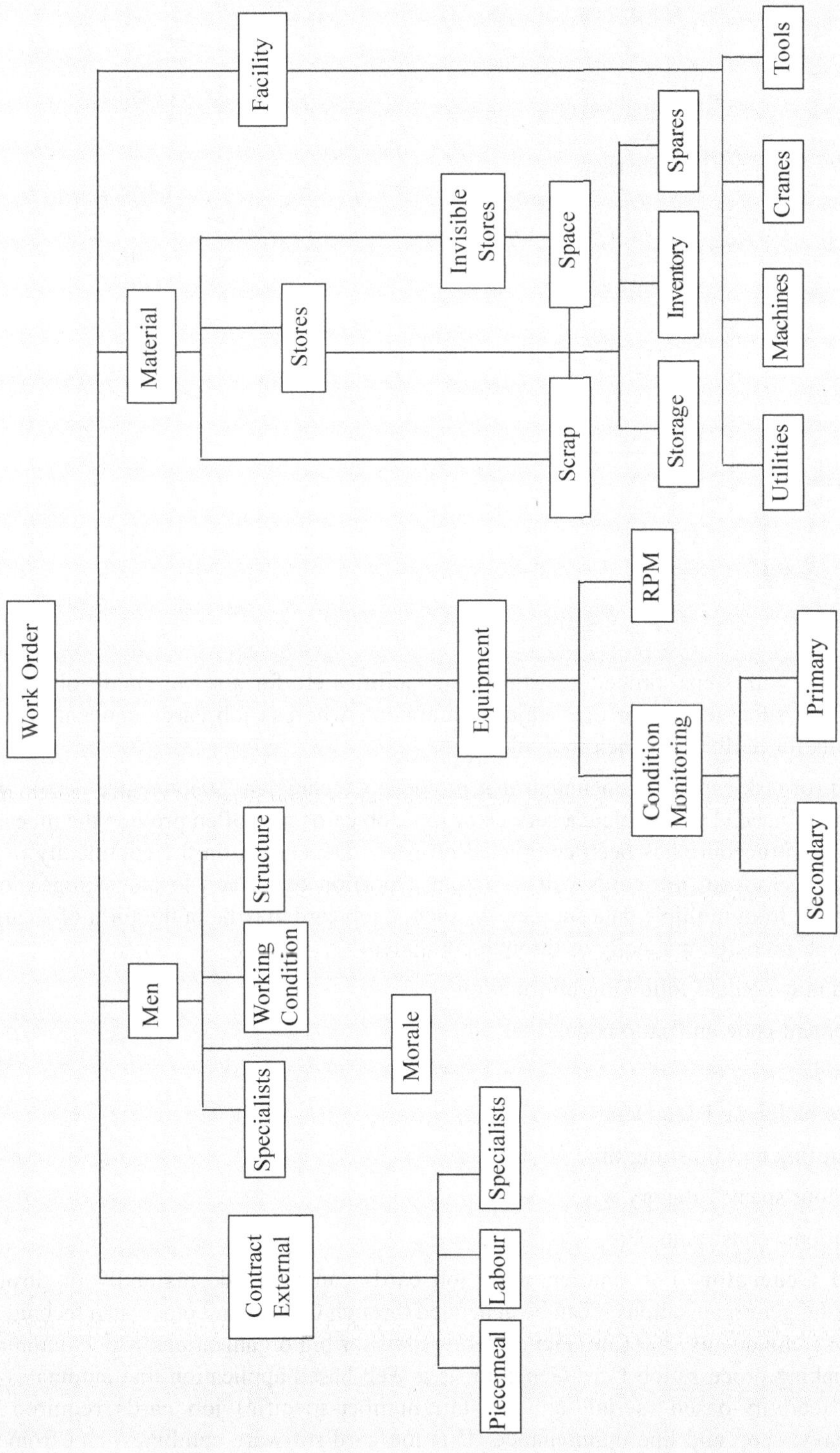

Fig. 8.5. An illustrative Work Book Order Tree.

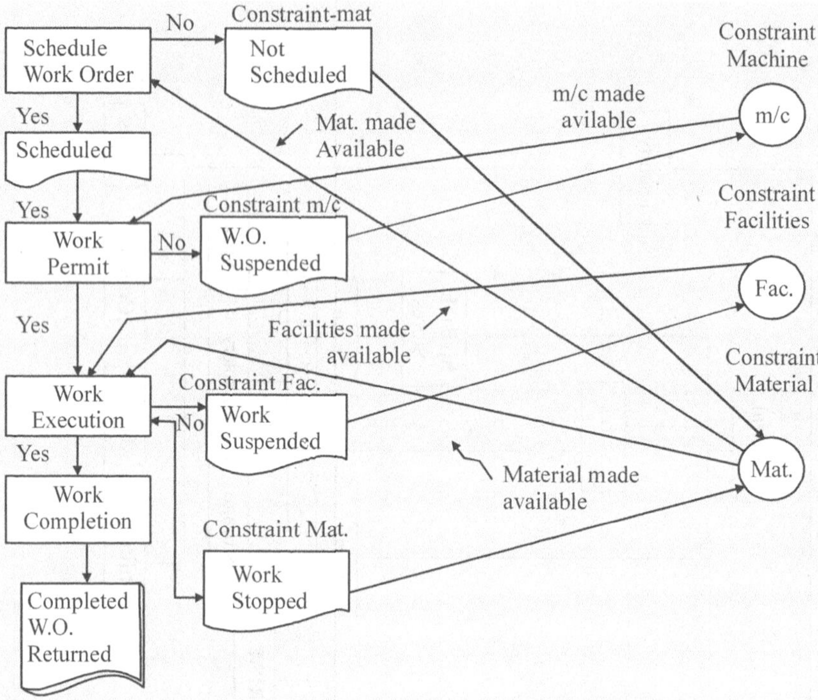

Fig. 8.6 Work Order/Work Permit Control

literature, dealing with steps, procedures, time and facilities etc for a set of small jobs of one equipment or one subassembly or one major component, whereas job cards concentrates for necessary details for performing each individual job.

A job card (or task card) is a document that provides a technician/ maintenance person with all the information needed to complete a service or job. Job cards also often provide the means of recording that the procedure has been completed properly. Depending on the complexity of the equipment being serviced, job cards can run from a portion of a page to many pages, with information drawn from multiple data sources. As such, a job card may be in the form of a card, a sheet or print-out, prepared manually or through computers.

A job card may contain following information—

- Equipment code and shop code,
- Job code,
- Nature of Job and Job Details—
- Job starting and finishing time,
- Man-hour spent; category wise,
- Deviations/ constraints; Etc.

Job Card Generator- For smaller units, job cards can be made manually or through computers. For bigger organizations it can be generated through CMMS and other such techniques. Recently, some technique as 'Job Card Generator' is used for big organizations which automates the job card making process. Job Card Generator is a Web-based application that automates the creation of affectivity-based (serial number/ tail number-specific) job cards required for performing heavy, shop and line maintenance. This job card software combines data from the maintenance and engineering system with the OEM content to produce a job card, even for unscheduled maintenance. The job card can be printed on paper or used electronically.

The completed and filled up job cards are signed by concerned maintenance personnel and sent to central planning and documentation section for necessary processing and compilation. The compiled information from job cards gives following feedback to maintenance engineer/ plant engineer—

- History of maintenance (History cards/sheets are updated as per need,
- Frequency of same job being done on same equipment,
- List of equipment where maximum resources were used,
- Job auditing and average maintenance time for various jobs,
- Cost of downtime,
- Category-wise manpower utilization; Etc.

Job Card Procedure- This means the procedures and steps associated with job card preparation and tracking/ follow-up, which are, generally the followings—

(*i*) Job lists/ requisitions received from control office or job planning office,

(*ii*) Job card preparation and issue for every job,

(*iii*) Job cards placed on loading board/ rack against the names of workers allocated or their supervisors,

(*iv*) Completed job card checked by issuing supervisor,

(*v*) Notes made for corrective works if any and fresh job cards made issued and placed in loading board/ rack,

(*vi*) Job list signed and summary of manpower completed,

(*vii*) Job list and job cards sent to control office or central planning office for documentation.

8.11 JOB EXECUTION, MONITORING, FEEDBACK AND CONTROL

After all the earlier mentioned steps have been taken, the jobs/ work are actually performed by different tradesmen/ workers in scheduled sequence and as per revised schedule, if any from time to time, till the jobs are completed.

8.11.1 Monitoring

Constant follow-up and monitoring, during the whole period of repair and maintenance, is very essential so that—

(*i*) Any deviation in time schedule can be immediately corrected by additional resources or facilities and, if absolutely essential, revised time schedule can be made and communicated,

(*ii*) Any deviation/ change in job content can immediately be incorporated in plans/ schedules/ charts and communicated to all concerned agencies,

(*iii*) Any constraint or deviation in technological methods can be corrected with re-thought and/ or improved technologies/ methods. Any improvement in technology/ method, over the scheduled ones, should also be recorded for future use.

Thus, constant follow-up and monitoring serves as an effective control for the total repair and maintenance jobs. Various methods/ techniques/ tools/ systems are used for such monitoring; a few of which are given below—

(a) **Daily brief meeting** (for small urgent jobs twice a day, if needed), at fixed time, with authorized representatives of all concerned agencies, to monitor deviations/ constraints over last meeting, anticipate/ visualize deviations/ constraints expected if any and accommodate necessary reschedule of resources of resources and facilities. Also decide priorities amongst various agencies for special facilities (mobile cranes, special machining facilities etc) so that waiting time is reduced and wrench time is increased, thus maintain or improve the time schedule.

(b) **A repair coordinator** to check and update/mark the progress in PERT, CPM and other networks or 'Wall mockups', Bar /Gantt charts etc, about twice a day and intimate the deviations to concerned agencies,

(c) **Line of Balance (LOB) method**-It is a method of showing the repetitive work, that may exist in any major repair or project programme, as a single line on a graph, rather than a series of activities on a bar chart. LOB charts can be used for any repair or project where there are a number of separate but similar activities or one activity with long duration. They are not well suited to individual activities, which have short duration and are undertaken in isolation.

Unlike a bar chart, which shows the duration of a particular activity, a LOB chart shows the rate at which the work that makes up all the activities, has to be undertaken to stay on schedule, the broad relationship of one group of activities to the subsequent group and if one group of activity is running behind schedule, how it will impact the following group.

(d) **Senior Level meetings**- Such monitoring meetings may be held weekly or fortnightly to assess and take major decisions about deviation and resources re-allocation etc. The level of participants depends of size of organization and also importance of job.

(e) **Progress/ monitoring reports**- Different progress and monitoring reports are issued at regular intervals (depending on the total job period) to all concerned agencies so that they can prepare and take necessary actions.

8.11.2 Feedback

It is slightly different from progress/monitoring report. Feedback is the inter-departmental or intra-departmental (within the maintenance department) communication on related topics/ matters. As one example, it is important for the maintenance planner to know (get feedback) if his planning and work package are providing trades/ workers with everything they need to perform the work requested on the work order or any improvement is needed. Feedback can be in the form of test reports, charts, graphs, cassettes or on-line on computer network (CMMS or EAM etc) or Internet web-network.

Though the feedback in an organization has a vast scope, here the term 'feedback' is limited to various information for proper control of maintenance and repair jobs. It is little more than 'job completion report' and may include –

- Job completion report (with constraints if any),
- Unscheduled/ pending job/work report,
- Ongoing work status report,
- Suspended or stopped work report, with reasons,
- Manpower engagement status report,
- Cost report; etc.

8.11.3 Job Planning Survey

This is a different type of feedback, which improves the quality of planning and scheduling. 'Job Planning Survey Form' should be completed by the supervisors, overseeing the work. The supervisor should seek input and assistance from appropriate technicians/workers. The completed job planning surveys should, then, be reviewed by planners, schedulers and concerned maintenance engineers to assess where the improvements are needed. Such job planning surveys, like any other surveys, often are voluntary and should include a set of questions, such as——

- Job instructions were—Clear/ vague/ incomplete/ misleading/ others;—Explain
- The estimated work force was— right size/ too small/ too big/ others— Explain
- Actual work performed, as compared job order work content was— less/ same/ more
- Are the work-orders/ job cards properly coded for the type of work and location—
- Etc, Etc.

Based on above mentioned monitoring and feedback etc, the total control of repair and maintenance jobs can be done effectively and maintenance time and cost can be optimized. Such control is done at various levels, within their span of control.

Fig. 8.7 shows one typical 'Maintenance Control System'. In Fig. 8.7, there is constant interaction between maintenance shop supervision and maintenance job executers and the unfinished or suspended jobs are again given for execution after the constraints are removed or the materials/ facilities are arranged.

The maintenance control functions embrace almost all the activities of maintenance and their implementation is achieved through suitable delegation of responsibilities at different levels of the organization. Of-course, budgetary control (refer chapter 14) is also a part of maintenance control.

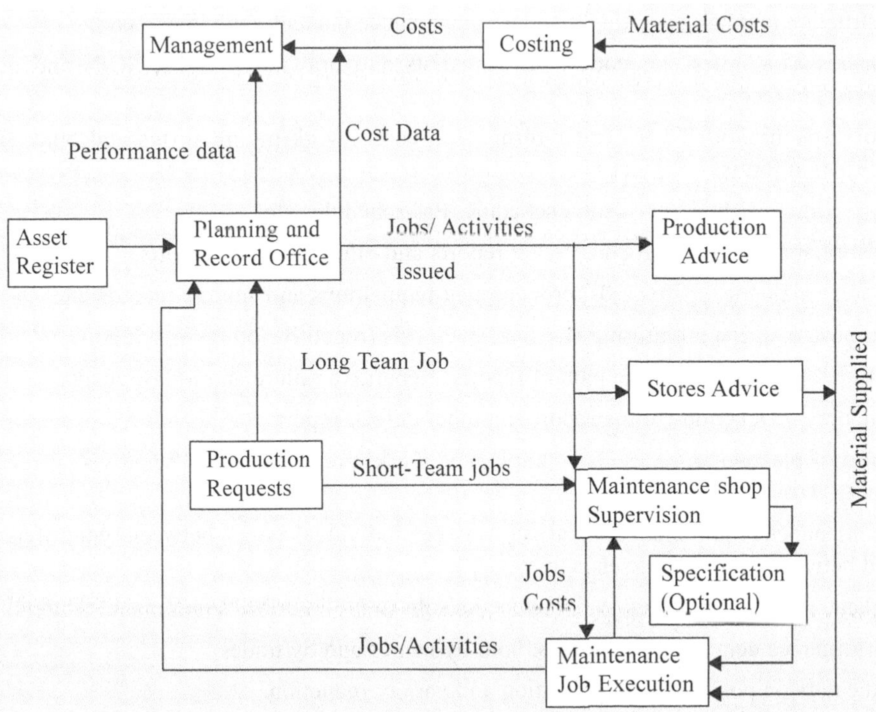

Fig. 8.7. Typical Maintenance Control System

8.12 Maintenance Records and Documentation

Maintenance records and documents are information/data pertaining to equipment's installation and subsequent maintenance done, condition/ defects observed and rectified, various plans/schedules implemented and also various plans/ schedules for future etc. The terms 'documents' and 'records' have little different meanings.

Documents are written policies (indicating what to do); process descriptions (indicating how it happens or works); work instructions, procedures (indicating how to do), and some blank forms or blank portions. Normally processes lead to procedures. These are mainly used to communicate information. Good documents are clear, concise and user-friendly.

Records are worksheets, registers, log books, forms, charts and labels etc and are used to capture information, activities or results, when performing a procedure. Records are essential for monitoring, tracking, failure identification, revisiting information and references etc.

These records/documents may be in different forms and are stored in a system in such a way that they can be retrieved expeditiously. Some data are of permanent nature, (e.g. instruction manuals, drawings etc) and some are to be continuously updated (e.g. history cards etc).

Recording maintenance information involves money to an organization and it is wasted if recorded information is not analyzed to aid in decision making process to control on maintenance cost, improve availability, reliability and overall efficiency. As such, management has to take a decision on the various records and documents that are to be used and how the actual recording will be done on those.

Some of the necessary maintenance records and documents have already been mentioned in this and previous chapters. There may be some repetition but a few are mentioned here–

(*i*) History Records/ cards,

(*ii*) Instruction manual, operating manual, maintenance manual, job manual etc,

(*iii*) Various catalogues and codes, drawing lists and drawings, SMPs, SOPs and other permanent records,

(*iv*) Work specification, work order forms, Job cards, work permit forms and other such instruction forms,

(*v*) Inspection schedule, lubrication schedule, PM schedules etc,

(*vi*) Performance reports, job completion reports and other feedback reports,

(*vii*) Spares catalogue, spare cards, procurement requisitions and storage records etc,

(*viii*) Various condition monitoring and fault diagnosis (repetitive/otherwise) reports,

(*ix*) Failure analysis and failure pattern reports; Repetitive failure and MTBF analysis, etc

(*x*) Production Delay/downtime report and other defect/ reports,

(*xi*) Ratio of planned work vis-à-vis unplanned work; preventive work vis-à-vis corrective work;

(*xii*) Maintenance requirement comparison between individual assets, between types of assets and between groups of assets, etc

(*xiii*) Indicators or reliability of equipments and subassemblies of different manufacturers,

(*xiv*) Performance details of individuals, both by names and by trades,

(*xv*) Indicators on possible standardization and variety reduction,

(*xvi*) Cost and Budget data; etc.

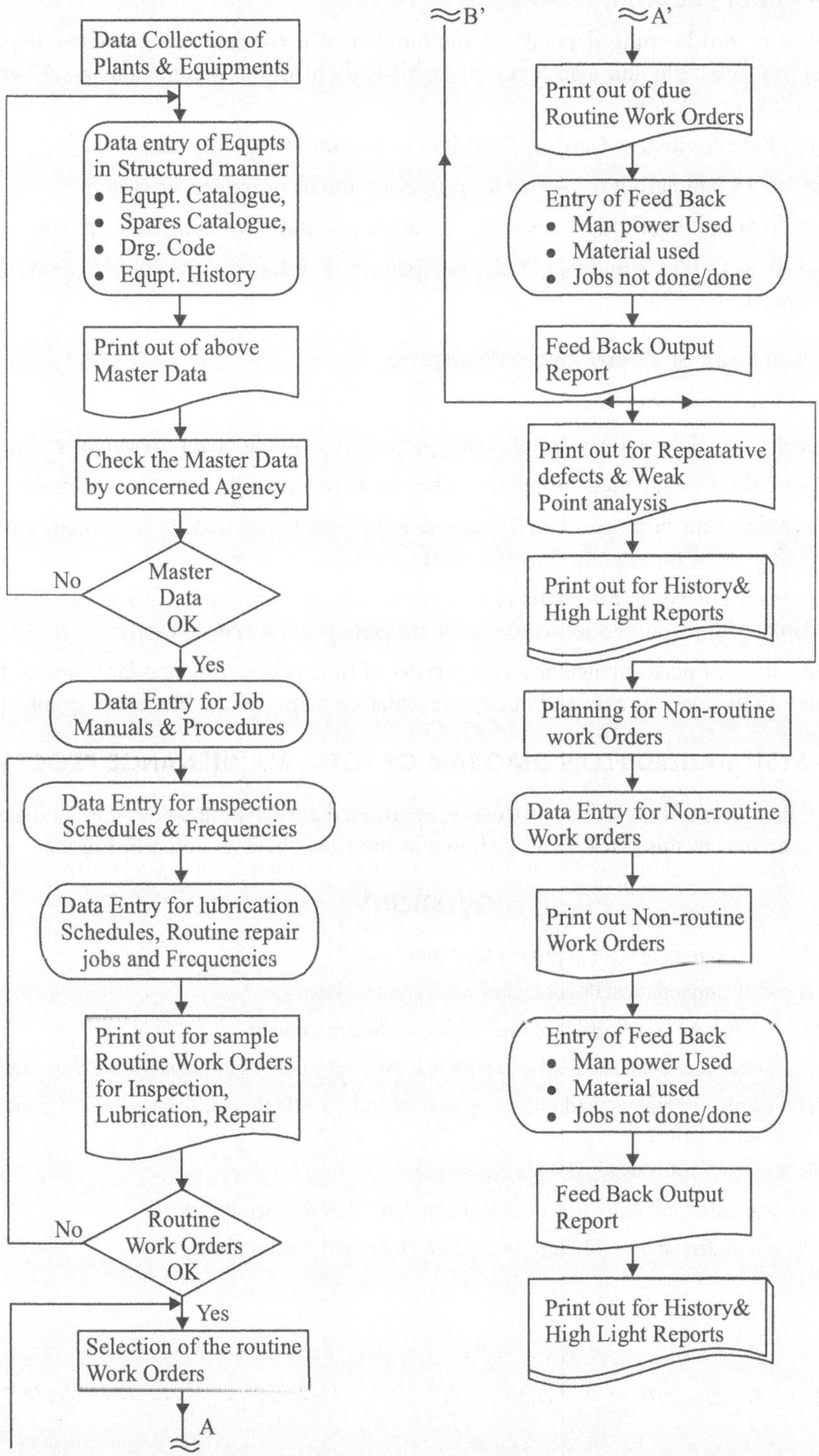

Fig. 8.8. Systematic Flow Diagram of Maintenance Functions.

8.12.1 Method of Record Keeping

Method of record keeping depends on the number of activities a maintenance department/ organization has to handle and also level of sophistication in that plant/ industry. A very rough guideline is mentioned below—

- 0 to 5000 activities— Cardex, Card-index, or computers (if in use),
- 1001 to 15,000 activities— Shorter/printer (manual) or computer system,
- 5000 to 50, 000 activities— Shorter/ printer (automatic) or computer system,
- More than 15,000 activities— Fully computerized system (CMMS/EAM) or Web-System (Internet).

8.12.2 Advantages of Good Recordkeeping

Though this is quite obvious, a few are mentioned below—

(*a*) It presents a clear and ready reference picture of maintenance programmes,

(*b*) Jobs held-up/ suspended/ stopped can be identified quickly and reprogrammed,

(c) Job cards/ work orders etc can be copied quickly and modified for next requirement, thus avoiding lot of typing and drawing works,

(*d*) Necessary records and formats can be retrieved easily and quickly and can be modified/ updated with respect to job contents or frequency or otherwise easily.

(*e*) Evaluation of performance and comparison of time taken with standard time or previous time can be done quickly and in any presentation form (text, table, chart, graph etc) ; Etc.

8.13 SYSTEMATIZED FLOW DIAGRAM OF TOTAL MAINTENANCE PROCESS

The various functions/ systems/ processes, mentioned earlier (and some may be discussed in subsequent chapters) in this book, can be shown in the flow sheet as given in Fig.8.8.

QUESTIONS

1. What do you mean by system approach to maintenance?
2. What is codification and cataloguing and what are its advantages?
3. Explain briefly, how codification of drawings are done in industries.
4. What is maintenance time standard and why is it difficult to arrive at maintenance time standard?
5. What are instruction manual and operating manual and how do they differ from maintenance manual and how do they differ?
6. What is History Card/ sheet and what are its uses?
7. What are Standard Operating Practices and Standard Maintenance Practices?
8. What do you understand by Maintenance Work Order and Work Permit?

CHAPTER

9 COMPUTERIZED MAINTENANCE MANAGEMENT SYSTEMS (CMMS)

Definition and Objectives

Before the introduction of computers, log-tables, slide rules and other such devices were use for calculating/ computing. When computers were introduced in industries, it was first adopted for production, planning, project and quality aspects and later it was adopted to help maintenance. Initially computers were used in maintenance for computing, report generation and support functions etc and that maintenance was termed as **Computer Managed Maintenance System** (CMMS). It can be said as transition from paper-based work-order system to electronic ones. With the further development of computers and the whole gambit of Information Technology (IT), all aspects of maintenance were also computerized and CMMS is, now, termed as **Computerized Maintenance Management System.**

As it works with different types of work-orders, CMMS is often referred as "**Work-order Software**" also.

Objective/Goal–Maintenance is more of an evolution than a 'goal'. The main objective/goal of CMMS can be said 'to move to a planned environment and allow for further progression along the maintenance evolutionary path'. Despite the advances in technology there are still a very large number of maintenance management areas that are extremely reactive in nature and these pose challenges for CMMS implementation.

Today, CMMS encompasses all aspects of maintenance including preventive and predictive maintenance tracking, inventory usage tracking, failure tracking, RCM, TPM, EAM, vendor management, documentation analysis and system integration etc. It will generate work orders automatically according to a schedule or manually from service request, can log authorized uses of a piece of equipment, and maintain history, employee time, downtime, parts inventory and purchase orders etc. All of this data, combined with built-in reporting tools, gives a powerful, efficient maintenance management system. A properly used and set up CMMS is a powerful tool for enhancing reliability efforts and improving Return on Investment (ROI). An improperly setup and or improperly used CMMS can have an equally negative impact on reliability and ROI.

The fundamental principle underlying a CMMS is "you can not manage if you can't measure".

9.1 IT VIS-À-VIS MAINTENANCE

Today IT is used extensively in all plants, industries, corporations and other establishments. Some of the commonly used areas are—

(*i*) CRM (Customer Relationship Management), to manage communications and information with their customers, (*ii*) SCM (Supply Chain Management), to manage their supliers,

(*iii*) ERP-(Enterprise Resourse Planning), for planning their resourse within the enterprise,

(*iv*) SDLC (Systems Development Life Cycle) or PLM / PLCM (Product Life-cycle Management),

(*v*) EAM (Enterprise Asset Management); etc.

PLM or SDLC and EAM, are both connected with maintenance of industries. **PLM** is the process of managing the entire lifecycle of a product from its conception, through design and manufacture, to service and disposal. PLM integrates people, data, processes and business systems and provides a product information backbone for companies and their extended enterprise.

Enterprise asset management **(EAM)** means the whole life optimal management of the physical assets of an organization to maximize value. It covers such things as the design, construction, commissioning, operations, maintenance and decommissioning/replacement of plant, equipments, machinaries and facilities. EAM systems are the family of systems that had their origins in extremely capital-intensive industries. EAM has been further discussed in chapter 11.

Some of the components, related to maintenance, of both, EAM and PLM are as below—

- MPM (Manufacturing Process Management),
- CAM (Computer Aided Manufacturing),
- CMMS,
- Generation of shopfloor work instructions,
- CAQ (Computer Aided Quality Assurance),
- ERP (Enterprise Resourse Planning), for communication with other systems,
- MOM (manufacturing Operations Management),
- PDM (Product Data Management),
- HMI (Human-machine Interface);

Thus, CMMS basically comes from two distinct IT families, ERP and EAM or PLM.

The various programmes and terminologies, mention above, is not to confuse maintenance engineers, but just to make them aware that they may, occasionally, come across these terms and they should not unnecessarily get perturbed about those.

CMMS software integration is the process of enabling two or more software systems to talk to each other by making common data available to each system and facilitating the exchange of this data between the systems. The data exchange can involve each system reading the other's data, writing to it or both. As example is the integration of CMMS systems with a purchasing system. Most companies already have a purchasing system but almost all CMMS systems now come with their own. So, they have to be suitably integrated. Similarly CMMS may be integrated with Accounts Department or other departments or Third-party systems.

9.2 INTRODUCTION/ DEVELOPMENT OF CMMS

CMMS essentially need to be developed and introduced when you can't keep track of all your information, when repairs and inspections are falling through the cracks, when you don't have the right spare parts in your storerooms, and when you realize you just aren't using your personnel effectively. But before implementing a CMMS, you have to do lot of preparations. You have to get back in control. You can't expect maintenance management software to produce order out of chaos. You have to clean up the chaos first; then a CMMS can help you stay organized.

Again, before developing and implementing CMMS, a **CMMS feasibility study** may be needed. Not all businesses may be effectively benefited from the introduction of CMMS software and system. **CMMS feasibility study** involves an audit of your existing maintenance practices and procedures. Should the results illustrate that it is feasible to proceed, you can then enter computerized maintenance audit process.

In any plant, industry or enterprise, CMMS can be introduced/ developed in following three ways—

1. In-house development–It can be developed, introduced and followed up by engineers, technicians and executives already available in that enterprise. This can be done section-by-section (area by area) or whole enterprise in one go. If you decide to implement CMMS area by area, you have to be careful in sequencing the areas. Phasing-in can also be done category-wise; e.g. first take up implementation of corrective work-orders, then take up PM work-order and then Inventory operations and so on. Up-gradation of existing maintenance system can also be considered.

2. Cooperation Development–The in-house (enterprise's own) engineers/ technicians/ executives can take help and guidance from engineers/ technicians/ executives from other enterprise already using CMMS or from any outside consultants to develop and introduce CMMS for their own enterprise. This can be also be done section-by-section (area by area) or category-wise phase-in or whole enterprise in one go. Up-gradation of existing maintenance system can also be considered.

3 Outright Purchase– Many CMMS programmes are available in the market from different vendors/ consultants/ providers for outright purchase. They are capable to fit or adjust their standard or tailor-made programmes to take care of the needs of individual enterprises. The provider studies the need, suitably adjust their available programme, implement in the enterprise and train the enterprise for satisfactory use. This category may also include **third party consultants** who may buy CMMS programme from any regular supplier and install in your plant/ industry.

Selection of the implementation of CMMS is governed by the level of expertise available in-house in the enterprise.

9.2.1 Outright Purchase Sources

There are very many sources (suppliers/ vendors/ providers/ consultants etc) available in the market today for selection/ procuring CMMS programmes. Plant Maintenance Resource Center website (www.plant-maintenance.com) conducted a survey, between May 2004 ~ July 2004, on worldwide CMMS implementation, by asking Voluntary (and confidential) responses and they received 105 valid responses (including 31 from USA and 4 from India) from a wide range of individuals working across a variety of industries. Some of the key findings of the survey are as below–

(*i*) The five most common CMMS in use were SAP, Maximo, MP2, MIMS/Ellipse and PMC. Few lesser used CMMS were from Mainsaver, MPACK, EPACK, AMMS, Avantis etc.

(*ii*) Most systems have been in place only for a few years, but a significant proportion has been in place for 5 years or more.

(*iii*) The most commonly stated reasons for selection of the current CMMS were General functionality and features and Integration with other commercial software.

(*iv*) CMMS are seen as being moderate-to hard to use, on average, with Maximo rating highest in terms of ease of use.

(*v*) CMMS are, on average, seen as being only moderately well endowed with functionality and features, although SAP and Maximo rate highly in this area.

(*vi*) CMMS implementations are generally seen as moderately successful, with little variation between ratings for the top 5 most popular CMMS packages.

(*vii*) Most of those respondents who rated their implementation as poor had implemented packages other than these most popular packages.

(*viii*) The most important factors in implementation success were obtaining Senior Management commitment, and effective training. The relative importance of selecting the correct CMMS appears to have declined over the last 4 years.

(*ix*) Most respondents reported that their CMMS implementation has led to some or significant benefits.

The findings of the survey is put here just as an indicator that the CMMS system is used throughout the world with reasonable success and benefits. The more used CMMS suppliers can easily be contacted. SAP is an well known portal (www.sap.com) and they can be contacted directly or through intermediate sources. For Maximo-CMMS programme you can contact www.gocfi.com/maximo.htm or write to CFI, 18000 West Nine Mile Road, Suite 550, South Field, Michigan-48075, USA. IBM also has Maximo programme. For MP 2, you can contact www.datastream.net or 13560 Morris Rd. Suite 4100, Alpharetta, GA 30004, USA. For MIMS-Ellipse, you can contact www.mincom.com. For PMC, you can contact DPSI, 4905 Koger Blvd, Suite 101, Greensboro, NC 27407, USA or sales@dpsi.com. Others can also be similarly contacted.

There are many companies/ firms developing or selling such programmes. Apart from SAP and IBM, others are Project Software and Development Inc. (PSDI), USA, FastMaint CMMS (SMGlobal Inc, North Carolina, USA), Q4 Power CMMS System (contact—info@engica.com), MainBoss CMMS (contact—info@mainboss.com), EZMaintenance CMMS, USA (contact—-sales@linkitsoftware.com) etc to name a few.

Of-course buying a CMMS programme can be tedious and costly affair. After doing the extensive research, choosing the most suitable package and seeing how the price-tag will effect the year's budget, many companies are tempted to cut cost—and one cost that often gets sacrificed is the cost of training. But you should guard against that as training can make huge difference between success and failure of CMMS.

9.3 FACTORS AFFECTING SELECTION OF CMMS

First and foremost criterion is what size of industry is yours; small size, mid size or large size? **Small organizations** or independent workgroups normally do not have any existing CMMS and often one or a couple of personnel are expected to handle all maintenance tracking and scheduling. If you have many types of equipment and several different types of maintenance tasks to do, the standard edition is a better choice. Alternatively, if you have fewer equipments, a small and simple CMMS will be OK. **Mid size organizations** may have maintenance personnel between 5 to 30 or where there are many people from multiple departments that need to help in maintenance. They may or may not be familiar with traditional maintenance software like CMMS. For these, standard CMMS programmes can be implemented that allows you to install the software on any number of workstations and create multiple user accounts. The only restriction is the number of users who may be logged in at the same time.

Large organizations have small divisions or workgroups within total organizations that need a maintenance solution that can be rolled out quickly and cheaply. There may be some need to integrate with enterprise level systems to provide maintenance data or reports. Some sort of traditional maintenance software like CMMS may or may not be there and if pre-existing, it may not be working very effectively. For those, full-fledged CMMS, with many workstations and multiple user accounts can be implemented where 10 or more users can log-in at the same time. Highly integrated large organizations with multiple locations may need an enterprise class system that will integrate with existing ERP systems and provide centralized control and audit of all maintenance activities.

Often the area which the senior plant manager feel is critical or crucial might be an area where the manufacturer's source of profit from the sales and service point of view. Therefore, the CMMS developer finds a potential **conflict of interest** between developing a maintenance program that is best for you, the end-user (maintenance men) or the one that is best as perceived by the senior plant manager. Here the Plant manager should be asked to have a focus on equipment reliability, as well as availability, a push towards zero downtime or zero in-service breakdown.

However, some of the basic points, which need to be considered for selection of CMMS in any type of industry, are as follows–

(*i*) Compatibility with previous/existing CMMS or other maintenance software,

(*ii*) Compatibility with existing hardware/operating system,

(*iii*) Ease of implementation and ease of use. In all, its flexibility, and ability to be configured to match our client's/ supplier's unique processes and existing documents/ systems.

(*iv*) General reputation of software and its vendor,

(*v*) Integration with other commercial and technical software,

(*vi*) Already being used in sister concern or other department of same organization,

(*vii*) Cost of implementation, including cost of purchase and consultants,

(*viii*) Availability of local support and training facility,

(*ix*) Availability in local language version;

(*x*) Extensive Management Reporting Capabilities,

(*xi*) Support Regulatory Compliance,

(*xii*) Multiple System Interfaces, Consistency between multiple sites; Etc.

9.3.1 Implementation Failure Syndrome

The fear or lack of confidence that the CMMS may not be subsequently followed up or may not give appropriate result or they would not be ably to justify the expenditure made on CMMS, leads to many smaller companies to develop or buy CMMS.

9.4 COST OF CMMS

When you want to determine the cost of a particular CMMS package, you have to look many points, e.g.—

(*i*) The actual price of the software, if it is purchased outright or the expenditure to be incurred for consultants and outside advisors if it is developed with their help,

(*ii*) The price of any hardware upgrades you may have to make to run the software,

(*iii*) The price of any software upgrades you may have to make to run the software (e.g. some packages only run on the most up-to-date version of Windows, so if you're running an older version, you have to upgrade your system before you can put in the CMMS)

(*iv*) The price of support for the CMMS.

(*v*) The cost of training for the CMMS (remember that training may be needed for management maintenance personnel, and/or non-maintenance personnel and support personnel).

(*vi*) The cost of implementation (which includes any costs in time that your personnel spend setting up the CMMS rather than doing their usual jobs)

(*vii*) The cost of the learning process—any slow-downs that might occur while people are getting used to working with CMMS; etc.

You have to be realistic. If you ignore ant possible cost, you may land in unpleasant surprises later on.

In addition to the money resources, you have estimate how much on the job time will be taken for the implementation and how many personnel will be involved in it full-time and how many part-time or squeeze in within their other duties.

9.5 FACTORS INFLUENCING SUCCESS OF IMPLEMENTATION

A CMMS is only as good as the organization's commitment of using it. For example, if maintenance personnel didn't record the exact labour and material cost of jobs done, that information won't be available later on and CMMS won't be beneficial to that extent.

Whatever the quality or reputation or cost of the CMMS may be its success of implementation and effective use depends on few other factors also, some of which are given below—

(*i*) Senior Management commitment,

(*ii*) Effective training and retraining,

(*iii*) Choosing/ developing the right CMMS,

(*iv*) Effective Change Management,

(*v*) Adequate budget,

(*vi*) Effective CMMS Vendor, consultant, supplier or third party Support,

(*vii*) Knowledgeable and dedicated end users; etc.

It is not unusual to find a company that owns several different maintenance software products. Although software is usually the first point of blame when implementation fails, humans are the real reason of implementations failure in most cases. CMMS will also try to integrate all those software.

9.6 MAIN BENEFITS

A properly used and set up CMMS is a powerful tool for enhancing reliability efforts and improving Return on Investment (ROI). An improperly setup and used CMMS can have an equally negative impact on reliability and ROI. Its one of the main benefit is "Improved Labour Productivity". Maintenance productivity can be raised considerably (up to 75%) by using CMMS. e.g.—

- Not much time is lost in locating and getting materials, as all information are available and materials are reserved for the job,
- Preparing work-orders and schedules take much less time as the previous one can be copied on computer, necessary modification done and printed.
- Less chance of adopting wrong or lengthy methods or duplicating as standard maintenance procedures are already known,
- More effective deployment of workmen as individuals capabilities etc are known, etc.

CMMS leads us from reactive to planned workflow status. Other benefits are as followings—

(*i*) Reduction in machine breakdowns and other machine outages and increased plant availability and reliability,

(*ii*) Reduction in overtime and optimum utilization of manpower; Lower maintenance operating cost, overall reduction in labour cost.

(*iii*) Reduction in other costs and improved cost control.

(*iv*) Improved equipment availability and reliability,

(*v*) Improved control over PM and PdM planning and scheduling,

(*vi*) Longer asset/ machine life cycle,

(*vii*) Better inventory management and reduced material cost,

(*viii*) Improved maintenance history and documentation,

(*ix*) Better trending and cost projections,

(*x*) Better yield and less scrap, waste and re-work etc,

(*xi*) Quicker access to Plant Maintenance or Facilities Maintenance Statistics,

(*xii*) ISO-9000 Accreditation,

(*xiii*) Conformity with Health and Safety Standards,

(*xiv*) Enhanced transparency and accountability,

(*xv*) Standardization between different maintenance departments and that of sister concerns,

(*xvi*) Compliance with Industry and statutory Regulatory Standards;

(*xviii*) Lower total cost of ownership of assets; etc.

The process of obtaining the benefits of CMMS to actual uses is shown in table (T-9.1) below—

Table T-9.1. Benefit Process of CMMS

Objectives	Strategies	Tasks	Working modules
Reduce Maintenance Cost	* Reduce labour cost	* Control jobs and schedules * Plan labour and materials	* Workorder * Schedules and charts * Employee, * Location
	* Reduce Inventory Carrying cost	* Analyze uses * Analyze investment / procurement	* Inventory
Increase Equipment Uptime	* Reduce MTTR	* Keep equipment data * Keep location data * Keep Bill of materials * Keep employee info. * Keep contractor info. * Keep supplier's info * Keep equipment history * Manage material stock	* Asset * Location * Employee * Requester * Workorder * Inventory
	* Increase MTBF	* Analyze reactive jobs * Schedule PM, PdM etc.	* Workorder, *Asset * PM, PdM, CBM

9.7 PRE-REQUISITES BEFORE IMPLEMENTATION

There are some pre-requisites or goals or projected achievements, which the organization has to consider and decide before actual implementation. A few of those may be as followings—

(*i*) Define the high level KPI's that the corporation wishes to use to monitor and control its performance— e.g. Unit cost of maintenance (by equipment, per employee or otherwise) and Overall Equipment Effectiveness (Availability and MTBE, MTTF, Utilization etc),

(*ii*) Defining what is capital works and what constitute operating costs; such as what criterion determines a plant improvement? It may include like for like replacement, Changes to process and instrumentation drawings, Changes to operating improvements etc.

(*iii*) Defining authorization levels in terms of money (Rupees) of each role within the maintenance organization; e.g. visualize and guard against bottlenecks that any levels may create.

(*iv*) Determining a prioritization system that will allow the best use of resources across the corporation. For example— are resources to be used across various plants, are resources to be used across various sites and what are the structures of the work groups or teams to be included etc.

(*v*) Determining what are the definitions of various types of work orders.—Safety, Capital or Maintenance jobs.

(*vi*) Determining what are the definitions of various types of maintenance and setting levels to be used as benchmarks— Refer point (C) of Sec. 9.8.4 for desired ratio of Preventive, Predictive, Corrective and Breakdown jobs for example.

(*vii*) Defining business processes and the KPI measurements required for controlling these. Examples may include Backlog management, Planning/ Scheduling systems, Execution and Data capture systems, engineering works, analysis and action loops, associated purchasing and inventory functions not covered within the etc.

This section may also be termed as "Defining Organization's Direction and Rules". Again, this may appear somewhat overlapping to Sec. 9.9. However, both are to be considered.

9.8 COMPONENTS/ FEATURES OF CMMS

Today, an integrated CMMS, in addition to other things, includes all the aspects mentioned in chapter 8 (system approach to maintenance) and most of the aspects mentioned in other chapters of this book.

In today's age of industrialization, there are different kind of industries— Process industries (including mining industries), Manufacturing industries, Transportation, Utilities, Supply chains and warehousing etc. Each of these industries can be further classified into different categories— Metallurgical industries, Chemical industries, Mechanical industries, Electrical industries, Pharmaceutical industries etc. Most of these industries face somewhat similar problems, such as—

- Huge volume of data,
- Monotonous type of job in data recording, storing, and retrieving etc manually or otherwise,
- Chances of human error during repetitions and recording similar information manually or otherwise,
- Difficulties in communication of information between different departments and facilities of the industry or sister concerns,

- Decision requiring checking and analysis of lot of statistical information and documents is not streamlined and takes lot of time and occasionally timely corrective action gets delayed or missed.
- Statutory and insurance aspects are often missed in view of pressures of other jobs,
- Correlating cause and effect (including RCA) for defects and the jobs done earlier is not done properly,
- Accurate and timely generation of information, reports and action plans etc;

CMMS attempts to solve all such problems. Again, using a CMMS is not just a matter of which button to click. It is deciding which information you need to record, and which is unimportant. It is making strategic choices about how to organize your data so that you quickly get the answer you need

Basic components/ features and scope of CMMS for any industry remains nearly the same and are suitably augmented/ modified or tailor made for any particular industry. The main components features or modules of a CMMS are as follows, through which you can carry out the required master data and transaction processes for the tasks you want to do.

9.8.1 Initial data, cataloguing and codification

Entering data is a key part of CMMS implementation. Then you have to assign codes for those data so that the data can be easily identified and read. Codification of drawings has already been discussed in chapter 8.

9.8.2 Equipment Classification and Management

It covers keeping track of the organization's machineries (manufacturing/ production equipments, assembly lines, vehicles and other associated equipments). Typical data to be stored would include asset number, equipment number, dept., asset name, model, serial number, drawing numbers, purchase price, location, supplier, equipment history and associated spares, etc. It also includes keeping records of equipment specifications, meter and other readings, warranties, service contracts, past repairs, necessary preventive maintenances, spare-parts needed to maintain the equipments, standard maintenance procedures, cost and replacement cost projections, MTBF etc. However, you should store only relevant data, which may be used now or in future. You should not populate all data fields just because they are available to you.

In addition to equipment codes and logic, you should also cover "Operating Location Codes and Logic". Every location that is to be recorded in CMMS has to be codified to satisfy a standardized format across of all locations.

In essence, a good CMMS can tell maintenance personnel anything they want to know about any equipment before they go for the job. It can also provide data that can lead detection of significant trends; e.g. if a number of machineries in the same building suffer electrical problems, may be there is some overall problem with building electrical system.

Equipment records can also simplify making cost-benefit analysis of maintenance activities. For example, it has been seen somewhere that amount spent on preventive maintenance of a set of electrical fans were much greater that the cost of just replacing the fans whenever they broke. This component may also be termed as "**Asset Registration, upkeep and Performance**". Equipment classification module (including codification) has been further discussed in some details in Sec. 9.10, which can be slightly modified with today's age of computerization and technology. It also includes—

(*i*) **Failure codes**–It builds and displays failure hierarchies. Different failures and defects are also codified. This has already been discussed in chapter-4.

(*ii*) **Condition Monitoring**– It creates and views measurement point records for your equipment (used for Predictive Maintenance or CBM). This has already been discussed in chapter-6. These are also suitably codified.

9.8.3 Work Request Management

It keeps track of problems received from various sections of maintenance and inspection departments and from complaints from other departments of organization. It also covers various maintenance job plans and schedules, Safety plans, Yearly or other periodical repair or overhaul plans and other action plans etc. This also has been discussed to some extent in chapter-4 and chapter-7.

9.8.4 Work-order Management (PM and Other Repair Module)

It keeping track of work-orders prepared by maintenance personnel. This includes (*i*) creation of work-order and (*ii*) recording the result of each job after it is finished (post-job information). It also includes automatic generation of work-orders for preventive maintenance.

(*i*) **Creation of Work-order (Open work-orders)**—It may include:

- Description of the job
- Date and time the job should be done,
- Estimated length of job,
- Assigning personnel and estimating labour cost,
- Locating and reserving materials expected to be needed for the work,
- Specifying accounting information for the job, e.g. account to which the cost should be charged,
- Recording other useful information for the workers, e.g. access information, contact people, applicable service contract, safety precautions etc.

(*ii*) **Post-job information (Closed Work-orders)**—It may include:

- Was the job actually completed?
- Actual start and end time of the job,
- Actual labour (or category-wise workmen) and actual material used,
- Length of downtime for any equipment involved,
- Actual cause or condition of the problems,
- Any other detail that might be relevant now or in future; etc.

(*iii*) **Generate PM work-orders**—It normally considers PM procedure library, PM schedule, PM periodicity, condition monitoring reports and built-in test equipment results etc. The CMMS should show "**Alerts**" in case of any deviation, missing or overdue jobs. The word **Maintenance Calendars** is also used in some CMMS, which not only produces print outs of all preventive maintenance schedules for all or any piece of equipment, but will produce actual monthly calendars showing all scheduled preventative maintenance, and the days on which it is to be done.

Work-orders or checklists for daily, weekly, fortnightly or other periodical checking of equipment (termed as Regular Inspection) are also issued here. These are carried out without isolating the equipment (i.e., while equipment is under operation).

As already mentioned in Sec.5.13, the percentage of different maintenance work-orders in one type of industry and reliability required, are as follows—

1. Preventative Maintenance work-orders .... 50 %
2. Predictive maintenance and CBM work-orders................ 30%
3. Corrective Maintenance work-orders......... ....15%
4. Breakdown.... .... Less than (<)- 5%

All these sub-sections also keep track of work-orders and issue necessary updates. These (A, B and C), including their cataloguing and codification) has been further discussed in some detail in Sec. 9.11, which can be slightly modified with today's age of computerization and technology.

9.8.5 Facility Management (Sometime Called as Property Management)

It keeps track of maintenance on an organization's premises/building (including its electrification, utilities supply, air-conditioning, security systems and landscaping etc), waste recovery and disposal and other such areas. These are also suitably catalogued and codifies. These statistics may also be used to set up maintenance Key Performance Indicators (KPIs).

9.8.6 Inventory Management

It Keeps track of materials (spares, consumables, tools, tackles, lubricants etc) to perform maintenance activities. This includes economic order quantity and reordering accordingly, reserving materials for use in up-coming jobs, indicating locations for easy availability by maintenance personnel, pointing unused or obsolete materials etc. Creates purchase requisition, Tenders (request for quotation), Purchase orders, Invoices, Currency management etc. When purchasing and receiving are also included in it, it may be named as **Material Management module**. This module is sometime called as **Resource Management Module.** It also includes information about different vendors and suppliers, service contracts, ancillary developments etc. These have been further discussed in some detail in Sec.9.12, which can be slightly modified with today's age of computerization and technology.

9.8.7 Personnel Management

It keeps track of staff cost and productivity; creates and works with craft/trade wise employees records, including pay-rates and special skill etc. A maintenance personnel database is required in your CMMS. Unlike the company personnel database, it would generally contain details of maintenance personnel only. Again requirements here will differ, but typical details will include name, trade, department, shifts worked, qualifications, special training received, authorization etc. CMMS would codify different craft/ trade designations, their grades/ skills, rates and range (Rs./ hour) etc.

9.8.8 Captive Engineering Shops Module

This has been discussed in some details in Sec. 9.13, which can be slightly modified with today's age of computerization and technology.

9.8.9 Training Module

Training will need to be developed, or adapted, in using the CMMS to accommodate the industry's rules and processes of the specific facility. The role specific focus of the training is required so that each of the employees know what their responsibilities, accountabilities and what the role relationships with others in the organizational structure are. This has been discussed in some detail in chapter-15.

9.8.10 Set-up and Evaluation Module

It covers preparation of various reports and charts, assessments and various documents. It creates ledger accounts and components vis-à-vis budget, maintains database and application security and performs related functions. It also includes User Evaluation (i.e. Evaluation sheets for production supervisor and that for production head, Evaluation sheet for Maintenance Engineer and that for Maintenance head) etc.

9.9 EMPHASIS FOR INTEGRATED CMMS

An integrated CMMS may be prepared and implemented by going through earlier mentioned components/ modules. However, there may be many specific points, which must be kept in mind; a few of which are mentioned below–

(*i*) Equipment Tracking and cost-roll-ups,

(*ii*) Hierarchical System Dependencies,

(*iii*) Fast Job Reporting, Viewing out-standing work-orders,

(*iv*) Failure Consequence and Cause Analysis,

(*v*) Risk Assessment and coordinated safety,

(*vi*) Production Outage Synchronization,

(*vii*) Resource Balancing,

(*viii*) Critical Path scheduling (refer chapter 7),

(*ix*) Synchronized Work Flow, work Packaging and Control,

(*x*) Supplier and Contract Vetting,

(*xi*) Mobile and Remote Work Access,

(*xii*) Continuous documentation and Drawing correction Management,

(*xiii*) Calibration Verification and Audit, Gauge and calibration management,

(*xiv*) Tool Management and Control, Portable equipment testing,

(xv) Expenditure Tracking and Forecasting and financial integration,

(*xvi*) Managerial Decision Analysis; Etc.

Important features of CMMS and their interdependences are shown in Fig F-9.1

9.10 EQUIPMENT CLASSIFICATION MODULE

As already discussed in previous chapter, equipment classification and codification is a pre-requisite for computerization or otherwise. All the equipments, assemblies and subassemblies etc are codified as numerical (sometime alphanumerical) identification so that the computer can take that up correctly. As already mentioned in Sec. 8.1, equipment classification and codification is very essential in CMMS, mainly for following two counts—

- Uniqueness of each component, subassembly, assembly and equipment,
- Huge number of components/ parts, subassemblies and assemblies and their dynamism (quick movement etc).

9.10.1 Classification Steps and Guidelines

A uniform system or method should be adopted for classifying components, subassemblies and assemblies of all the departments of a plant/ industry to avoid duplication, confusion and wrong actions. Following guidelines should be used as rules for such classification—

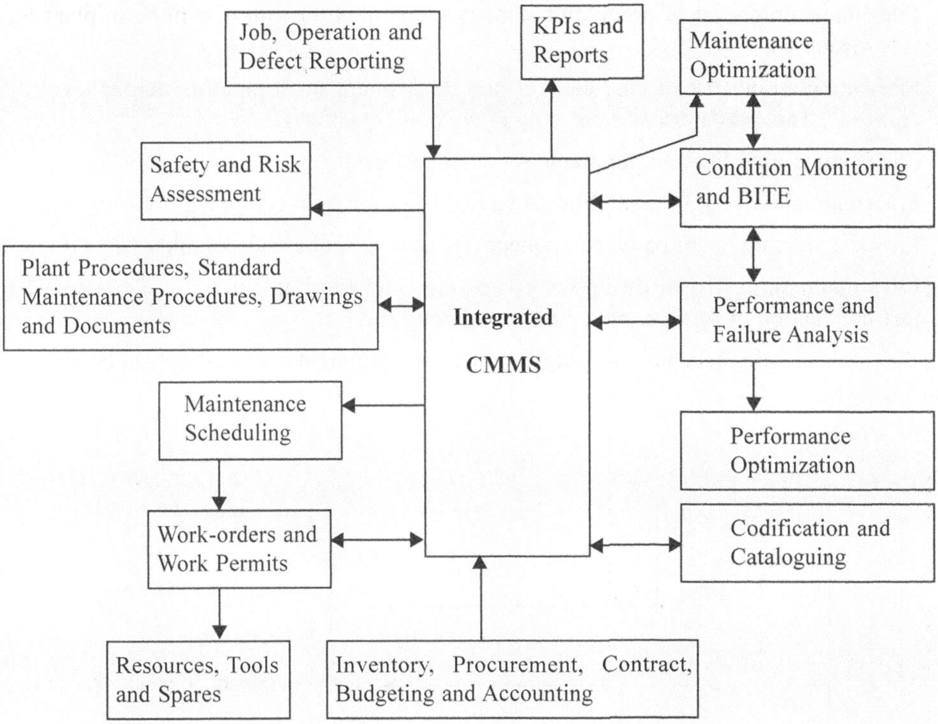

Fig. F-9.1 Integrated CMMS Features

(*i*) Division of plant/ industry, upto each part of its equipment must be done stepwise, i.e. in sequence of Department, equipment, assembly, subassembly and part etc,

(*ii*) Each level division must follow a physical justification or logical justification. Both justifications should not be followed together at one level. Physical justification means dividing departments that are physical in nature, i.e. come one after another. Logical justification means discipline-wise, i.e. mechanical department, electrical department etc. Of-course, occasionally different justification may be used at different levels.

(*iii*) Number of steps or levels should be a judicious decision so that uniqueness is brought out easily and also tracing the location and address of the item is simple. This level classification is determined as per the largeness and complexity of the industry. Having too many levels to reach a single part will make the classification very large unnecessarily. Too less level, on the other hand, will cause problem in tracing.

(*iv*) At each level of division, a larger slot of number range should be allotted than actually needed at the time of classification. This would help in maintaining the same logic of classification and not disturbing existing data, as and when additional equipments/ components are procured in future. This, of-course is a judicious judgment. If, at a level, there are 99 items, we should allot 3 digits (***) for that.

9.10.2 Classification Records

The classification of equipments, decided as per above guidelines, must be suitably recorded in computers. Often a backup printout record is also maintained. The records should preferably be separate at each level of classification or combination of more than one level, as per suitability. Following different files may be maintained–

(*i*) Files maintaining list of production centers (departments) with discipline in plant level (say Group E),

(*ii*) Files maintaining list of cost centers in a department or production center level. Say Group D. This may not be same as mentioned at (1) above,

(*iii*) Files maintaining list of equipments for each cost center (say Group C),

(*iv*) Files maintaining list of assemblies/subassemblies for each equipment (Group B),

(v) Files maintaining list of parts/ components for each assembly/subassembly (say Group A),

(*vi*) Files maintaining list of discipline codes (e.g.1 for mechanical, 2 for electrical, 3 for instrumentation, 4 for operation etc), (say Group A).

By adopting generally acceptable 11 digits code, the equipment codification can be shown as below in Fig. 9.1.

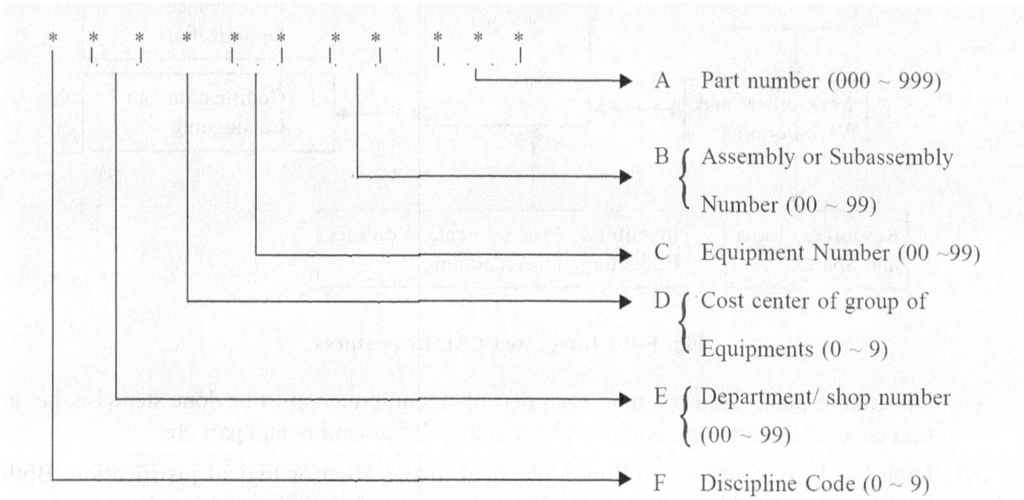

Fig. 9.2. Equipment Codification

While making such codification, following points should be noted–

- The discipline code (Group F) should, logically, come either between group B and group C or between group C and Group D, as each major equipment or group of equipment may have some mechanical, electrical and instrumentation assemblies/ components. But in many cases, the discipline code is used first digit in integrated plants for better centralized monitoring.
- Occasionally maintenance activities, that are carried out at part level (Group A) are also common at assembly level (Group B), i.e. assemblies are also treated just like parts. Hence, when we mean the item as assembly, we identify the two digits at assembly level (Group B) and put three zeros (000) at part level (Group A), but when we mean the same item as part, we put the same two digits at assembly level (Group B) and put a separate three digit part number at Group A. As an example, the last five digits may be 65000, if referred as subassembly or 65031, if referred as part.
- Level at Group D may be used as group of equipments in a department/ production center or may be used as cost center in the department. If it is not possible to accommodate numbers of cost centers in one department within one digit (0 ~ 9), we may devise ways to use levels (Groups) D and E together to describe number of departments and also cost

centers within the departments. If it is still not possible, we can add one more digit i.e. the system would be 12 digit system and not 11 digit system.

- In a big industry, some cost centers are common to more than one department/ production center. Compressed air lines and gas lines (which are generated centrally and distributed throughout the industry) are two examples of such cost centers. In such cases total cost incurred is calculated at plant/ industry level and, then, apportioned to different departments/ production centers. To account for such cost center, we can reserve (allocate a particular range) within the cost center level (Group D). This would help in calculating in maintenance cost department/ production center-wise.

Table 9.3 gives one illustrative example of equipment classification (List of equipments) and Table 9.4 gives illustrative example of equipment classification (List of assemblies and parts).

Table 9.3 Equipment Codification (List of Equipments)

Code	Full Description	Short Description
151	Blooming Mill (Mechanical..........	BLM (M)
251	Blooming Mill (Electrical)	BLM(E)
351	Blooming Mill (Instrumentation)..........	BLM (I)
152	Billet Mill (Mechanical)..........	BLT (M)
252	Billet Mill (Electrical)..........	BLT (E)
161	Bar Mill (Mechanical)..........	BM (M)
694	Traffic (Operation)..........	TR (O)
etc	etc..........	etc.......

In Table 9.1, discipline codes 1, 2, 3 and 6 are taken as Mechanical, Electrical, Instrumentation and Operation respectively. Department codes 51, 52, 61 and 94 have been taken as Blooming Mill, Billet Mill, Bar Mill and Traffic Departments respectively.

Table 9.4 Equipment Classification- List of Assemblies and Parts)

Department	-	Blooming Mill Mechanical	1	5	1
Cost Centre	-	900 Mill Complex			3
Equipment	-	900 Mill Manipulator Drive Mech................		4	0
Assembly	-	Gear Box-Entry side			1

Assembly	Description	Draw -ing No.	Part No.	Material (Kg.)	Weight	Number installed	Part No.	Store Cat. No.
01/000	Output Shaft Assembly	****			156	2	****	
01/001	Output Gear	"	001	IS-1030	23	2	****	
01/002	Pinion Shaft	"	002	C-45	15	2	****	
01/003	Jib Key	"	003	M.S.	1	4	****	
02/000	Input Shaft Assembly	****			45	2	****	
02/001	Input Shaft	"	004	IS-1030	40	2	****	
02/002 etc	Input Gear	"	005	C-45	20	2		

Table 9.4 also leads to formation of equipment catalogues. Drawing numbers and stores catalogue numbers are indicated as ***** and other figures are taken only for illustration.

Table 9.5 illustrates total classification (Code Group wise) of three specific parts.

Table 9.5 Equipment Part Classification

Code Group	Example-1	Example-2	Example-3
A	Sliding Shoes	Armature	Bevel Gear
B	Wepuko Pump-5	Motor No.1	Roller Assembly No.9
C	Press Hydraulic System	Lathe No.14	Front Working Table
D	2000 T Press Complex	Lathes	700 Mill Complex
E	Forge Shop (Mechanical)	Mechanical Repair Shop (Electrical)	Bar Mill (Mechanical)
F	Mechanical	Electrical	Mechanical

For a very big industry, like integrated steel plants, often 13 digit codification is used to pinpoint the location and interchangeability etc of each equipment part. In such industries, many similar assemblies are installed at different locations (either in different shops or as parallel lines or standby units of the same shop/department). Similarity may start from group of equipment level or equipment level or assembly level etc. In such cases, link data set is introduced so that manual feeding of entries in computer is done only in one case and computer will "assume" similar entries upto part level in all 'subsequent' files.

9.10.3 Another way of Equipment Codification

Today, composite alphanumeric codification is more common as it takes less space (i.e. less number of characters). As an example, in one medium size industry, using Maximo-CMMS, they use seven character alphanumeric codes for equipments. The first alphabet designates the discipline; i.e. M for mechanical, E for electrical, I for instrumentation, etc. Nest two alphabet say the equipment of that discipline. The 4th character is digit, signifying the unit/ shop/ department, assuming maximum nine units/shops in that industry. (In case there are more shops using similar equipments, we may use alphabet as 4th character also.) The next three digits say the serial number. Table 9.5.1 shows few of such classification.

Table 9.5.1. Another way of Equipment codification

Sl.	Discipline	Equipment	Code
1	Mechanical	Ball Mill	MBM
2	"	Blender	MBL
3	"	Centrifugal Pumps	MCP
4	"	Furnace/ Firing Chamber/ Burner	MFC
5	"	Gas cutter/ drilling machine/ Bench grinder/ Vice	MMC
6	Electrical	Circuit Breaker	ECB
7	"	Meter/ Killowatthour meter/ Energy meter	EMR
8	Instruments	Flowmeter/ Mass flowmeter/ Rotameter/ Vortex	IFL
9	"	Sensor/ RTD/ Thermocouple	ISE

An example— Code MCP2121 means (Mechanical) Centrifugal Pump, at Unit/shop 2, and serial number is 121

9.10.4 Reports/ Forms

Reports can be obtained in any suitable forms, as desired by the user. In addition to master data file, other types of reports can also be generated such as assembly details, linking information of assemblies, group-wise or assembly-wise report of any department/ production center etc.

9.11 PM AND OTHER REPAIR MODULE (WORK-ORDER MANAGEMENT)

Preventive maintenance and repair planning etc have been discussed in some details in previous chapters. Various elements of preventive maintenance/ planned maintenance are again given herewith–

(*i*) Fixing of maintenance standards/ types for every equipment- If Fixed Time Maintenance (FTM) or Condition Based Maintenance (CBM); As few CBM jobs calls for periodic inspection (except few cases of continuous on-line monitoring), such inspection jobs can be taken as FTM jobs and included in PM schedules,

(*ii*) Fixing inspection schedules at predetermined frequency for all equipments- Manufacturer's recommendation, our past experience and equipment history records etc can help in determining frequency,

(*iii*) Checking and recording defects/ faults, detected during different inspections,

(*iv*) Fixing standard time periods (MTS etc) for execution of each maintenance job,

(*v*) Preparation of job plan, i.e. shutdown planning- This includes making a job list (based on inspection reports, defect lists and other requisitions) for all jobs needed to be done during shutdown, computing the total shutdown period for each job and arranging for spares, tools and other facilities. Network analysis techniques (PERT, CPM) and Bar Charts etc are used.

(*vi*) Recording the actual jobs done and deviations made for each job and equipment and updating of history of jobs and equipments; necessary documentation.

(*vii*) Analysis and evaluation of performance of equipments and checking equipment's availability and reliability.

All the above mentioned jobs and other maintenance planning jobs call storage of huge data and, then, use the data to draw out various job plans. For a big industry, doing this job manually is almost impossible, and, as such, CMMS (or EAM or other such techniques) have been adopted. Apart from such data, innovative thinking can also be accommodated in CMMS or EAM etc.

Codification of equipment has already been discussed in this chapter. Same equipment classification is used for identifying equipments for maintenance planning. Some other codification for CMMS is discussed later in this chapter.

9.11.1 Job Codification/ Catalogue

This mainly includes—

- Defining each job as unique,
- Codifying each job numerically (alpha-numerically if needed).

Unique definition of each job is essential so that computer does not make mistake and picks up correct job every time. 'Carry out bolt tightening' and 'tighten bolts' may mean the same for human but computer will always take these two different jobs. Jobs can be defined at three levels— (a) equipment level, (b) assembly/ subassembly level and (c) Part level. In a standard gearbox changing, the jobs at three different levels are as follows—

- Gearbox changing—(equipment level),
- Input gear shaft assembly changing——(assembly/subassembly level),
- Bearing No.(—) changing or bevel gear changing— (part level)

Having defined jobs properly, each job has to be allotted a numerical code. In an industry, normally, three digit codes (000 ~ 999) can accommodate all maintenance jobs. Depending on this type of system adopted, the jobs can basically be classified in following two categories—

(*i*) Frequency based (FTM etc)- This may include jobs like checking, adjustment, lubrication, replacement, minor repair etc,

(*ii*) Condition Based (Predictive, CBM etc)- These may include other jobs like replacement, repairs, overhauling etc.

In order to differentiate and identify the jobs as per above classification, the same three digit code (000 ~ 999) can suitably divided in such a way that a particular range is allocated to a particular type of job. One such division is illustrated below in Fig. 9.3

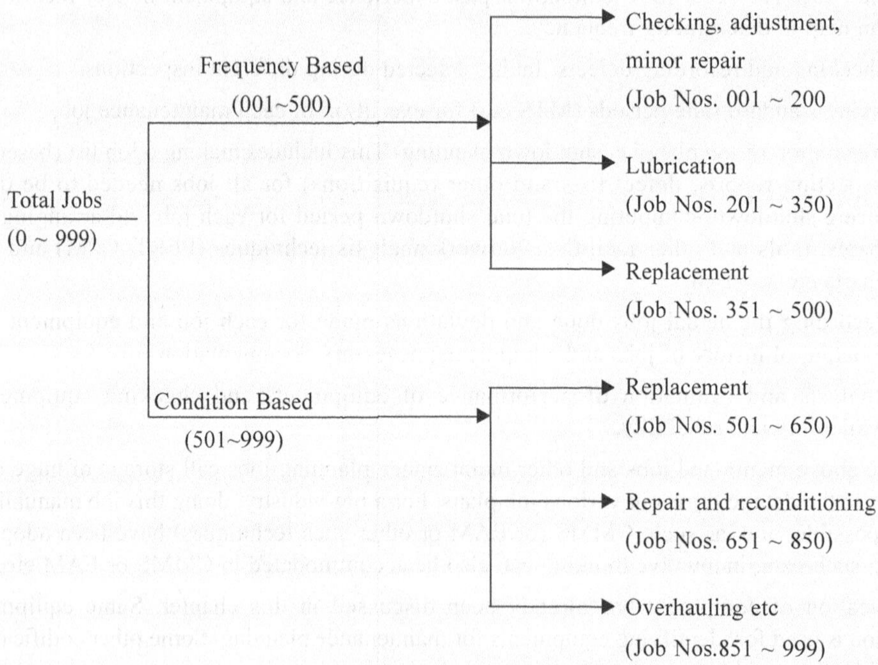

Fig. 9.3 Job Classification

The frequency based replacement and CBM based replacement have been given different code ranges in above classification. Enough blank code numbers are to be made available for future jobs which are not in the job catalogue at the time of initial preparation. In a still bigger industry, either more digits or alpha-numeric system can be adopted.

Thus for each job catalogue, equipment code number (indicating job location) and job code number (defining what to do), together makes the complete description of the job. Again, definition of the job should be standardized so that so that the same job details (not necessarily of similar in size or volume of job) can be used for similar type of equipments in different shops.

Following guidelines should be considered while making job list and job catalogue—

- One format for every equipment/assembly, indicating all jobs on that, with frequency time duration and expected or standard man-hour etc,
- One format for every job, indicating detailed manpower and time duration,
- One format for every job, indicating all other details (materials, tools, special facilities and special instructions, if any. For smaller organizations, one format can be used to indicate these with manpower and time duration details.

The system gives following advantages—

(*i*) Each job is standardized and the same description is used by everyone,

(*ii*) Quick analysis of one type of job (like motor changing or gearbox changing etc) done by a particular department, is possible,

(*iii*) While entering the data, text/ description of the job is keyed only once.

9.11.2 Codification of Breakdowns

Different numerical codes can be adopted for —

- Different types of breakdown,
- Different causes of breakdowns.

Generally two digit codes (01~99) are adopted for different types of breakdowns on each type of equipments. As an example, for all the mills of a steel plant, code 01 may be taken as top roll breakage, 03 may be entry manipulator shaft breakage, 12 may be taken as hydraulic system failure and so in. Again one digit code (0~9) or two digit code (00~99) may be taken to indicate the causes of failures.

An illustrative breakdown data set may have following information—

1. Department Number,
2. Equipment number,
3. Date and time of breakdown occurrence,
4. Date and time of handing over of equipment after repairs,
5. Code for breakdown type,
6. Code for causes of failures/breakdowns,
7. Code for affecting/ not-affecting production,
8. Codes for materials used; etc.

9.11.3 Codification for materials/services/facilities needed

A three digit code (001~999) may be used to indicate materials, services, facilities etc needed to do the job. For differentiating and identifying materials, services, facilities etc needed, the three digit system can be divided into following ranges—

- Range 001~200 for – Materials,
- Range 201~300 for – Transports, cranes etc,
- Range 301~500 for – Special tools, fixtures, facilities
- Range 501~600 for – Safety and environmental instructions/ precautions,
- Range 601~700 for – Information to other units; etc.

9.11.4 Job Sequencing File/ Record

Certain jobs have to be done in a certain sequence as doing otherwise may not be possible or may cause catastrophic damage or may take unduly very long time. The data about such sequence of jobs must be stored in job sequencing file/ record, so that correct execution plan may be taken out. This information is needed to take out job schedule and Bar/ Gantt chart etc.

As an example, for a coke oven gas line job, blanking, purging with nitrogen gas and testing for no presence of gas is essential before cutting that line. In manual scheduling, human brain automatically thinks and writes in a correct sequence, but for CMMS and other computerized scheduling, such sequencing file has to be maintained.

9.11.5 Frequency Determination

For determining the frequency of a job, many factors are to be considered, some of which are given below—

(*i*) Type of job, i.e. – inspection and adjustment jobs,
– Repair jobs,
– Lubrication jobs,
– Reconditioning and overhauling jobs, etc.

(*ii*) If the job would affect the production or if standby equipments available,

(*iii*) Condition as per inspection report, i.e.

- If vibration, temperature or other parameters are within limits, or
- If vibration, temperature or other parameters are in caution zone, or
- If vibration, temperature or other parameters are in danger zone, etc.

(*iv*) Availability of resources (spares, tools, tackles etc) and facilities (cranes, transports, in-situ testing and machining facilities etc,

(*v*) Urgency of production, etc.

9.11.6 Another way of PM Job Codification

A medium sized industry, using Maximo-CMMS, codifies all PM and other jobs using 12 digit alphanumeric code. First two alphabets say the type pf job (e.g. PM means Preventive Maintenance, RI means Regular Inspection etc). Next three digits designate the frequency (e.g. 015 means 15 days or fortnightly, 030 means 30 days, 180 means six-monthly, 360 means yearly etc) and last 7 alphanumeric digits says the equipment number, as indicated earlier in Sec.9.10.3. As an example, code PM030MCP2121 means monthly (after 30 days) PM of mechanical centrifugal pump of Unit/Shop No.2 (Sl. 121).

Job Plan Codification- For codifying the PM and other job plans, the same industry adds two alphabet (say JP) just before the said job code to make a 14 character job plan for that job. As an example, for the PM job, PM030MCP2121, the code for Job Plan would be JPPM030MCP2121.

9.11.7 Preparation of Work-Order and Job Schedule

Before preparation of any work-order or job schedule, all the files, records, data-sheet, as mentioned earlier or otherwise, must be handy. Data creation, data updating and data retrieval must, almost, be a continuous process.

Following steps (briefly mentioned earlier also) are generally taken for making work-order and job schedule–

(*i*) Making monthly job schedules, matching with the earlier made yearly maintenance plan and adding as per inspection reports and defect lists etc,

(*ii*) Making daily inspection schedules,

(*iii*) Making daily job schedule (crew-wise or worker-wise,

(*iv*) Making job plans for particular shutdown; this should include—

- List (compile) all jobs to be carried out, in the form of activities,
- Arrange all the activities (jobs) in logical sequence (considering preceding, succeeding and concurrent activities,
- Compute the total shutdown period, basing on the standard time for individual activities (jobs) and manpower resources available,
- In case of prefixed time duration for shutdown, arrange necessary resources and facilities to accommodate all jobs during that period,
- Arrange spares, tools, facilities etc, before start of jobs,
- Fix-up proper monitoring process and personnel,

(*v*) Make separate work-order for each major or medium job, indicating components of job (what to do), time schedule (when to do), manpower plan (who to do), materials/ facilities needed and special instructions if any.

Table 9.6 shows illustrative examples of Daily Inspection Few schedules are already shown in chapter 7. Basing on those, other schedules can be drawn-out.

Table 9.6 Daily Inspection Schedule

	Preventive Maintenance and Repair Planning (Daily Inspection Schedule)					
	Department: .... Date.................. Cost Center: .... Equipment: .... Assembly					
Sl. No.	Job No.	Job Description	Frequency	EOP	I/R	Action
1	......	......	......	......	......	......
2	......	......	......	......	......	......
3						
4						

Notes: EOP means Effect on production (Answer Y for yes and N for no)

I/R means inspection to be done in idle or running condition of equipment,

Action—Put Y for done and N for not done (with reason)

9.11.8 Follow-up and Documentation

The observations of defect lists, failure analysis reports, inspection schedules (both manual and online etc), lubrication schedules, records of repair schedules, PM schedules and other schedules made and their compliance and pending jobs etc should all be properly documented and necessary follow up is initiated. Proper documentation at every level is the lifeline for CMMS and any other such programmes.

For checking the effectiveness of total process, many methods/parameters are mentioned in chapter 13. However, two most commonly used ones are as below—

- Equipment availability reports (text, graphical and other forms),
- MTBF- Mean Time Between Failure reports (text, graphical and other forms),

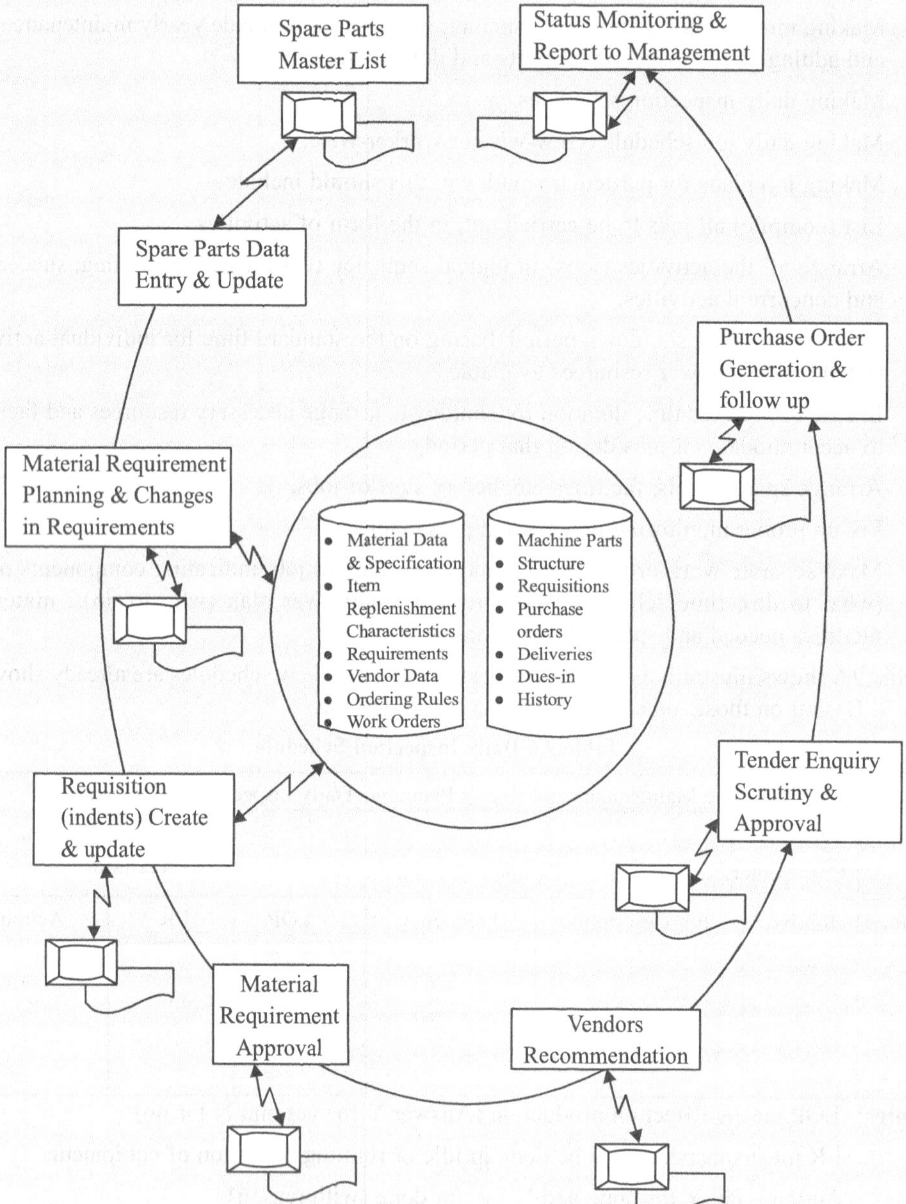

Fig. 9.4. Material Management Module

9.12 MATERIAL MANAGEMENT MODULE

Though material management is the responsibility of Material Management Department in any big plant, it is equally important for maintenance personnel as availability of equipments depend substantially on the timely availability of quality spares, tools, tackles, consumables etc. As such 'Material Management Module' is included in CMMS and works in conjunction with plant's material management department, with more focus on material planning activities.

Following areas are emphasized in this module—

(*i*) Cataloguing of spares and consumables,

(*ii*) Compilation of basic details and specifications about materials,

(*iii*) Classification and codification of spares/components,

(*iv*) Requirement planning and indent (requisition) generation,

(*v*) Tenders/offers evaluation and recommendation,

(*vi*) Monitoring of purchase indents (requisitions) and purchase orders,

(*vii*) Necessary help to inspection cell of Material Management Department for specialized technical inspection,

(*viii*) Inventory monitoring,

(*ix*) Standardization and rationalization,

(*x*) Import substitution, indigenous development and development; etc.

Overall working of material management module of CMMS is roughly shown in Fig.9.4. The advantages of such modules are as followings—

- Timely availability of materials in right quantity and right quality,
- Reduction in lead time for procurement (both internal and external),
- Budgetary control through quick and good information system,
- Proper identification and location of spares and other materials,
- Consumption control,
- Inventory control; etc.

9.12.1 Cataloguing and Codification of Spares and Consumables

Codification of spares is a scientific record keeping and information system, which is essential for proper identification, reducing duplication, material planning, procurement, storage, use and inventory control. Like other codification, this also adopts a scientific numbering system for each spares, components and consumables. The complete code for each of such item is also known as catalogue number for that item.

One example of a 10 digit material codification (catalogue number) is given in Fig. 9.5. In this codification, the 'unit of quantity code' (A) may be 0 to 9, indicating numbers, sets, pairs, meters, sq. meters, kilograms, tons etc. Other codes are self explanatory. In a big industry, if necessary, we may increase the number of digits or use alphabets at some places (alpha-numeric codification) to accommodate more items in the same category.

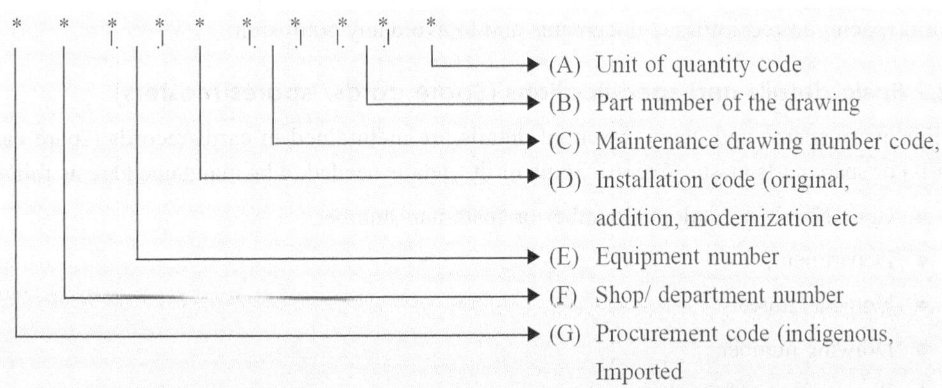

Fig. 9.5. Spares codification—Catalogue number

In any industry, there are normally two types of spares——

(*i*) Unique spares, which may be used at only one place, and

(*ii*) Common or interchangeable spares, which may be used at more than one place in the same industry,

(*iii*) Consumables.

The codification shown at Fig. 9.5 is for unique spares. For common spares, the

Codification (catalogue number) of the first shop/place may be repeated at other shops/ places and intimation may be made in the Equipment catalogue accordingly.

In addition to such spares, lot of consumables, such as cotton wastes, lubricating oils, common abrasives (paper/tapes, pastes or small wheels etc), common welding electrodes, common nuts/ bolts/screws/washers etc, common small tools (files, pliers, etc), common GI and PVC pipes and fittings etc are used almost by all shops/areas of maintenance. As such, codification and cataloguing of these are done differently. Fig.9.6 shows a typical codification of such consumables, again taking 10 digits codification.

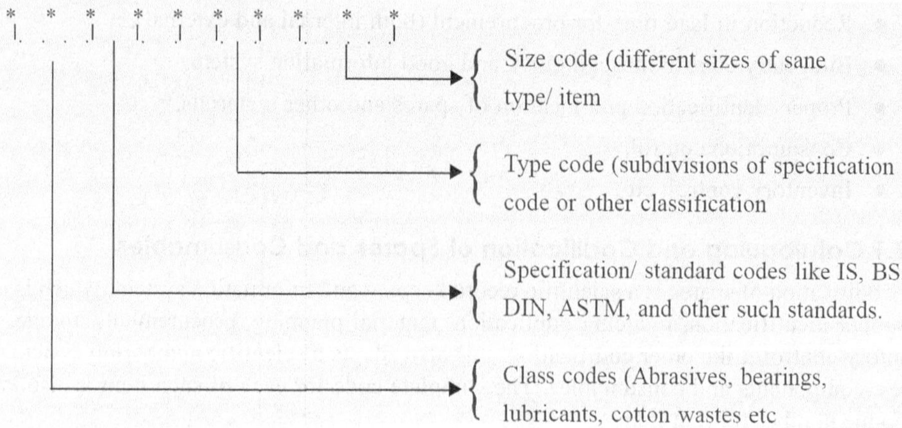

Fig. 9.6. A Typical consumables codification/ catalogue numbering.

The above spares codifications have been shown mainly for items of mechanical nature. The codification for items of electrical, instrumentation natures etc may also be done in the same way, but, may be adopting some different logics. The total number of digits for all the spares (unique or common) and consumables, used throughout the industry should preferably the same, to have uniform spacing in receipt/issue documents and to avoid any confusion.

9.12.2 Basic details and specifications (Spare cards/ spares masters)

For every spares and items, essential details are maintained in cards/records (spare cards/ records) or spare parts master/ record. Some of the details needed to be maintained are as follows–

- Code Number/ catalogue number or spare part number,
- Department,
- Nomenclature,
- Drawing number,
- Part number (of the drawing),
- Installed quantity,

- Weight/ volume (for bigger items),
- Criticality; etc.

Fig. 9.7 shows a typical spare part card/ record. It may be in the form of card or computer record or in any other form. The detailed specification of each part/ item is also maintained (and is periodically updated as per need), to enable procurement of correct item.

SPARE PART CARD (MECHANICAL)				INFORMATION RECORD				
DEPARTMENT :				PRODUCTION				
COST CENTRE :				CRITICALITY	VITAL	ESSENTIAL		DESIRABLE
GROUP OF EQUIPMENT				PROCUREMENT	SCARCE	DIFFICULT		EASY
SUB GROUP OF EQUIPMENT				CRITICALY				
DRAWING NUMBER				MOVEMENT	F Fast	S Slow	N Now	C Capital Repair	I Insuranc
PART NUMBER (DRG):				SCALING				
PATTERN/TEMPLATE NO.	STD NO.			AVERAGE DEMAND	NORMAL MAINT SPECIAL REPAIR			
		Rough	Original	Qty. in stock as on				
Import	Installed Quantity			Safety Stock (Minimum) Re Order Level (ROL)				
Indian	Operation	Finish	Substitute	Rate Contract (RC)				
Internal	Circulation			Number of Deliveries (No) Qty. Per Delivery (Q)				
SPARE PART / STOCK NUMBER				UNIT OF ISSUE				
				CODE	DESC	DESCRIPTION OF THE PART		

Fig. 9.7 Typical Spare Part Card

9.12.3 CLASSIFICATION OF SPARES AND CONSUMABLES

In addition to the classification of spares and consumables mentioned earlier (based on drawings etc), another way of classification of these can be based on its movement pattern. Fig. 9.8 shows one such classification.

There may be two other ways of classifying spares–VED Analysis based and SDE Analysis based.

(*i*) **VED Analysis based classification**–

- Vital (V) spares, non-availability of which makes the equipment total inoperative or Unsafe,
- Essential (E) spares, which reduces the equipment's performance and efficiency, but do not make the equipment totally inoperative immediately,
- Desirable (D) spares, which are non-functional and do not affect the performance of equipments, but gives better look or better ease of operation and maintenance

(*ii*) **SDE-Analysis based classification–**

- Scares (S) spares are those which are not easily available, may have to be imported or mar require source development,
- Difficult (D) spares are those which are not readily available, or can be made with difficulty within captive shops or through outside sources,
- Easy (E) spares are those which are readily available off-the-shelf or can be made easily, in captive shops within short notice.

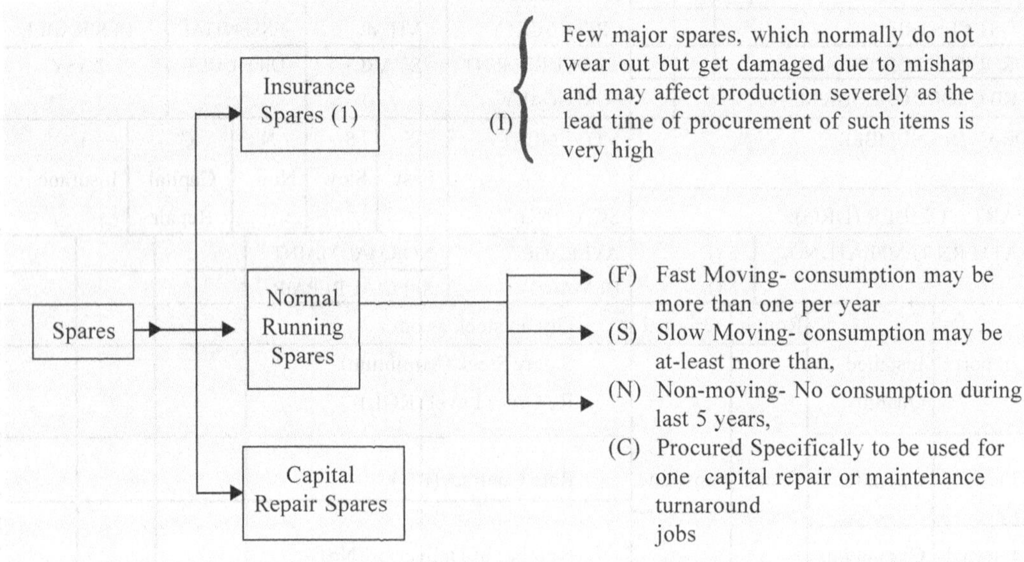

Fig. 9.8 Movement-based spares classification

These classifications are used more by material management department, but some indication should also be made in material management module of CMMS for help in spares planning.

9.12.4 Requirement Planning and Indent Generation

Requirement planning job is very challenging for maintenance as it should not lead to overstocking (inventory built-up) or under-stocking, i.e. equipment and plant should not suffer for want of spare/item in time. CMMS provides following data to help in planning optimum requirements—

(*i*) Consumption data and pattern for individual items,

(*ii*) Detailed specifications, drawings and codes for items,

(*iii*) Performance and reliability of parts and machines/equipments,

(*iv*) Lead time (specifically for imported or difficult or scares items, needing more time),

(*v*) Information about economic repair/ salvaging of old spares/ components,

(*vi*) Obsolescence of items,

(*vii*) Planned capital repair or major repair jobs,

(*viii*) Stock available in stores (central and departmental) at the time of planning; etc.

After requirement planning, next step in generation of 'purchase indent/ requisition'. Utmost care should be exercised at this stage to supply all necessary information and nothing should be taken as "assumed". Any mistake or omission at this stage may lead to getting incorrect material. Further jobs are the responsibility of plant's material management department. Of-course help of maintenance personnel may be required and corresponding CMMS module may be linked with Material Department's computers.

9.12.5 Standardization and Varity Reduction

Standardization and rationalization of spares, as applicable in CMMS, calls for following steps and considerations—

(*i*) **Using standard items/ components**– As much as possible, we should select spares and components from standard components from reputed manufacturers/ suppliers. We may also use prevailing plant or interplant standards or national/ international standards, even though the selected component may appear to little over-rated.

- Equipments such as pumps, compressors, electrical motors, contactors etc can be selected from the standard catalogued products of reputed manufacturers,
- Components, such as gear-boxes, valves, couplings, carbon-brushes, limit-switches etc, again as standard catalogued products of reputed manufacturers,
- Consumables, e.g. general purpose welding electrodes, oils, lubricants, cutting tools, packing and jointing, belts and common bolts etc directly from reputed manufactures,
- Tools and tackles, e.g. hydraulic and screw jacks, chain-blocks, pull-lifts etc, again as standard products of reputed manufacturers; Etc.

(*ii*) **Reducing Tailor-made items**- Specially tailor-made items should be reduced where-ever possible, even-though it may lead to using little over-rated items,

(*iii*) **Introducing Interchangeability**- As much as possible, spares should be made interchangeable at different equipments within the shop, plant, or sister-concern, even though it may need little over rated component at few places.

(*iv*) **Variety Reduction** i.e. reducing the number of varieties in sizes and types etc. Such standardization, rationalization and variety reduction gives following advantages–

- Reduction in inventory of spares and components and thus, less material cost,
- Interchangeability of spares and components, also leading to much less chances for stock-out of those items and less chances of equipment downtime for materials,
- More reliability by using standard and proven components from reputed suppliers,
- Speedy procurement of proper spares and other materials,
- Less inventory storage and carrying cost; Etc.

9.12.6 Alphanumeric Codification

As composite alphanumeric codification is more in use today, the spares and assemblies can also be codified in line with equipment classification, mentioned in Sec.9.10.3, by adding few characters (alphabets and digits). Again suitable alphabets can be added to indicate the exact location of the spares/ subassemblies (e.g. in central store-No.1 or No.2 or No.3 or in shop stores or sub-stores etc).

9.13 CAPTIVE ENGINEERING SHOPS MODULE

In most developed countries (and lately in few industries in India), plant maintenance can afford to depend fully on outside sources for their spares (including making new or reconditioning old ones) because of existence of versatile and capable engineering concerns nearby to supply the item in time. But in India and many other developing countries, the small and medium ancillary industries formed at the outskirt of big industries (like steel plants) have limitation in taking up large and intricate spares manufacturing or reconditioning. As such most of the steel plants and few other big industries are provided with some manufacturing facilities to make few small and big spare parts and also to repair/ recondition old components. Such facilities are called captive engineering shops.

Earlier such captive shops were merely considered as 'repair shops' and were not given due but now, with the introduction of CMMS in such shops also, these shops have got due importance and, by making and reconditioning more spares and by making few other components for other shops, such captive shops are turning themselves as "Profit Centered Shops".

Captive shops in a big industry (like integrated steel plants), normally under chief of maintenance, includes following sections/shops—

1. Machine shop (normally light or medium),
2. Fabrication shop and special welding shop,
3. Assembly and overhauling shops,
4. Forge shops (normally light or medium) and smithy,
5. Small foundry facility/section,
6. Small heat treatment section,
7. Wagon Repair shop; Etc.

For evaluating the performance of captive shops, one has to know 'shop capacity', 'work center' and 'job sequencing'.

9.13.1 Shop Capacity

Shop capacity is computed from normal/ standard capacity of each machine/ facility. Units of capacity should be easy measurable. For machine shops, measuring units may be one or combination of many parameters, such as–

- Number of hours machine actually ran per day or per month, as compared to the hours scheduled, or
- Volume of material removed per shift, or
- Value of items produced per day or per month, including cost of manpower; Etc.

Again for welding and fabrication shops, the measuring units may be one or combination from many parameters, such as,

- Man-hours actually worked on the job, or
- Volume of weld-metal deposited or welding electrode consumed per job (for big jobs) or per day/ per month, or
- Value of items produced per day or per month, including cost of manpower; Etc.

For assembly jobs, which are only manpower oriented, the measuring units may be man-hours actually worked per day or per month, including some element of cost. Accordingly, measuring unit of all other shops can be worked out.

However, actual measuring unit in a plant is decided on the basis of nature of business and other local conditions. After knowing the measuring units of each machine/ facility and knowing the number of machines and facilities and also number of workmen, shop capacity can be worked out.

9.13.2 Work Centers and Job Sequencing

A big captive shop complex is divided into many work centers or group of machines for uniformity and ease of evaluation. Each work center may consist a number of similar type and near similar capacity machines of machines. As an example, all lathes of same or nearly same "center distance" and "swing-over bed" may form one work center. For assembly shops, each group of workmen or crews may form one work center.

Proper and logical sequencing is very essential for effective and economic utilization of machines and facilities, to reduce wastages and to reduce waiting time. Fig 9.9 shows a simple and typical example of job sequencing, where one part (P_3) is bought-out item (say bearing), one part (P_1) is made from raw material R_1 and another part (P_2) is made from raw material R_2, but with the help of fixture F_1, which has to be made from raw material R_3. All the three parts are then assembled. Levels 1, 2, 3 and 4 are different working levels in such a way that Level 3 follows after completion of Level 4, level 2 follows after completion of level 2 etc. OPR 1, 2, 3 and 4 indicates different types of operation (e.g. gas cutting/welding, machining, fitting, testing, painting etc).

- Make or buy decision,
- Process planning,
- Material planning and procurement,
- Estimation of delivery period,
- Dispatch; Etc.

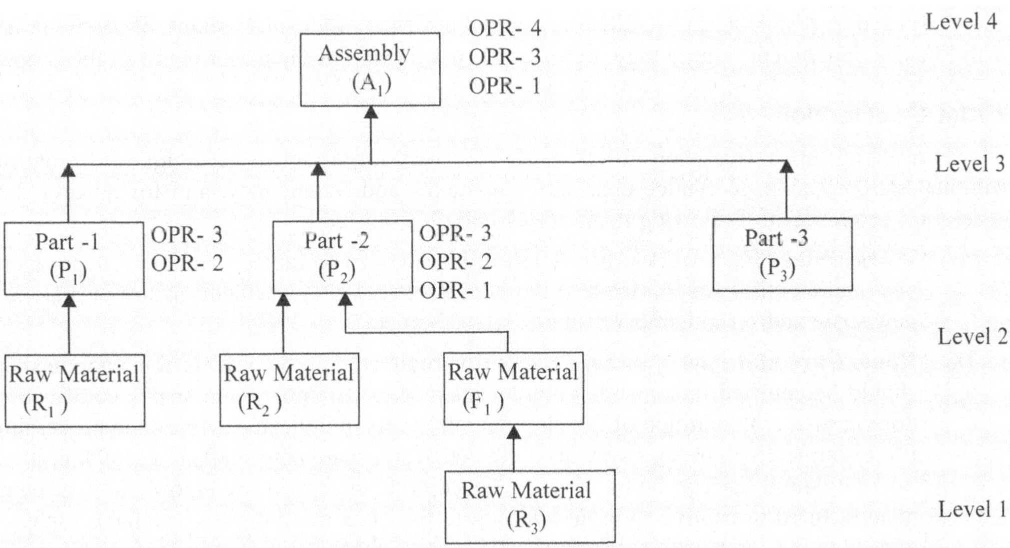

Fig. 9.9. Sequencing of Operations

9.13.3 Process Planning

Process planning is, probably, the most important step in captive shop utilization. It involves preparation of followings, in addition to other jobs—

- Route Card or Job Card or Technology Card,

- Machine Loading Chart,
- Cost Card/ Cost data

Route card (also known as job card or technology card at different places) is a carefully designed form, indicating total operations and sequences for easy follow-up by shop floor operators. One 'route card' is made for each of the element. For example, for sequences shown in Fig. 9.9, separate route card should be made for A_1, P_1, P_2 and F_1. There are no separate route cards for bought-out items or raw materials, but these can be mentioned in connected route cards. A route card should include following information—

- Work-order number and operation number,
- Work center,
- Estimated operation time per piece or per item,
- Set-up (SU) time, i.e. time for studying the and initial minor adjustment,
- Move Time (or additional set-up time), i.e. time to be spend before starting the next operation,
- Narration of the execution procedure,
- Number of pieces for the operation; Etc.

Fig.9.10 and Fig.9.11 shows the two sides of a typical Route Card, being followed in a small plant in SAIL (manual system). However, in a good CMMS, this can be fully computerized and changed suitably.

Machine Loading Chart is made for each machine from available and in-process work-orders so that waiting time (unutilized time) for each machine is reduced to minimum and priority work orders are also taken care-off.

Cost card or cost data can be suitably designed in CMMS for captive shops, of-course in line with Finance department terminology, for effective monitoring of the performance of the shop.

9.13.4 Codification

Codification in captive shop utilization is also very important to get unique information, without any confusion, to reduce chances of mistakes and redundancy, and for quicker data generation and retrieval. Following items are essentially codified—

(*i*) **Work-order**– Each work-order or job-order is coded to indicate initiating department, year and month it was initiated, if the item is of repair/replacement nature or making a new spare and serial number of the captive shop etc.

(*ii*) **Route Card and Part Number**– Every part/route card, to be processed in captive shop, should be codified in somewhat similar manner as equipment and spares codification, done earlier, with addition of few digits or alphabets to indicate captive shop processing. For example, in eight digit system, if the code for a spare (say shaft) is 12300456, it can be 12300456R if it is being repaired in captive shaft or 12300456M, if it is being manufactured in captive shop. In addition to this, few more digits or alphabets may be added to indicate the sub-sections of a big captive shop complex processing that spare or route card.

(*iii*) **Material**– For captive shops, materials may be parts to be processed, bought-out items or raw materials. Codification of parts has been discussed earlier. We can add fewdigits/ alphabets to any part code to identify it for captive shop; say we may add 00 to indicate bought-out items, CA for Casting, FO for forging etc. Considering 8 digit codification for parts, we can identify as followings—

x x x x x x x x 00 (Bought-out items for captive shop)

x x x x x x x x CA (casting processed in captive shop)

x x x x x x x x FO (Forging processed in captive shop) ; etc.

(*iv*) **Machines**– Every machine and facility of the captive shop should be codified in similar way as equipment codification, discussed earlier, to indicate location, type of machine, serial no./ model number of the machine etc.

(*v*) **Operations**– Every operation or process of the captive shop must also be codified and each route card, work-order etc must indicate that code number.

History cards, containing both— permanent files/ records like machine data and dynamic (temporary) files/records, and many other records procedures, discussed earlier, to be maintained for captive shops also.

CMMS also deals with centralized facilities (e.g. compressed air and steam generation and distribution, fuel oil and gas procurement and distribution, water (drinking and industrial) supply systems etc) in similar way with suitable addition or identifications.

9.14 OTHER CODIFICATIONS

In CMMS all other aspects/ activities, concerning maintenance and connecting departments are also properly codified for smooth and correct information flow through computers and various reports etc. In almost all industries of today, different failures, all the crafts/ trades, different designations, their pay scales/grades, performance evaluation methods etc etc are codified and would not be discussed further here.

9.15 FUTURE OF CMMS

CMMS has been in use for nearly a quarter century. It has become an essential part of any maintenance organization. Most CMMS packages deliver on the value proposition that maintenance software is designed to help stay organized, reduce maintenance costs and increase overall margins. However, one thing has been missing with many of the CMMS packages: they are not often, web-based.

With the advancement of Internet technology, business is now conducted much differently than earlier. CMMS is also moving towards web-based. Today, maintenance managers from locations around the globe can share maintenance information using web-based maintenance management software. The maintenance and operations departments of schools, universities and hospitals are recognizing the deep benefits associated with investing in web-based technologies.

With web-based software, you can access your maintenance data securely from anywhere in the world. Minimal investment is required to get up and running, which leads to a lower overall cost of ownership and quicker realization of the return on your investment. Typically, there have been high costs associated with installing software on every computer. Now, all you need is a web browser and an Internet connection to use maintenance management software on a day-to-day basis.

There are quite a few vendors/ providers who have built such software solutions that run through the Internet. These solutions are called **'Web-based CMMS'**, which have put a whole new twist on maintenance management. Maintenance organizations can now much more easily gain access to their maintenance data 24 hours per day via the Internet. It is not that the traditional client-server model of CMMS does not work, but rather that the Web-based CMMS just add value to ultimate CMMS software package.

<table>
<tr><td colspan="6">Technology Card (ROUTE CARD) MED/ MRS/ JM/ 09
MRS/ MED/ ASP Work Order No. — **/****/S/ ***</td></tr>
<tr><td colspan="6">Department : Order By: Phone No :
Equipment : Drawing No :
Catalogue/Code No :
Part Description :
Job Description : Quantity Location :
Remark :
Detailed Material Description :</td></tr>
<tr><td>Item</td><td>Material and Description</td><td>Pcs. Input : Output Size</td><td>Input Size</td><td>Output Weight</td><td>Time Weight</td></tr>
<tr><td>01</td><td></td><td>0</td><td></td><td>.000</td><td>.000</td></tr>
<tr><td>02</td><td></td><td>0</td><td></td><td>.000</td><td>.000</td></tr>
<tr><td colspan="4">Total Weight:</td><td>.000</td><td>.000</td></tr>
<tr><td colspan="6">Detail Process/ operation Description</td></tr>
<tr><td>Item</td><td>Sl.No.</td><td>Element of Operation and Operation/ process code</td><td>Machine</td><td colspan="2">Remark</td></tr>
<tr><td>01</td><td>01</td><td></td><td></td><td colspan="2"></td></tr>
<tr><td>01</td><td>02</td><td></td><td></td><td colspan="2"></td></tr>
<tr><td>01</td><td>03</td><td></td><td></td><td colspan="2"></td></tr>
<tr><td>02</td><td>01</td><td></td><td></td><td colspan="2"></td></tr>
<tr><td colspan="6">Additional Information/ Tools/ Tackles</td></tr>
<tr><td colspan="6">Special Note (If any)</td></tr>
<tr><td colspan="6">Prepared By Approved By</td></tr>
</table>

Fig. 9.10. Front Side of one Typical Route Card

ASP/ MED/ MRS		ROUTE CARD			(2/2)
DATE	SHIFT NO./CODE	MACHINE	JOB PROGRESS	STAGE INSPECTION	DEVIATION

INSPECTION				JOB COMPLETION REPORT		
DATE	SHIFT	INSPECTED	REJECTED	DISPATCHED	SIGNATURE	COMMENTS (IF ANY)
Total						Shift In-charge

Fig. 9.11. Back side of same route card

QUESTIONS

1. What is CMMS and what are its benefits?
2. What are the scope of computerization and IT in maintenance?
3. How do you classify and codify equipments in an industry?
4. How do you codify and catalogue spares and consumables?
5. What are captive engineering shops and how CMMS can be used there?
6. Explain briefly different components/ features of CMMS?
7. What are the different elements in Preventive Maintenance and Repair Module In Computer Managed Maintenance System?
8. What are the factors you should consider while selecting a CMMS for your plant?

CHAPTER

10 TOTAL PRODUCTIVE MAINTENANCE (TPM)

TPM is a philosophy of continuous improvement that creates a sense of ownership in the operators and other connected persons for machine as well as in their supervisor. We all reach a new attitude of higher responsibility and commitment. Its focus is on "Ease the operation", "Improve the availability and productivity", "Ease the maintainability" and "Make the operation and area Safer" etc. Once implemented, TPM has just begun as it is an ongoing process of continued improvement supported by all the team. *It can be considered as the medical science of machines.*

10.1 DEFINITION

The three different words of TPM can be defined as following—

Total

(*i*) All employees are involved in it,

(*ii*) It aims to eliminate all breakdowns, defects and accidents;

Productive

(*i*) Many actions are performed while production goes on,

(*ii*) Troubles for production are minimized, if not eliminated.

Maintenance

(*i*) Check, clean, lubricate and repair,

(*ii*) Keep machine in good condition.

In nutshell, TPM combines the traditional practices of preventive and predictive maintenance with total quality control and total employees involvement/ participation, to create a culture where operators develop ownership of their equipment and becomes full partners with maintenance, engineering and management etc to ensure that equipments operate properly everyday at required time.

10.2 ORIGIN OF TPM

In earlier days, job of maintenance was supposed to be dull and donkey's job. Planned maintenance can be traced back to 1951 when preventive maintenance was introduced in Japan. However the concept of preventive maintenance was taken from USA for broader use.

As the operating staff did not involve themselves in maintenance and did not try to know the mechanism, very often some safety aspects were missed, leading to somewhat unsafe working environment and practices. Further, it is seen that when any equipment problem develops, it is much easier and less costly to attend to that by the operating staff that to keep running for some more time or waste time in calling maintenance personnel, leading to problem getting aggravated (i.e. a stitch in time saves nine).

All these led to the development of "**Productive Maintenance**' concept, where the operating personnel are also involved in maintenance and upkeep of their equipments. Just as a man looks after his own health and tries to prevent minor sickness and calls doctor only for major problem or periodic checking, the operating personnel looks after daily checking, lubrication and routine maintenance of their equipments and specialized maintenance groups attend to major problems/ breakdowns and major PM jobs etc. This concept was first introduced by M/s General Electric of USA in year 1960, but gained momentum much later.

Later Dr. Deming introduced statistical analysis and used the resulting data to control quality during manufacturing (TQM). As some general concepts of TQM did not work straightway in maintenance environment, efforts went on to use somewhat similar principle in maintenance. However, maintenance became an integral part of TQM in early 90s.

Meanwhile, the concept of Productive Maintenance was extended to all the employees of the plant/ industry, including material men, quality men and others and all the employees worked as one team, under consent. This was called "Total Productive Maintenance" (TPM). This concept was first introduced by M/s Nihon Denso Co. of Japan in year 1971 and was subsequently promoted to national level in Japan in 1983, by Japanese Institute of Plant Maintenance.

10.3 FEATURES/ PRINCIPLES OF TPM

TPM has led to better maintenance and upkeep of the equipment and resulted in enhanced reliability of equipments and quality of products, thus leading towards economic efficiency excellence. Some of the features/ principles considered for successful TPM are as followings—

1. Increase of Overall Equipment Effectiveness (OEE),
2. The operator is the best condition monitor,
3. Involves all departments, such as operation, maintenance, connected planning groups, material management, quality, finance and administration (of-course the first three departments are totally involved) and utilizes cross-functional teamwork.
4. Involves the participation (up-to required level) of members of all groups from top management level to shop-floor worker level.
5. Provides training to upgrade operation and maintenance skills,
6. Generally promotes small autonomous group concept for better motivation and coordination;
7. Establishes a total system of complete maintenance and upkeep of the equipments, generally covering the entire plant lifecycle; etc.

The difference between preventive maintenance, productive maintenance and TPM is shown in Fig. 10.1

		Total Productive Maintenance	Productive Maintenance	Preventive Maintenance
Employees Involvement for the plant	All employees of the plant	0		
	Operation and Maintenance personnel only		0	
	Maintenance Dept. personnel only			0
Characteristics of system	Economic Efficiency Excellence	0	0	0
	Life Cycle approach for equipments	0	0	
	Small Autonomous Group approach	0		

Fig. 10.1. Comparison of TPM with other Maintenance groups

10.4 TPM IS NOT DECENTRALIZED MAINTENANCE

In decentralized maintenance, the maintenance and operation personnel may stay together at work spot but their involvement and functional relationship remains completely different whereas in TPM the involvement is complete.

10.5 TPM VIS-À-VIS TERRO-TECHNOLOGY

The purpose of TPM and Terro-Technology (developed in UK in 1970) is somewhat similar, i.e. to pursue economically efficient life cycle cost; but the scope of Terro-Technology is much larger, as it involves designers, manufacturers, erectors and other outside agencies, in addition to user plant personnel, and as such, it does not concentrate much on maintenance. But the TPM is connected with only user plant personnel. There are differences in approach also as TPM calls for direct and total involvement of all connected members for the same goal, whereas in Terro-Technology, goals of all groups/ agencies may not be the same. Fig. 10.2 (Relationship between TPM and Terro-Technology) explains the relationship between the two where it is shown than user (TPM) is a small component of whole Terro-Technology.

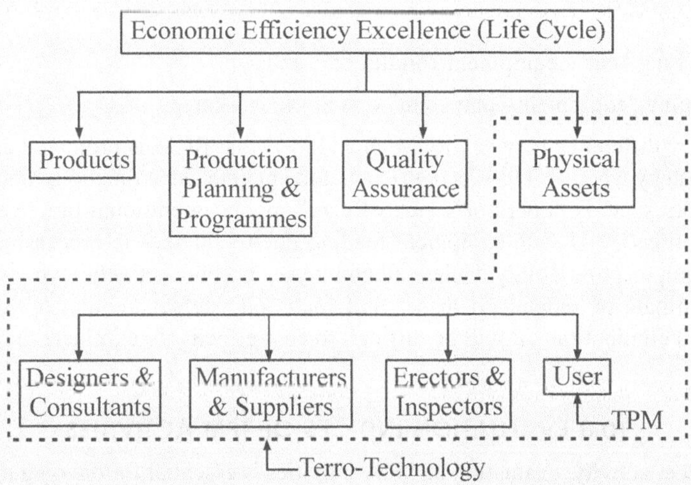

Fig.10.2 Relationship between TPM and Terro-Technology

10.6 TPM VIS-À-VIS TQM

TPM resembles to Total Quality Management (TQM) or its extended version Total Quality Process (TQP) to the extent that both are small group activity, both employs tools such as employee empowerment and commitment, benchmarking, documentation etc, there are still many differences between the two. Some of the differences between TPM and TQM (or TQP) are given below:

Category	TQM (or TQP)	TPM
Object	Quality (Output and Effect)	Equipment (Input and Cause)
Means of attaining goal	Systematize the management. It is software oriented. It centers, normally, on improvement of systems and standards	Employees participation. It is hardware oriented. It centers on improvement of equipments and their performance
Mode of selection of teams	Its small group activities or quality circles are generally voluntary	Its small group activities are company-led and are integral part of workplace and organization
Target	Quality for PPM (Parts per million)	Elimination of losses and wastes
Training of Employees	Centers on management and inspection technology (quality circles etc)	Centers on maintenance technology, e.g. knowledge and skills peculiar to maintenance and up-keep of plants and machineries

10.7 TPM VIS-À-VIS RCM

The original precept of RCM was developed for aircraft industry where "basic equipment condition" (no looseness or contamination or lubrication problems etc) are mandatory and where operator's (pilot's) skill level, behaviour and training is of high standard. Unfortunately, in most manufacturing and other industries, these "basic equipment conditions" and operator's skill and behaviour levels do not exist to that level, thus undermining the basis of RCM applications.

For this reason, application of TPM is used in most industries for wide improvement strategy, as it helps in:

- Establishing "basic equipment conditions", and
- Developing "equipment-competent" operators/ workmen.

The other key difference between RCM and TPM is that RCM is promoted as a maintenance improvement strategy whereas TPM recognizes that the maintenance functions alone can't improve reliability. Factors such as "operator's lack of care", poor "operational practices", poor "basic equipment conditions", "adverse equipment loading due to changes in processing requirements" etc also effect equipment reliability. Unless all employees become actively involved in recognizing the need to eliminate or reduce all losses and focus on "defect avoidance" or "early defect identification and elimination", it will be difficult to cost-effectively eliminate the failures in most industries.

10.8 EVALUATION INDICES OF TPM ACTIVITIES

As in any other activity, evaluation of TPM activities is essential for ongoing monitoring. Such indices may be following three categories

1. Economic Indices, e.g. maintenance cost as % of total cost, etc
2. Performance Indices, e.g. % Downtime, % Availability, % Jobs outsourced, Overall Equipment Effectiveness (OEE) etc,
3. Quality Indices, e.g. % Repetitive breakdown, Reliability etc

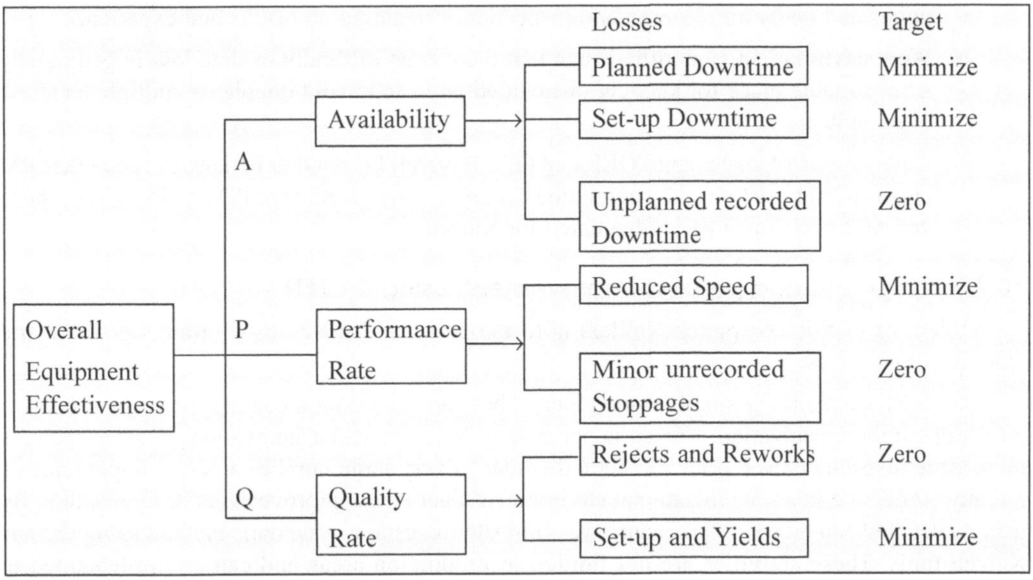

Fig.10.3. OEE Model for TPM

Evaluation Indices are discussed in more details in Chapter 13. Fig.10.3 (OEE Model of For TPM), given here, also helps in its working and evaluation.

10.9 COMPONENTS/ PILLARS OF TPM

Followings are eight basic components or pillars of TPM

10.9.1 Autonomous Maintenance (Jishu Hozen)

It promotes small autonomous group concept, involving operation, maintenance and all connected disciplines to required degrees. In order to progressively increase operator's knowledge, participation and responsibility for their equipment, following steps are taken—

1. Train the operators to close the gap between them and the maintenance staff, making it easier for both to work as one team (flexible operators),
2. Change the equipment so the operator can identify any abnormal conditions and measure deterioration before it affects the process or leads to a failure,
3. Operator to perform initial cleaning and inspection; take countermeasures for the causes and effects of dirt and dust, provide covers/ bellows and other protective or self-cleaning devices.
4. Other countermeasures– Inaccessible regions had to be reached easily. If there are many screw to open a fly wheel door, hinge door can be used. Instead of opening a door for inspecting the machine, acrylic sheets can be used

5. Establish cleaning and lubrication standards and operator to carryout minor lubrication and monitor lubrication gadgets,
6. Eliminating the defects at source through active employee participation,
7. Autonomous Inspection, including continuous monitoring and changing frequencies and points and items for cleaning and inspection, depending on results and experience.
8. Work environment is modified such that there is no difficulty or time loss in getting any item; suitable place for keeping dismantled parts and avoid double or multiple handling of same item.
9. Autonomous Management– OEE and OPE (Overall Production Efficiency) and other PM targets must be achieved by continuous improve through Kaizen; PDCA (Plan, Do, Check and Act) cycle must be implemented for Kaizen.

10.9.2 Equipment and process improvement, using KAIZEN

Objective of this component (pillar) is to maximize efficiency by eliminating waste and manufacturing losses.

Basically kaizen is for small improvements, but carried out on a continual basis and involve all people in the organization. Kaizen is opposite to big spectacular innovations. Kaizen requires no or little investment. The principle behind is that "a very large number of small improvements are move effective in an organizational environment than a few improvements of large value. By using a detailed and thorough procedure we eliminate losses in a systematic method using various Kaizen tools. These activities are not limited to production areas and can be implemented in administrative areas as well.

TPM identifies following major sources of losses and then works systematically to eliminate them

(A) Losses that impede equipment efficiency (Equipment Losses)

(*i*) Downtime Loss

1. Equipment failure/ breakdowns
2. Set-ups and adjustments,
3. Scheduled downtime loss,
4. Consumable breakage/change loss (cutting blade loss etc)

(*ii*) Speed Loss

5. Minor stopping/ idling
6. Reduced speed,

(*iii*) Quality Loss

7. Process errors
8. Reworks/ scraps

(B) Losses that impede human efficiency (Manpower Loss)

9. Cleaning and checking,
10. Waiting for instructions,
11. Waiting for materials
12. Waiting quality confirmation,
13. Measurement and adjustment loss,

(C) Loses that impede effective use of production resources (Material Loss)

14. Material Yield Loss,
15. Energy Loss,
16. Consumable material loss (Die, jig and tool breakage loss etc).

All these losses can be classified as below for further action towards their elimination

Classification of Losses/ wastes

Aspect	Sporadic Loss	Chronic Loss
Causation	Causes for this failure can be easily traced. Cause-effect relationship is simple to trace.	This loss cannot be easily identified and solved. Even if various counter measures are applied
Remedy	Easy to establish a remedial measure	These types of losses are caused because of hidden defects in machine, equipment and methods.
Impact / Loss	A single loss can be costly	A single cause is rare - combination of causes trends to be a rule
Frequency of occurrence	The frequency of occurrence is low and occasional.	The frequency of loss is more.
Corrective action	Usually the line personnel in the production can attend to this problem.	Specialists in process engineering, quality assurance and maintenance people are required.

TPM attempts to reduce substantially, if not eliminate, by using KAIZEN tools, such as PM analysis, Why-Why analysis, Summary of losses, Kaizen register and Kaizen summary sheet etc.

10.9.3 Planned Maintenance

Objective of this component (Pillar) is to establish Preventative, Predictive Maintenance systems for equipment and tooling.

Natural life cycle of individual machine elements must be achieved for trouble-free working, producing defect-free products for total customer satisfaction, by planned maintenance, which can be grouped in following categories

- Corrective Maintenance,
- Preventive Maintenance,
- Predictive and CBM Maintenance
- Maintenance Prevention.

All these systems have been discussed in chapter 5. Correct operation, Correct set-up, Proper cleaning, Proper lubrication, Necessary retightening, Feedback and repair of minor defects, and Quality spare parts etc are specifically taken care off.

Following are the six steps for effective Planned Maintenance

1. Equipment evaluation and recoding present status.
2. Restore deterioration and improve weakness.
3. Building up information management system.

4. Prepare time based information system, select equipment, parts and members and map out plan.
5. Prepare predictive maintenance system by introducing equipment diagnostic techniques and
6. Evaluation of planned maintenance

10.9.4 Early Management of new equipment

The objective of this component (pillar) is

1. To establish a system to shorten/ minimize the time for new equipment or product development,
2. System shorten the start-up, commissioning and stabilization time for quality and efficiency,
3. New equipment needs to be
   - easy to operate
   - easy to clean
   - reliable and easy to maintain
   - have quick set-up times
   - have easily visual and approachable built-in-test equipment and condition monitoring equipments,
   - operate at the lowest life cycle cost.

10.9.5 Process Quality Management and Quality Maintenance

This component (pillar) calls for a process for controlling the condition of equipment and machineries that affect variability in product quality and dives defect-free products.

Its objective is to set and maintain conditions of equipments to accomplish zero defects as the quality rate of any product has a direct correlation with followings

- Material condition,
- Equipment condition and precision,
- Production methods,
- Process parameters

We gain understanding of what parts of the equipment affect product quality and begin to eliminate current quality concerns, then move to potential quality concerns. Transition is from reactive to proactive (Quality Control to Quality Assurance). It aims to

- In-line detection and segregation of defects
- Focus of prevention of defects at source
- Focus on poka-yoke. (fool proof system), etc.

10.9.6 Education and Training

TPM is a continuous learning process. It is aimed to have multi-skilled revitalized employees whose morale is high and who is eager to come to work and perform all required functions effectively and independently. Education and training is given to operators/workmen to upgrade their skill. It is not sufficient to know only "Know-How" but they should also learn "Know-why".

By experience they gain, "Know-How" to overcome a problem what is to be done. This they do without knowing the root cause of the problem and why they are doing so. Hence it become necessary to train them on knowing "Know-why". There are following four knowledge stages of a workman about a job

Stage 1: Do not know

Stage 2: Know the theory, but can not do,

Stage 3: Can do the job but can not teach,

Stage 4: Can do the job and can teach also.

Aims and Targets of Education/ Training in TPM

It should aim for the followings

1. Improvement of knowledge, skills and techniques—The two major components of the Skills and techniques are

   Soft kills training: How to work as teams, diversity training and communication skills etc,

   Technical training: Upgrading problem solving and equipment related skills etc.
2. Creating a training environment for **self-learning** based on needs felt,
3. Training curriculum/ tools/ assessments etc conducive to workmen's revitalization,
4. Training to remove workmen fatigue and make work enjoyable,
5. Achieve and sustain zero losses due to lack of knowledge / skills / techniques,
6. 100 % participation in suggestion schemes.

The pre-requisite for this component (Pillar) is checking present status education, training and skills of the available men and setting priorities and policies for improvement.

10.9.7 TPM in administrative and support departments (Office TPM)

Administrative and support departments can be seen as process plants whose principal tasks are to collect, process, and distribute information. Its aim is to improve productivity, efficiency in the administrative and support functions and identify and eliminate losses. This includes analyzing processes and procedures towards increased office automation. Office TPM addresses many major losses, few of those are:

1. Processing Loss,
2. Cost loss including in areas such as procurement, accounts, marketing, sales leading to high inventories
3. Communication loss
4. Idle loss and set-up loss
5. Office equipment breakdown,
6. Communication channel breakdown, e.g. telephone and fax lines etc
7. Time spent on retrieval of information,
8. Non availability of correct on line stock status,
9. Customer complaints due to logistics
10. Expenses on emergency dispatches/purchases, etc.

The Office TPM group or sub-committee should preferably be headed by Finance or Purchase or Management Information departments, with members from all support functions, and representatives from production, quality control and maintenance etc. The Plant TPM group should give the Office TPM group/ subcommittee necessary guidance etc.

The Office TPM should specifically examine and improve the **P, Q, C, D and S, i.e.**

P Production output lost due to want of material, Manpower productivity, Production output lost due to want of tools.

Q Mistakes in preparation of cheques, bills, invoices, payroll, Customer returns/ warranty attributable to BOPs, Rejection/ rework in BOP's/ job work, Office area rework.

C Buying cost/unit produced, Cost of logistics - inbound/outbound, Cost of carrying inventory, Cost of communication, Demurrage costs.

D Logistics losses (Delay in loading/unloading)

- Delay in delivery due to any support function,
- Delay in payments to suppliers,
- Delay in information,
- Other losses in office or other support areas.

S Safety in material handling/stores/logistics, Safety of soft and hard data.

Using these P, Q, C, D, S, TPM should concentrate on the followings——

1. Inventory reduction,
2. Lead time reduction of critical processes,
3. Motion and space losses
4. Retrieval time reduction,
5. Equalizing the work load
6. Improving the office efficiency by eliminating the time loss on retrieval of information, by achieving zero breakdown of office equipment and communication tools (telephone and fax lines, internets etc.

Extension of Office TPM to Suppliers and Distributors: After pursuing satisfactorily internally (within plant), Office TPM can be extended to suppliers and distributors also. With suppliers it will lead to on-time delivery, improved 'in-coming' quality and cost reduction. With distributors it will lead to accurate demand generation, improved secondary distribution and reduction in damages during storage and handling. In case of some of the larger companies, they have started to support clusters of suppliers.

10.9.8 Safety and Environmental Management

Assuring safety and preventing adverse environmental impacts are important priorities of ant TPM efforts. In this area focus is on to create a safe workplace and a surrounding area that is not damaged by the processes or procedures. This component (pillar) will play an active role in each of the other pillars on a regular basis. A committee is constituted for this pillar which comprises representative of officers as well as workers. The committee may be headed by Senior Vice President or I/c (Technical). Utmost importance to Safety is given in the plant. Manager (Safety) is looking after functions related to safety. To create awareness among employees various competitions like safety slogans, Quiz, Drama, Posters, etc. related to safety can be organized at regular intervals.

Aims and Targets:

Zero accident

Zero health hazard/ damage, and

Zero fire, etc.

The '5-S Concept' can also be considered as an important component (pillar) of TPM and it is almost embedded in the pillars mentioned above. The '5-S Concept' is discussed in some detail in chapter 11 (Sec. 11.6)

10.10 TPM IMPLEMENTATION

Followings are the phases/stages and steps for implementation in any industry—

10.10.1 Preparatory Phase

This can also be said 'pre-requisites'. It include steps like-

1. Discussions with worker's union representatives
2. Announcement by management, through company's communication channels and newsletters, to introduce TPM, highlighting TPM objectives and gains.
3. Introductory education campaign for the workforce; This can be through seminars for executives and managers and through slide presentation and letters to all workmen.
4. TPM Promotion- Special committees at every level to promote TPM,
5. Establish Basic TPM Policies and goals—Analyze the existing condition, set goals/ targets and predict results.
6. Preparation and formulation of master plan—This step includes—
    - A master plan for you goal, what will you do to achieve them and when will you achieve them,
    - Detailed plan for each pillar has to be prepared
    - Necessary segregation of jobs—Jobs essentially for operation, maintenance, planning and other groups.

10.10.2 TPM Take-off Phase

At specified time, TPM Take-off should take the form of a formal presentation by top executive, with all employees attending. This opportunity can be used to gain the full support of all employees. We can also invite our external customers, suppliers, contractors and affiliated companies in order to apprise them and to get their support and they can, later-on, be the part of extended TPM. This phase can also be called **Introduction stage**.

10.10.3 Implementation Phase

This phase may include following steps—

1. Develop an equipment management programme—This step includes followings—
    - The tools of TQM and continuous improvement (Kaizen) are applied for management and improvement of equipments,
    - Form project teams—A typical team may have 6-7 persons, consisting of 4-5 operators, 1~2 maintenance men, and one technical expert; each team should have a leader.

- Identify and analyze equipment problems and develop solutions and proposals for Improvements,
- Tools used may be Cause and Effect Diagram, Pareto, RCA, Methods analysis etc.

2. Develop a Planned Maintenance Programme—This step includes the followings—
   - Make plans and schedules to carry out work on equipments before it breaks down, in order to extend the life of equipment,
   - Plans and schedules for periodic, predictive and condition based maintenance,
   - Management of spare parts, tools and consumables for timely availability,
3. Develop an autonomous maintenance programme- This step includes the followings-
   - Operator should be flexible operator, taking some of the maintenance tasks from specialized maintenance group,
   - Operator should take up routine jobs like cleaning, inspection, lubrication, set-up and adjustments etc,
   - Operator should feel responsible for total care of equipments,
4. Increase skills of production and maintenance personnel- This step includes—
   - Training sessions should start immediately after Take-off. The training should include both, the soft skill training and technical training,
   - Train leaders together with workmen,
   - Leaders should share information with group members,
5. Develop early equipment management programme—Jobs in line with Sec. 10.9.4;

The principle of designing for maintenance prevention should be applied to new products and to new and existing machines.

10.10.4 Stabilization Phase

Once the TPM has been implemented, it should be properly monitored and evaluated and some reward system can be introduced for further encouragement. The Japanese Institute for Productive Maintenance runs an annual PM Excellence Award. They provide a checklist for companies or groups applying for the award.

Subsequently, higher goals may be set and TPM levels can be raised.

10.11 TPM BENEFITS

The benefits of TPM may be vast. A few of them are given below—

1. Increased equipment productivity,
2. Increased plant capacity,
3. Reduced equipment downtime,
4. Lower maintenance and production costs,
5. Immediate attention of smaller problems (nip-in-bud) to prevent its aggravation.
6. Approaching zero equipment caused defects,
7. Reduced inventory levels and inventory carrying costs in all parts of the supply chain
8. Better utilization of work area and improved productivity of employees,
9. Enhanced Job-satisfaction, and
10. Increased Return on Investment (ROI).

QUESTIONS

1. What is TPM and how does it differ from decentralized maintenance?
2. How does TPM differ from TQP?
3. What is the basic difference between TPM and RCM?
4. Explain briefly the components (pillars) of TPM on which it works?
5. Explain the stages for implementing TPM in an industry.
6. Explain briefly the basic system and features of working of TPM in any industry.
7. Can TPM be used for clerical staff ? If so, how?
8. Is TPM a component of Terro-technology? What are the benefits of TPM?

CHAPTER

11 OTHER CONCEPTS OF MAINTENANCE SYSTEMS/ STRATEGIS

This chapter includes few modern maintenance practices which are not any specific systems or strategies, but are amalgamation of different strategies and systems, with need based improvements and innovations. These have come up because of increased competition and the concept of lower cost per unit of production etc. This called for improved availability and reliability of equipments, which needed substantive changes in performance of maintenance and optimization of maintenance.

11.1 RELIABILITY CENTERED MAINTENANCE (RCM)

RCM's roots trace back to the 1960s when it was advanced to improve the safety and reliability of commercial aircraft. Since then, it has begun to move into the industrial sector as a result of work conducted by several authors. RCM is a procedure for determining maintenance strategies based on reliability techniques and encompasses condition monitoring and well-known analysis methods such as Failure Mode Effects and Criticality Analysis (FMECA). The primary objective of the RCM process is to identify ways to avoid or reduce the consequences of failures which, if allowed to occur, will adversely impact personnel safety, environmental health, mission accomplishment, or economics. Today, RCM) is an advanced maintenance strategy for increasing asset availability by minimizing downtime caused by failures, reducing an asset's total cost of ownership and increasing Overall Equipment Effectiveness (OEE). It is the optimum mix of reactive, preventive, predictive (CBM) and proactive maintenance practices. It involves some design/ redesign and redundancy also.

To re-emphasize the point, RCM can be said a methodology to balance the resources being used with the required inherent reliability based on the following precepts—

1. A failure is an unsatisfactory condition and maintenance attempts to prevent such condition from arising,
2. The consequences of failure determine the priority of maintenance efforts,
3. Equipment redundancy should be eliminated, where appropriate.
4. Condition based or predictive maintenance tactics are favoured over time based methods and
5. Run to failure is acceptable where warranted.

RCM can be conducted and implemented in many ways. One way can be based on rigorous FMEA and FMECA, complete with mathematically-calculated probabilities of failure based on design or historical data, intuition or common-sense, and/or experimental data and modeling. Such approaches may be called Classical or Rigorous RCM. Other way may use more of CBM, PM optimization and some FMEA and RCA but less of analysis and calculations and such approach is generally called Streamlined RCM. However, the RCM approach in industries may base on Consequences of failure, Probability of failures, Historical data available and Risk tolerances etc. Thus, along with other techniques, RCM uses—

1. RCA, RCFA, FMEA and FMECA etc,
2. Multi-channel vibration analysis, including Real Time Analyzer (RTA) and sophisticated data-logger,
3. Sophisticated Lubricant and oil analysis,
4. Interface with equipment manufacturers and using RCM group's condition monitoring tools to cross-check manufacturer's monitoring while making or rebuilding equipments; especially checking tolerances, mis-alignment, dynamic balancing and vibration signatures etc,
5. Develop occasional calibration check programme for different monitoring equipments,
6. Occasional use of miniature video cameras into maintenance programmes; etc.

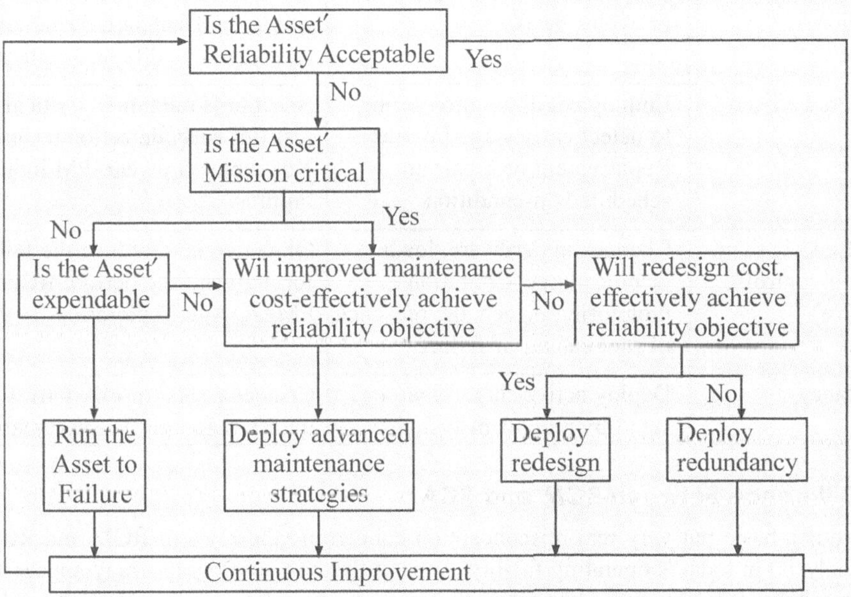

Fig. 11.1 RCM Flow Diagram

11.1.1 Few Attributes of RCM

Followings are some of the attributes/ principles of RCM

- RCM is function oriented,
- RCM is system focused,
- RCM is reliability centered,
- RCM acknowledges design limitation,
- RCM is driven by safety and economics,

- ***RCM Defines Failure as "Any Unsatisfactory Condition***
- RCM Uses a Logic Tree to Screen Maintenance Tasks
- RCM Tasks Must Be Applicable
- RCM Tasks Must Be Effective
- ***RCM is a Living System; etc.***

As shown in Fig. 11.1, RCM uses different strategies, as per need; i.e. run to failure, Fixed-time PM (Time directed maintenance), CBM, minor redesign, failure analysis (FMEA and RCA etc) and proactive maintenance etc. This is also illustrated in Table T-1.

Table T-11.1 (RCM Application)

Maintenance Strategy	Action Required	RCM based Application
Run to Failure (Reactive)	Repair or replace upon Failure	Non-critical and small items; Costs to control or detect failure exceeds benefits (Not cost-effective)
Scheduled change or restoration (preventive)	Repair or replace on fixed time or cycles basis.	Asset has a well documented MTBF and a small standard deviation; subject to wear-out and failure pattern known.
CBM (Predictive)	Employ condition monitoring to detect early stage failures. Replacement or repair are scheduled on-condition	Asset fails randomly. Critical nature justifies early detection techniques; Not subject to wear, PM induced failures.
Minor Redesign and condition-control (proactive)	Changes in hardware, loading or procedures. Condition monitoring detects the presence of root causes of failure.	Objective is to reduce the failure rate for a given time period; RCFA, FMEA, Age Exploration.
Redundancy	Deploy active shared-load or stand-by redundant systems	Critical assets (or mission) for which no other approach is acceptable

11.1.2 Difference Between RCM and RCA

There is a basic but very real misconception concerning the roles of RCM and RCA (Root Cause Analysis) in today's operating facility. Apparently the two programs may appear to do the same thing, but that is not true. Although both programs are extremely beneficial if implemented properly, the purpose of each is entirely different. When implemented together they compliment each other and provide the greatest overall benefit to the facility. The purpose of RCM is "to determine the maintenance requirements of any physical assets (**Equipment**) in its operating context." In RCM, all of the potential modes of failure are uncovered and a predictive maintenance strategy is devised to mitigate the consequences of the failure based on the criticality of the failure mode. In RCM, these failure modes are identified as the root cause(s) of the failure.

The purpose of RCA is "to uncover the underlying reasons (root causes) why an event (not just equipment related events, but any type of event) is occurring so that the necessary steps can be taken to eliminate the event in its entirety." This is accomplished by analyzing the modes (the point at which RCM stops). RCA uses a logic tree that stresses verification at every level. The advantage is that the actual root causes that are uncovered are facts that have been derived from the verification process. The comparison between the two programs is striking – RCM is driven by preventive maintenance strategies while RCA is driven by maintenance prevention strategies. It should be clear that the difference between RCM and RCA is that RCM treats the symptom while RCA finds and corrects the cause. Both techniques would solve the immediate problem of equipment, but only RCA would uncover and eliminate the actual cause of the problem.

RCM is also called Reliability Based Maintenance (RBM) by few users/ consultants.

11.1.3 The RCM Process

The most common reason for achieving disappointing result from RCM programme seems to be the failure to connect and integrate them fully and completely with overall asset management strategies and supporting information. Modern CMMS or EAM can be used to prioritize and target RCM programmes, to provide the data required to support the questions generated by the RCM process and to act on the results of the analysis.

The followings are the seven key steps of the RCM process

1. Overall RCM Programme Development. At this stage, it is essential to select analysis equipments for the area in which the greatest benefits can be achieved. Poor targeting will not give desired result. It is essential to understand priorities and to focus on the areas that exhibit the greatest problem in terms of cost, reliability, downtime, environmental risk and their impact on business performance. This can be presented in the graph form.

For example, in one industry, taking the contract, labour, material and service costs together, drying, centrifuging, site services, conveyors, treatment, packaging, compaction, boilers, evaporators and power plant were the top 10 maintenance costs in order of severity. KPIs, such as MTBF and MTTR can be easily accessed together with other associated cost information for those top 10 aspects.

Other important aspects of the plan, at this stage, are implementation of the RCM analysis results and the different ways the in which the outputs of the RCM programmes will be sustained.

2. Equipment Classification and Analysis. To prepare for the detailed RCM analysis, important decisions are to be made regarding the way equipments are classified from main system to subsystem and within the unit levels. This must be done in the way that is consistent with the existing maintenance strategy, tasks, routines and levels of unit replacement. This step is nearly similar to equipment classification discussed in chapter 9.

Connectivity with specific RCM tools is also important in order to reduce workload and error resulting from manual data exchange.

3. Failure Modes, Effects and Criticality Analysis (FMECA). FMECA isolates ways in which the equipment can fail and seeks to identify the significance of these failures in terms of their impact on safety, environmental risk, operations and costs. In doing this FMECA evaluates both, probability and consequence of failures, providing the basis for making risk based decisions to optimize maintenance activities.

This can be given a graphical display. For example, in one industry, the problem codes for failures on centrifugal were— Leak- 46%, Stopped- 27%, Low Volume- 18% and Low Pressure- 9%.

4. Criticality Ranking of Failure Modes by Probability and Consequences. This step requires that the failure modes be ranked accordingly to their probability and the severity of consequence for the organization. It should, then, be ascertained whether the loss of equipment function resulting from each failure is significant or not and hence whether further consideration of the failure by RCM is justified. This is a key step in determining how effective RCM can be in excluding non-productive work activities.

By utilizing the analytical tools and historical data, as outlined in step (2), priorities can be set for the subsequent stages of the RCM process.

5. Cost Effective Maintenance Task Selection Linked to Criticality Ranking. The earlier RCM process suggested listing of all the options and then selecting the most cost-effective mix from those options. However, now, most of the organizations have some sort of CBM practices and procedures already in operation and with the help of that, selection of cost-effective tasks becomes easy and less time consuming. This strategy also takes advantage of the skills, technology and practices already working successfully within the organization.

Software/ programmes can be developed to integrate RCM and CBM practices and direct interfacing with plant based data acquisition systems and hardware such as programmable logic controller (PLC) and supervisory control and data acquisition (SCADA) systems.

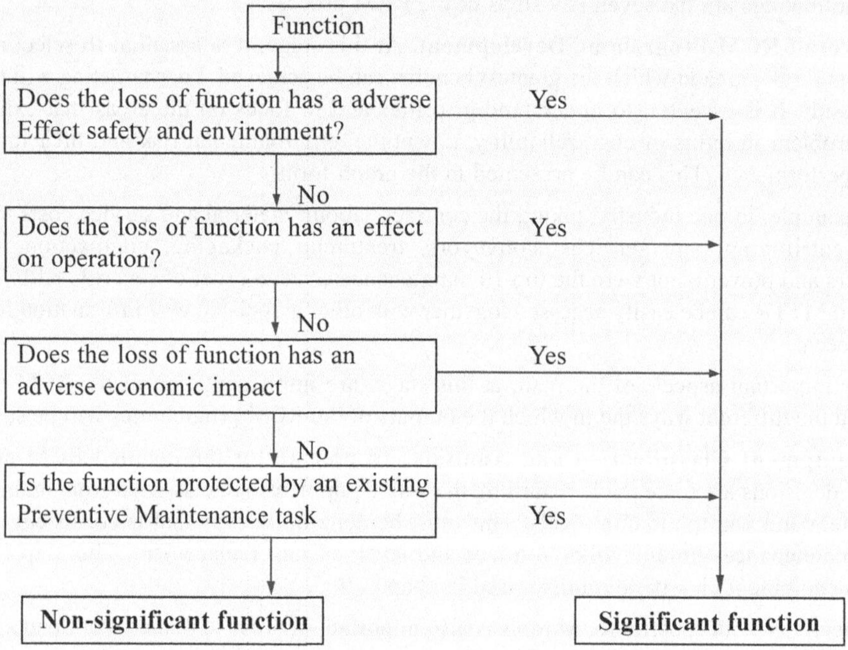

Fig.11.2: Criticality Ranking

6. Task Packaging, Implementation and Incorporation into overall Strategy. Organization adopting RCM process achieves no real tangible benefits until the results are implemented as part of the existing asset maintenance approach (CMMS etc).

As such, individual tasks, determined as a result of the analysis, need to be packaged and integrated within existing maintenance programmes, supported by local skills, resources, knowledge or culture. The existing asset management programme of maintenance software should provide traceability and audit trails linking the packaged and integrated work tasks back to the RCM decision-making process that created them.

7. Ongoing Monitoring and Continuous Improvement. One important step in establishing successful on-going monitoring and continuous improvement process is to understand the "as-in" situation, which can be established at the Step (1) of the process by setting out the underlying logic for the areas targeted by the programme in terms of their impact on key factors, such as, security, health and safety, environmental risk, operation and cost etc. Implementation results of RCM programme can also be monitored by utilizing the powerful and user-configurable KPIs and Dashboards etc.

11.1.4. Advantages and Disadvantages of RCM

The advantages and disadvantages of RCM can be summed-up as followins

Advantages (RCM)	Disadvantages
1. Can be the most efficient maintenance program. 2. Lower costs by eliminating unnecessary maintenance or overhauls, 3. Minimize frequency of overhauls, 4. Reduced probability of sudden equipment failures, 5. Able to focus maintenance activities on critical components, 6. Incorporates root cause analysis,7- Increased component reliability.	1. Can have significant startup cost, training, equipment, etc, 2. Savings potential not readily seen by management.

11.2 ENTERPRISE ASSET MANAGEMENT (EAM)

EAM is an information management system that connects all departments and disciplines within a big company, making them an integrated unit. It is the organized and systematic tracking of an organization's physical assets—its plant, equipment and facilities etc. It aims that the machinery you use to manufacture your product has to be reliable, the electricity must flow, your computers must remain on line, and your delivery trucks must be operational etc. Although most companies have some type of EAM system in place, many rely on low-tech, largely paper-based communications that significantly underutilize the resources dedicated to asset management. Adding a wireless infrastructure plus rugged mobile computers to your EAM system allows you to fully utilize your resources and improve the efficiency of personnel as well as the performance of critical assets, while reducing operational cost.

The EAM system is also, sometimes, referred to as enhanced computerized maintenance management systems (Enhanced CMMS). By using batch and radio-enabled mobile computing hardware, application software and communication infrastructure and using other IT techniques, EAM ensures generation of quality data and timely flow of required data throughout the organization. Maintenance workers carry rugged, handheld mobile computers equipped with maintenance management applications that easily integrate into dedicated EAM software. Thus it reduces paperwork, improves the quantity, quality and timeliness of information and provides information to technicians at the point of performance and gives workers access to job-specific information at the worksite.

EAM system can help, qualitatively and quickly most of the activities, e.g.

- Workorder generation and communication,
- Easily find and view parts and sub-assemblies on equipment,
- Analyze failures by equipment, type of equipment, system, building, manufacturer, model number,
- Inventory management,
- PM; collects key parameters, history and verifies actual jobs to be done,
- Corrective maintenance; real time assessment of repairs, communicates work status immediately,
- Inspection; Records inspection readings and failure codes etc,
- Tracking job times; pinpoints workers activity by scanning equipment and location etc,

- Asset Audits; conducts physical counts, by using simple scanning, also validates and collect other information,
- Regulatory and QA compliance;
- Displays your facility asset hierarchy in an easy to interpret graphical tree-list while providing drill-down capability at all level; etc

However, EAM is more useful in big and modern industries. Many EAM or similar software are available in the market from reputed suppliers [MRO (maintenance repair overhaul) Software, Data-stream and Indus, as well as systems offered by vendors including Siebel, SAP, Oracle, JD Edwards, Intentia and Great Plains etc].

Pursuant to discussions in chapter 9, in addition to maintenance of equipments and machineries, the EAM also encompasses all other assets of the enterprise; e.g. all fixed assets, mobile and continuous assets, accounts payable, financials, human resources, pay-rolls, times and attendances, inventory, warehouses, waste recovery and waste disposal, engineering, process automation, external warranty services and statutory obligations etc. Companies selecting a new ERP system should include a review of the EAM function, particularly in process and asset intensive industries where good EAM is frequently seen as being critical to the business. However in most cases, after acquiring EAM software, interfacing with existing systems and assets has to be done.

Many regard EAM as CMMS on steroid, which does not paint the true picture. Typically, CMMS deals strictly within the confines of the work order and preventive maintenance activity, but the EAM software, in addition to those, offers many features that can provide additional capabilities, value added functionality, and savings to your company. As an example, a CMMS solution usually includes purchasing and procurement, inventory management, as well as equipment, parts, and asset tracking. However, CMMS applications typically do not have financial and accounting (other than mere cost recording) or *human resource* (HR) management capabilities (other than basic staffing needs recognition), i.e. EAM support financial and HR management more deeply. Technically, the EAM applications are also designed to scale to larger numbers of users and facilitate running at multiple sites from a single central database, thereby catering better to entire enterprises, rather than departmental or individual plant's needs.

11.2.1 Comparison- CMMS vis-à-vis EAM

A rough comparison between CMMS and EAM may be as shown in Table T-11.2,

Table T-11.2 (Difference between CMMS and TPM)

Features/ Components	**Typically Found In**	
Functions and Features	**CMMS**	**EAM**
Database structure and hierarchy	✓	✓
Repair parts availability	✓	✓
Manpower resource availability	✓	✓
Purchase requisition	✓	✓
Preventive maintenance scheduling	✓	✓
Cost accumulation and tracking	✓	✓
Inception recording and tracking	✓	✓
Standard and exception reporting	✓	✓

Whole life asset care	✓	✓
Maintenance administration		✓
Predictive maintenance analysis		✓
Maintenance alternatives analysis		✓
Physical asset risk management		✓
Reliability-centered maintenance		✓
Root cause analysis		✓
Financial cost/life analysis		✓
Technical document change management		✓
Strategic usage analysis		✓
Strategic planning for asset management		✓

11.3 SIX SIGMA MAINTENANCE

It is the application of six sigma principles in maintenance. Six Sigma is a maintenance process that focuses on reducing the variation in business production processes. By reducing variation, a business can achieve tighter control over its operational systems, increasing their cost effectiveness and encouraging productivity breakthrough.

Six Sigma is a term created at Motorola engineers, during the mid 1980, to describe the goal and process used to achieve breakthrough levels of quality improvement. Sigma is the Greek symbol used by statisticians to refer to the six standard deviations. The term Six Sigma refers to a measure of process variation (six standard deviations) that translates into an error or defect rate of 3.4 parts per million. To achieve quality performance of six sigma level, special sets of quality improvement methodologies and statistical tools were developed. These improvement methods and statistical tools are taught to a small group of workmen known as Six Sigma champions or Black Belts who are assigned full-time responsibility to define, measure, analyze, improve and control process quality. They also facilitate the improvement process by removing the organizational roadblocks encountered by the improvement teams The Black Belts lead cross-functional teams of employees (individually known as Six Sigma Green Belts) to achieve breakthrough levels of process. Today Six Sigma can be applied to many fields such as operation/ production, services (including maintenance), medical and insurance procedures, call centres etc.

Six Sigma methodology improves any existing business process by constantly reviewing and re-tuning the process. To achieve this, Six Sigma uses a methodology known as **DMAIC** (**D**efine opportunities, **M**easure performance, **A**nalyze opportunity, **I**mprove performance, **C**ontrol performance). This six sigma process is also called DMAIC process.

Six Sigma relies heavily on statistical techniques to reduce defects/ failures and it incorporates the basic principles and techniques used in Business, Statistics, and Engineering.

Six Sigma methodology can also be used to create a brand new business process from ground up using **DFSS** (*Design For Six Sigma*) principles.

11.3.1 Six Sigma Maintenance Process

Some of these steps closely resemble RCM processes, although the overall goal of Six Sigma maintenance is very different. The steps of six sigma maintenance are same as DMAIC process. To apply Six Sigma in maintenance, the work groups that have a good understanding of preventive maintenance techniques in addition to a strong leadership commitment. Six sigma helps in two principal inputs to the maintenance cost equation: Reduce or eliminate the need to do maintenance (reliability of equipment), and improve the effectiveness of the resources needed to accomplish maintenance (people, parts, and outside services).

Fig. 11.3 shows the six sigma maintenance process flow diagram.

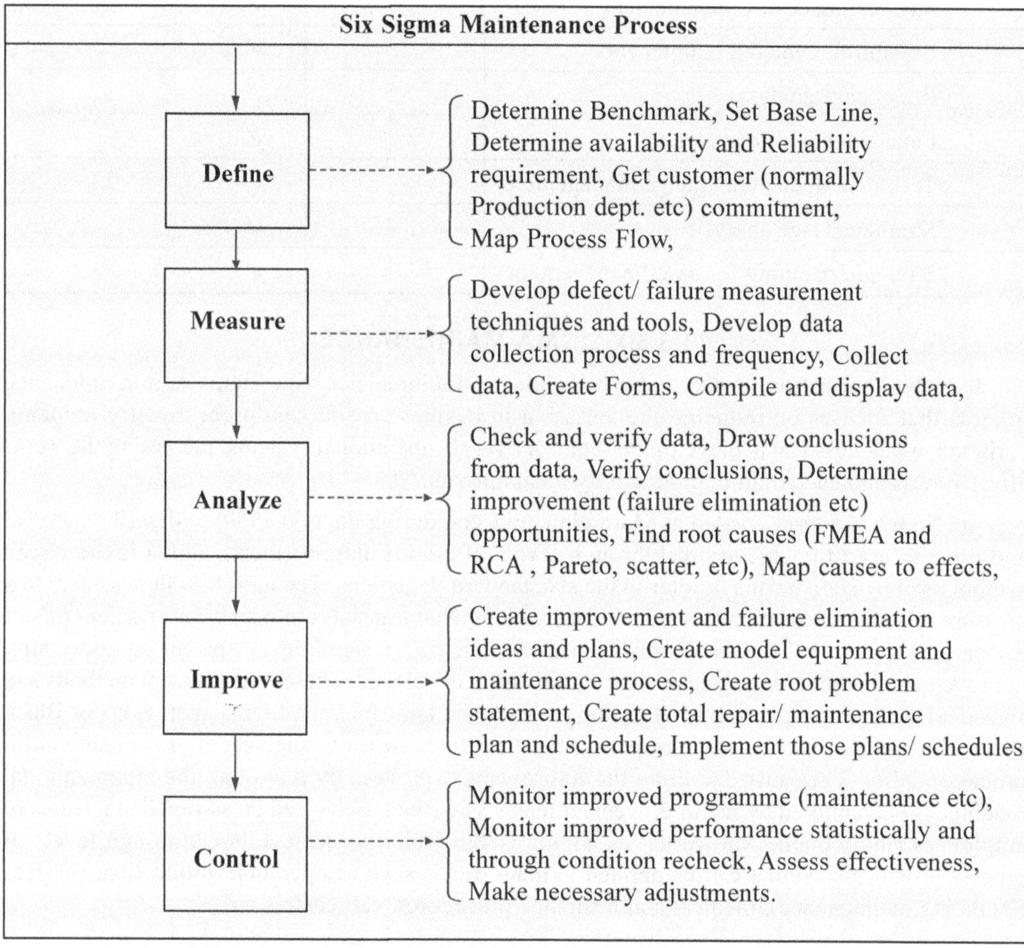

Fig. 11.3: (Six Sigma Maintenance Process

Of the steps mentioned above, measuring becomes a key element of the effort to improve. How is the current process performing? Where are the wastes? How often do these wastes occur (frequency, or MTBF)? How severe are the waste events (time wasted, or MTTR)? The total process also considers "What outputs/ actions are really important to the maintenance and production process (in addition to wrench time i.e. actual working time)?", "Analyze and mitigate the failure modes of key inputs", "Improve the process by mapping the should-be process" and "Validate the improvements and measure the new process capability (sigma, defects per million opportunities/ operations" etc.

11.3.2 Six Sigma Control Levels

Basing on the six Sigma concept, following six sigma control levels can be used and each sigma control builds on the previous level, until the near zero maintenance failure culture is achieved.

1. One Sigma Control (Reacting): One sigma is set in the era of 3 Es of maintenance, e.g. engineer, educate and enforce. The tools for this level control are work orders, schedules, standard maintenance practices, compliance reports etc. The first level represents the ability to sustain repairs and maintenance. As with one sigma in quality, the effective compliance should be at least 69%.

2. Two Sigma Control (What we see): The tools for two sigma control include observation programmes, e.g. inspection various condition monitoring gadget's reports, failure reports and failure analysis etc. At this level, the awareness and analysis tools are applied to reach a two sigma level (around 98.5%) of maintenance compliance and objective.

3. Three Sigma Control (What we do): Three Sigma control requires well defined controls, responsibilities and accountabilities (including individual's accountability) to provide predictable results on regular basis. Generally, three sigma control is quite good and it achieves around 99.7% result.

4. Four Sigma Control (What we believe): This level concentrates on the non-observable "what we believe" for the maintenance and up-keep. Proper maintenance audits and surveys are periodically conducted and compared with best known results and standards to identify shortcomings and implement continuous improvement. This level aims for 99.97% success.

5. Five and Six Sigma Controls (How we engage and how we lead): The next challenge is to utilize the data achieved at the previous four control levels; i.e. the fundamentals, observable processes, accountabilities of what we do and information on what we believe from maintenance perception survey and aim for 99.997% success. Though this level controls are not common, followings can be done to attain zero failure maintenance culture—

- A regular sanctioned meeting system with actionable rules and mechanism and trained leaders to manage the continuous improvements,
- Continuous use of Six Sigma analytical techniques/ tools for maintenance and project-able data.

11.4. LEAN MAINTENANCE

Lean Maintenance is relatively new term, but but its principles are well established in TPM. In the late 1980s, the term 'lean' was coined by researcher John Krafcik, who said that, "Lean production is 'lean' because it uses less of everything". Integrating "Lean" methods with legacy systems and organizations becomes real challenge. The Toyata Motor Company was credited with pioneering the concept of 'Lean Production' that was copied by other Japanese companies and later by others. The root of 'lean' in Toyata production system was, "Eliminate waste to reduce manufacturing cost".

Though lean maintenance took its concept from lean manufacturing, lean maintenance is neither a subset nor a spinoff of lean manufacturing. It is instead a prerequisite for success as a lean manufacturer.

11.4.1 Definition and Concepts

As a broader definition, including most of the key concepts/ elements, lean maintenance is a proactive maintenance operation employing planned and scheduled maintenance activities through total productive maintenance (**TPM**) practices using maintenance strategies developed through application of reliability centered maintenance (**RCM**) decision logic and practiced by empowered (self-directed) action teams using the **5S process**, weekly **Kaizen** improvement events, and autonomous maintenance together with multi-skilled, maintenance technician-performed maintenance through the committed use of their work order system and their computerised maintenance management system (**CMMS**) or enterprise asset management (**EAM**) system, and also with 'just-in-time'(**JIT**) material availability.

As proactive, it is the opposite of reactive where the maintenance operation reacts to equipment failures by performing repairs. In the proactive maintenance operation the prevention of equipment failures through performance of preventive and predictive maintenance actions is the objective. Repair is not equivalent to maintenance.

As empowered (self-directed) action teams, its activities are task-oriented and designed with a strong performance focus. The team is organized to perform whole and integrated tasks, hence requiring multi-department membership. The team should have defined autonomy (that is, control over many of its own administrative functions such as self-evaluation and self-regulation—all with limits defined). Furthermore, members should participate in the selection of new team members. Multiple skills are valued. Benchmarking (assessment and comparing with known best) are also considered.

Kaizen is the philosophy of continuous improvement, that every process can and should be continually evaluated and improved in terms of time required, resources used, resultant quality, and other aspects relevant to the process. Kaizen improvement events are often referred to as a Kaizen blitz—a fast turnaround (1 week or less) application of Kaizen improvement tools to realize quick results.

Other techniques/ terms mentioned in the definition, e.g. 'planned andscheduled', 'RCM', '5-S process' and 'autonomous maintenance', 'TPM', 'CMMS', 'EAM', 'Multi-skilled', 'Work-order system' etc are explained elsewhere in this book.

Other concepts, often used, as per need, in lean maintenance are **Bench Marking** (comparing with best in the world and setting target accordingly), **Hoshin Planning** (where you want to be in future), **Kanban** (one machine at a time), **PDSA** (Plan, do, study, act— specifically one machine basis) and **Poka-Yoke** (a mistake proofing device or procedure to prevent a defect during order taking or manufacture) etc.

Making 'lean maintenance' work also means making 'maintenance' work in lean organizations. It focuses on followings:

1. Standardized Work: Drive-out human variations, procedure based maintenance and operation,
2. Training and Qualification: focus on task-specific training on fast and sustainable results,
3. Avoid un-essentials— Avoid the 'Activity' or 'programmatic' approach of implementing 'lean' tools in hope of improving performance (it seldom works)
6. Focus on results— Address the compelling business case for very specific improvements in revenue, throughput, costs and wastes etc; See business KPIs etc.

11.4.2 Challenges of Lean

Lean maintenance is becoming more desirable in today's scenario, because of following reasons

1. Declining state of maintenance and industries in industries,
2. Retiring good employees and shortage of labour, skills and knowledge
3. Lack of vocational-technical trainees and programmes,
4. Traditional operation and maintenance methods tend to be out of date and inefficient, and
5. Competitive pressure etc..

11.4.3 Lean and Waste Control

like lean manufacturing, lean maintenance also identifies different forms of waste, waste processes and waste practices and includes a process for continuous improvement. In a lean maintenance system, sources of waste usually consist of outdated procedures and overstocked and underused inventory of equipment, material, and parts, as well as wasted labor, time, and transportation etc and it continuously works for elimination of those.

Lean Maintenance recognizes following seven forms of Waste in Maintenance

1. Overproduction: mainly of un-necessary PM and other maintenance activities,
2. Waiting: maintenance-related waiting includes the production department waiting for maintenance personnel to perform a service and waiting for tools, parts documentation, transportation, clearances and other items also is wasteful
3. Transportation: wasting time in repetitive walking to different places etc,
4. Process Waste: in reactive maintenance, repairs are typically conducted to achieve operation as soon as possible, sometimes eliminating an opportunity to perform a longer term or higher quality repair.
5. Inventory: carrying obsolete or rarely used material and creating bottlenecks,
6. Motion: wasted motion/unnecessary process in the maintenance operation usually revolves around PM tasks that do not add value to the output,
7. Defects: job recurrence because of improperly performed repair work is a huge source of waste.

Lean Maintenance is basically equipment reliability focused and reduces need for maintenance troubleshooting and repairs. Any maintenance engineer can begin Lean Maintenance by protecting automation, electronics, hydraulics and computer-controlled equipment from the root causes of malfunctions, failures, and downtime-chronic stress discussed above. From the above sources of wastes, uptime is to be improved and cost should be lowered by shedding the following costs—

1. Onsite Spare Cost,
2. Onsite board or component repair, and
3. Onsite technicians, maintenance engineers etc.
4. Downtime from Operator or Programmer Error
5. Downtime from inadequate PM procedure or performance,
6. Downtime from chronic wear and stress to circuit boards, hydraulic components and other system components, etc.

On the basis of these wastes we can arrive at six simple ways to lean, e.g.-

(*i*) Improve tool storage using strong visual cues such as painted outlines or foam cut outs to show where tools should be stored,

(*ii*) Improve documentation storage. Make sure it is neat, organized, and accessible by those who need it. Have a system that clearly shows when a file is out or missing.

(*iii*) Use visual cues/ displays to show how things should be in normal state.

(*iv*) Use visual cues/ displays to indicate status, next inspection due date, etc.

(*v*) Create a simple and uniform work request/ job order system that provides visual cues for operators, maintenance, and supervisors.

(*vi*) Use root cause analysis techniques to solve problems permanently.

Though there may be some duplication, various losses, mentioned in chapter 10 (TPM) are also considered as wastes and to be taken care of in lean maintenance.

11.4.4 Lean Maintenance Process

The lean maintenance process is explained in Fig. 11.3 (Typical Lean Maintenance and TPM Flow Chart).

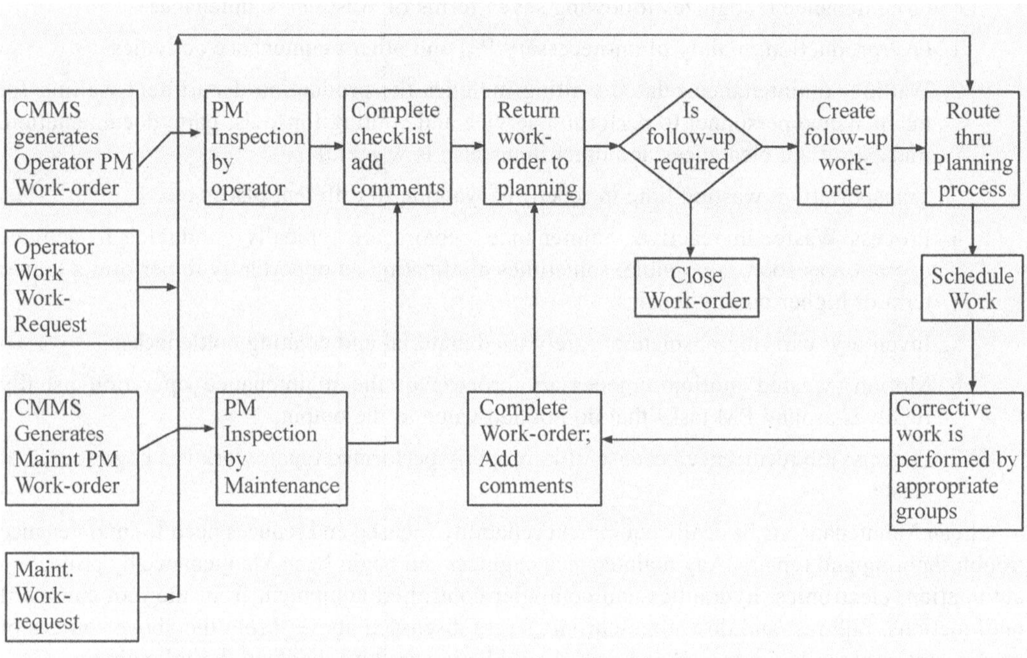

Fig.11.4: (Typical Lean Maintenance and TPM process Flow chart)

11.4.5 Lean Maintenance vis-à-vis TPM

Though both are essential for good result, lean maintenance employs TPM heavily but TPM may not employ lean maintenance concepts and strategies. Again, TPM is more equipment focused but lean maintenance also concentrates on maintenance management processes.

Lean Maintenance is often used in conjunction with six sigma maintenance or DMAIC process.

11.5 FIVE ZERO (5-ZERO) MAINTENANCE CONCEPT

This concept was introduced by Toyota Motor Works in Japan. It calls five "Zeros" as **"0-Breakdown", "0-Fault", "0-Delays", "0-Stock" and "0-Paper work".**

These "Zeros" do no more than represent "ideals" or "aims" to be achieved. It is dramatically opposite to what happens in many places where we have 'Records', Reports' and 'Reviews' but no 'Results'. Let us examine the individual zero concepts in some details-

11.5.1 Zero Breakdown

This implies a very high degree of reliability, combined with very good maintenance capability. The reliability falls within the competence of manufacturers. The maintenance capability, to some extent, falls within the competence of manufacturer (providing in-built monitoring and maintenance facilities) and more within the competence of maintenance and operation personnel for sound maintenance and operation practices. As such both, manufacturers and maintenance personnel, should associate and aim for zero breakdown. As productivity improves progressively, suitable training and dialogue should be a continuous process so that breakdowns are nearly eliminated.

11.5.2 Zero Fault

This actually means that fault should not be allowed to occur but that should be detected much earlier and corrected at initial stage. A policy of zero fault implies a maintenance policy with a high technical capability, using aids/ facilities like most suitable vibration and lubricant monitoring gadgets and other elements for prognosis of equipment behaviour. Action should be taken so that wear and tear are not induced and operational speed ought to be sufficiently below the maximum speed, thus necessitating the provision of safety margin. Continuous online monitoring for most equipments is first step.

11.5.3 Zero Delay

This concept calls for planned and rapid response to equipment outage. Points requiring attention must be correctly known so that no time is lost for fault diagnosis or repair. Availability and assimilation of operating and maintenance manuals, standard operating and maintenance practices, timely relaying and interpretation of condition monitoring data and alarm and effective communication are essential to reduce any possible delay on that account. It also calls for planning and scheduling essential maintenance jobs with available shutdowns in such a way that actual delay is almost nil and no separate shutdown is needed exclusively for maintenance.

11.5.4 Zero Stock

This concept means that spares and other materials are procured in such a way that those are received only when those are actually needed for consumption. Thus the materials are not unnecessarily stocked in stores and inventory carrying cost is almost nil. Zero stock concept is possible only when

- Failure do not occur (zero fault) or, if at all occurs, it should occur in predictable manner,
- Spares part, which are needed, are available at known locations on an assured basis (specifically involving zero delay),
- Command procedures, appropriate to the situation, are arranged.

The concept of zero stock and zero inventory carrying cost tries to ensure that some spares, which are essential to be stocked, are stocked at nearby manufacturer's godown with arrangement that no time will be lost (zero delay) when the spares are actually needed.

11.5.5 Zero Paper Work

Maintenance engineers/ personnel are, generally, poor in paper work. Zero paper work concept tries to eliminate the paper work or, at least, reduce that to barest minimum. This involves extensive use of computers, PCs, palmtops and other handheld devices and integrates the same with system computers with necessary local area network (LAN) and customized software. This enables the maintenance personnel to immediately know what, how, when and where of the job and intimate feedback. Good CMMS procedures and software is also helpful.

Thus the 5-zero concept covers almost all areas with specific aim of high availability and reliability.

11.6 5-S CONCEPT IN MAINTENANCE

This concept/ method, developed in Japan, assume no effective and quality job can be done without clean and safe environment and without behavioral rules. The 5S are five action verbs (**Sort, Clean, Set in order, Standardize** and **Sustain/ Progress**), all starting with an **S** in Japanese (Seiri, Seiton, Seiso, Seiketsu, Shitsuke). "Five S or 5-S" concept has come from those.

Table T-11.3, below, explains the meanings of terminologies

Table T-11.3 5-S Concept

Japenease Term	Translation and Meaning	Equivalent English Term
Seiri	Sort-out and organize items as critical, important, frequently used, useless or not needed at present.	Sort
Seiton	Tidiness; The concept here is that "*Each items has a place, and only one place*". The items should be placed back after usage at the same place.	Systematize(Set in order or Simplify)
Seiso	Cleaning; it involves cleaning the work place free of burrs, grease, oil, waste, scrap etc. No loosely hanging wires or oil leakage from machines	Sweep(Shine the workplace or Scrub)
Seiketsu	Standardisation; Discuss together and decide on standards for keeping the work place / Machines/ pathways neat and clean. These standards are implemented for whole organization and are tested/ Inspected randomly.	Standardize
Shitsuke	Self-discipline; Considering 5S as a way of life and bring about self-discipline among the employees of the organization. This includes wearing badges, following work procedures, punctuality, dedication to the organization etc.	Self-discipline (Sustain)

The Terms mentioned in the brackets in above table are the terms advocated by different persons/ consultants as equivalent of Japanese terms. Table T-11.4 roughly shows how the sorting can be done.

Table T-11.4 Typical Sorting-out procedure

Priority	Frequency of use	How to use
Low	About than once per year, or may be once in more than six months.	Throw away, Store away from the workplace
Average	Once per week to once in six months	Store together but offline
High	Daily to once per week	Locate at the workplace

5-S deployment is to be done in a logical and progressive way. The 3 first **S**s are shop floor actions, while the two last ones are sustaining and progress actions. Initiated and successfully deployed in the workshops, 5-Ss are being used effectively in every other department in the company, including maintenance and offices. 5-Ss are the solid base, the foundation on which to build continuous improvement, install lean manufacturing and lean maintenance tools and methods. They are also an effective lever for change management and empowerment.

Sort also means to separate needed tools, parts, and instructions from unneeded materials and to remove the latter. Standardize, again, means to conduct Sort, Simplify, and Scrub at frequent, indeed daily, intervals to maintain a workplace in perfect condition. Sustain means to form the habit of always following the first four Ss. Some other experts designate 5S as **S**ort through and sort out, **S**et in order and set limits, **S**hine and inspect through cleaning, **S**tandardize and **S**ustain. However, the aim of all sets of terms 5-S is same and gives good results. The benefits are many, e.g.

1. Improves on-time delivery or time-completion,
2. Improves quality and reduces defects/ failures,
3. Reduces waste in materials, space and time,
4. Reduces lead times,
5. Reduces inventory and storage costs,
6. Reduces changeover time and equipment downtime and
7. Increased safety and profitability etc.

Of the five Ss, the first one, i.e. Sort appears relatively more important and will influence other Ss. Fig. 11.3 shows how sorting can be done. While sorting, different techniques like VED (Vital, Essential, Desirable) and ABC sorting (A is to be kept at hand; B is to be stored closely i.e. work zone, within workshop, office, in the same room... and C is remote place i.e. Archives, shack, long storage...) etc can be used. Similarly, many techniques can be used for other Ss.

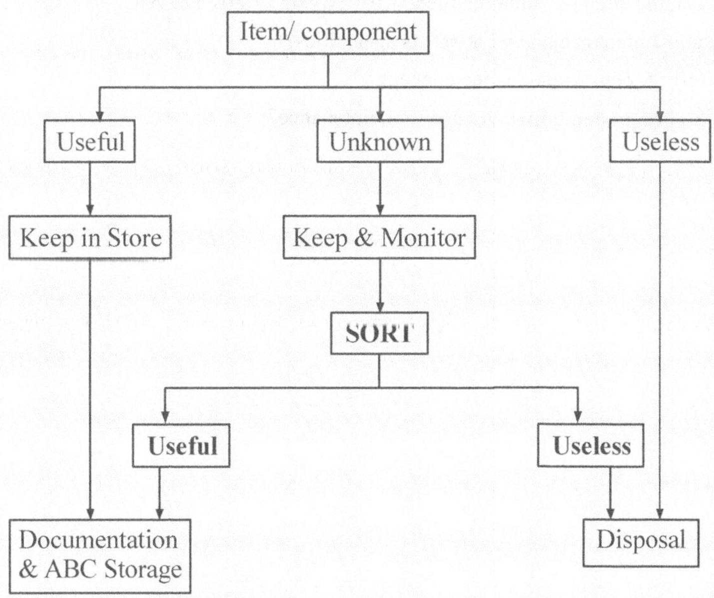

Fig. 11.4: (Process of sorting in 5-S)

Instituting the 5 S's is usually one of the first step companies take in implementing TPM, Lean Manufacturing, Lean maintenance or Six Sigma etc. To introduce 5-S concept or method, top management support is more important than hiring any outside consultant. To introduce and practice this concept call for few simple series of actions, such as:

1. Training in 5-S to all personnel. All those persons, who will never participate as auditors, will always be audited.
2. Select a representative of each department (if possible) to be part of the 5-S team, which will conduct periodical assessments of all types of premises, machines and facilities etc.
3. Conduct audit using a format, making a numerical measurement and taking note of all non conformed situation in order to make the resolution of those.
4. Make a report. Give it to the responsible of audited area. Demand corrective actions. Communicate results to all personnel.
5. A common effort of improvement (including a friendly competence, to be the best or the less worst).
6. And finally, the most important: The top management has the compromise to allow the 5-S team audit his own offices. In several cases, the only places "skipped" by 5-S team are the big shot ones, because "they suppose they always have a neat place.

QUESTIONS

1. What is RCM? Explain its working with a flow or block diagram.
2. How does RCM differ from RCA? Mention few attributes of RCM.
3. Explain briefly the EAM in maintenance.
4. How EAM differs with CMMS?
5. What is six sigma and six sigma maintenance? Explain briefly DMAIC.
6. What is lean maintenance and what are its specialties?
7. Explain briefly 5-Zero Maintenance concept.
8. What is 5-S concept and how is it used in maintenance?

CHAPTER

12 MAINTENANCE ORGANIZATION

DEFINITION AND PURPOSE

Organization means a group of people who work together in a structured way for a shared purpose. In any industry, the various maintenance functions are planned and executed by workmen of different trades/ disciplines/skills and of different levels (workmen/ employees, supervisors, executives and engineers etc) and they have to be arranged/ grouped/ organized in such a way that they have to work as a team towards the common goal of improving maintenance effectiveness/ efficiency and for improving equipment availability and reliability. Such grouping and arranging of maintenance personnel, with their interlinking and relationships, is generally termed as maintenance organization. Maintenance organization, as in the case of other organization, requires delegation of authority and span of control, with necessary feedback and control system.

There are almost infinite numbers of organizational structures in use but many of those are not configured to provide effective utilization of workforce and have unbalanced indirect (clerical, administrative and other support) personnel. Maintenance organization (or for that reason, even corporate organization) is both science and art. The science of the organization lies in different dimensions (e.g. different categories of workmen, supervisors, staff and support personnel, contractors, vendors, sophistication, prevailing cultures and regulations etc) on which they are designed. How these dimensions are coordinated, interlinked and governed is more of an art than science. Again, the total number of workforce in an organization may not be a mere addition of persons needed for different trades or dimensions but a judicious balance is to be arrived. Organization, like individual, must preserve their integrity. It must be a balance between policy and practice; between philosophy and performance; between decisions and deeds.

In 1986, the Society of Manufacturing Engineers, USA, examined the balance between the current level of manufacturing technology and of company's organization and the conclusion of the study was that, 'American industries, in its drive to become more competitive, is attempting to put fifth generation technology into second generation plant organization'. Thus the maintenance organization should also correspond to the advancement of maintenance practices in use.

Employees' involvement, to a reasonable extent, may also be considered in deciding maintenance organization through suggestion system, information sharing and survey feedback etc. However, while doing this, care should be taken not to increase the staff un-necessarily or not to create additional management structure.

12.1 BASIC CONCEPTS FOR MAINTENANCE ORGANIZATION

Followings are the basic concepts

(*a*) Establish reasonably clear division of authority, with minimum overlap—Authority can be divided functionally, geographically, on the basis of expediency or it can be on some combination of all the three. But there must always be clear definition of the line of demarcation to avoid confusion and conflict. The over lap should be bare minimum.

(*b*) Keep vertical line of authority and responsibility as small as possible— Unclear or many levels of intermediate supervision or over-application of specialized functional employees must be minimized. Whenever such practices are felt to be necessary, clear division of duties must be established.

(*c*) Maintain an optimum number of people reporting to one individual— In a good organization, number of people reporting to one individual may vary from 3 to 6, depending on the type of job. When a job requires a fairly small amount of supervision, one man can direct the activities of 12 or more persons.

(*d*) The total maintenance workload should be reasonably distributed amongst all concerned persons to avoid conflicts amongst workers.

(*e*) Maintenance is not subordinate to operation. Again, difference between 'supportive service' and 'subordinate service' should be kept in mind.

12.1.1 Factors governing Maintenance Organization

Even with the above concepts in mind, decision of maintenance organization will vary as per the following local factors of the organization

1. Type of operation: e.g. buildings, machine tools, process equipments, manufacturing facilities, piping, mining machineries, electrical equipments etc—each will affect the character of organization and supervision required.

2. Continuity of operation: e.g. if the operation is 5/6 days a week, single shift one or a 7 days a week, round the clock (3 shift) one or others— considerable differences may be in these on how the maintenance engineering department may be structured and how many personnel may be needed.

3. Geographical location: The maintenance that may be needed in a compact plant will vary considerably from the one that is dispersed through several buildings or over a larger area. The later often leads to area maintenance shops and additional layers of intermediate supervision at local centres.

4. Equipment's age and condition: Older the equipment, more maintenance and supervision may be needed (for older plants).

5. Size of plant/ industry: Here also the actual size of plant/ industry and its volume of business will dictate the number of maintenance employees needed and amount of supervision for those. Many more subdivisions in both, the line and staff personnel, can be justified since the overheads can be distributed amongst more departments.

6. Scope of Plant maintenance engineering department: This scope is a direct relationship of management policy. Inclusion of responsibilities for a number of secondary functions means additional manpower and supervision.

7. Employees level of training and reliability: This highly variable factor has a strong impact on the maintenance organization because it dictates how much work can be done and how well it can be performed. In industries where sophisticated equipments predominates, with a high wear or failure incidences, more monitoring systems and more employees and supervisors may be needed. Higher need of reliability of equipments may need bigger maintenance organization.

8. Complexity of business and machines and extent of automation and built-in test and monitoring equipments provided: This may lead to less frequent physical check and correspondingly less manpower.

9. Prevailing maintenance types/ systems of the industry (for older plants).

10. Extent of out-sourcing available or permissible: Availability of highly competent maintenance contractors around the industry may dictate out-sourcing some maintenance jobs (intermittent or continuous) and correspondingly reduce the industry's maintenance organization.

11. Local labour laws and prevailing industrial culture and practices.

12.2 AIMS/ OBJECTIVES OF MAINTENANCE ORGANIZATION

Much of these have been discussed along with its purpose in previous section. The basic objective/ aim of any maintenance organization or department is to ensure that production plants and equipments are available, at minimum cost, for production for scheduled hours and quantity, operating to agreed standards, safely and with minimum waste.

There must be recognition at the top management level and at all other levels of the vital need for maintenance organization and an understanding of its aims and objectives. It should be clear that maintenance department/ organization exists only by virtue of it being necessary for plant's operation and upkeep. It is not a self sufficient or self contented unit. It is a part of team that can perform satisfactorily only when firmly cooperating with others.

In order to develop an effective maintenance organization, following features must be kept in mind:

1. Maintenance is basically a teamwork activity; still the individuals should be able to work reasonable freedom for the specific/ defined job.
2. The jobs/ trades/ crafts of individuals should be so interlinked that they are supporting to each other and pulling towards a common goal. This is very essential as maintenance department consists workers of very many trades/ crafts; few of those are given below.

Main disciplines	Trades/ crafts	
Mechanical	1. Fitters,	2. Welder/ Gas-cutters
	3. Riggers	4. Plumbers/ pipe fitters
	5. Lubricators	6. Mechanics (esp. automobile mechanics)
	7. Machinists	7. Blacksmiths
	8. Markers	9. Millwrights (combined fitter, rigger, welder)
	10. Utility operator (for running pumps, compressors, boilers etc) Etc.	
Electrical	1. Armature winder	2. Electrician
	3. Switchmen	4. Electrician technicians
	5. Electrical inspectors	6. Electrician and cable jointer; Etc
Instrumentation	1. Instrumentation technician	
	2. I and C (Instrumentation and controls) technician	
	3. I and E (instrumentation and electrical) technician; Etc.	
Civil	1. Mason,	2. Surveyor
	3. Carpenter	4. Structural fabricator
	5. Overseers	6. Landscaper; Etc.
Others	1. AC and Refrigerator mechanic	2. Draftsmen (Drawing Office)
	3. Tool and Die technician; Etc.	

In many organizations, many of these trades/ crafts are clubbed together to make fewer trades/ crafts, which help in better utilization of manpower and leads to lesser maintenance personnel.

12.3 TYPES OF MAINTENANCE ORGANIZATIONS

One way of classifying any organization is

- Formal and
- Informal

Though this is not a popular way of classifying, with respect to maintenance organization, it can be said that 'Formal Maintenance Organization' means a separate distinctive maintenance department within an industry or plant, whereas 'Informal Maintenance Organization' means that maintenance jobs are clubbed with operation jobs and same personnel do the operation as well as maintenance jobs as per need and help of external agencies may be taken for heavier jobs, as per need.

As a more popular way, maintenance organization may be classified in following ways—

- Line and staff organizations,
- Functional organizations
- Centralized/ decentralized organizations

12.3.1 Line and Staff Organization

In the line organization, top management has complete control, and the chain of command is clear and simple. Examples of line organizations are small businesses in which the top manager, often the owner, is positioned at the top of the organizational structure and has clear "lines" of distinction between him and his subordinates. A line position is directly involved in the day-to-day operations of the organization, such as producing or selling a product or service. Line positions are occupied by line personnel and line managers. Line personnel carry out the primary activities of a business and are considered essential to the basic functioning of the organization. Line managers make the majority of the decisions and direct line personnel to achieve company goals. This is called '**Line Authority**'. However, two important clarifications should be considered, however, when discussing line authority

(1) line authority does not ensure effective performance, and

(2) line authority is not restricted to line personnel. The head of a staff department has 'line authority' over his or her employees by virtue of authority relationships between the department head and his or her directly reporting employees.

Staff positions serve the organization by indirectly supporting line functions. Staff positions consist of staff personnel and staff managers. Staff personnel use their technical expertise to assist line personnel and aid top management in various business activities. Staff managers provide support, advice, and knowledge to other individuals in the chain of command. Although staff managers are not part of the chain of command related to direct production of products or services, they do have some authority over personnel. '**Staff authority**' has the right to advise or counsel those with line authority. For example, a human resource development manager advises and aids a production manager or maintenance manager by training their workmen to improve their skill and capability. Therefore, staff authority gives staff personnel the right to offer advice in an effort to improve line operation.

The line-and-staff organization combines the line organization with staff departments that support the line departments. Most medium and large-sized firms exhibit line-and-staff organizational structures. The distinguishing characteristic between simple line organizations and line-and-staff organizations is the multiple layers of management within line-and-staff organizations.

An advantage of a line-and-staff organization is the availability of technical specialists. Staff experts in specific areas are incorporated into the formal chain of command. A disadvantage of a line-and-staff organization is conflict between line and staff personnel. Often, due to different positions and types of authority within a line-and-staff organization, conflict between line and staff personnel do occur, but a clear chain of command, which is a consistent characteristic among line-and-staff organizational structures, help minimize such conflicts.

Though maintenance as a whole is, often, considered as a staff organization (with production personnel as line organization), within the maintenance department, we have line organization and staff organization, thus terming the whole maintenance as a line and staff organization. Fig. 12.1 shows a typical line and staff organization of Mechanical maintenance department of Alloy Steels Plant (SAIL), Durgapur, W.Bengal, India. It shows the line and staff functions at two different levels. At first level, I/c Bar Mill Maintenance, I/c Sheet Mill Maintenance and I/c of other mills are line functions and Office staff, I/c Repair and Machine shop and I/c Hydraulic Cell are staff functions. At next level, Shift/ crew in-charge, along with crew workmen are line staff and store keepers and I/c Planning cell with material chaser etc are staff organization.

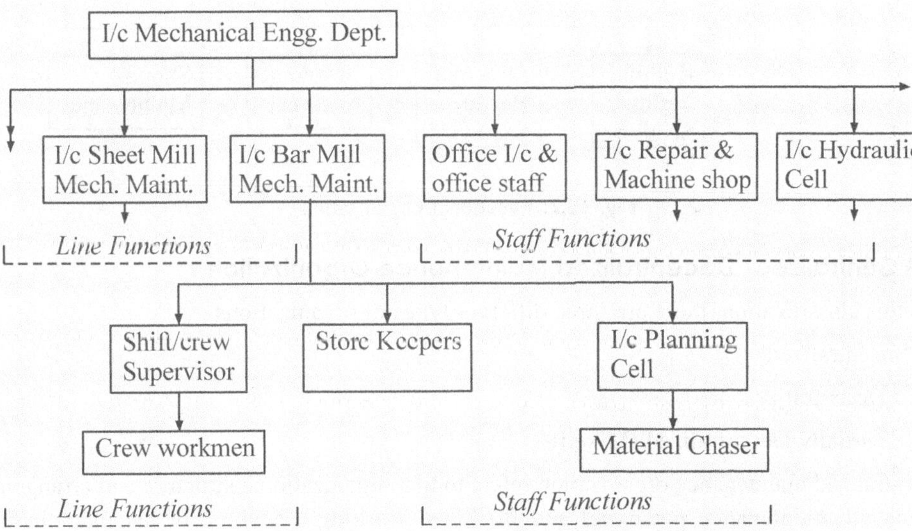

Fig. 12.1: Line and Staff Organization

12.3.2 Functional Organization

In this type of organization, the arrangement is not made by level, designation or hierarchy of maintenance personnel but by the type of functions they perform. Such organization is basically not individual based but is based on the functional areas, shops and sections. This only highlights the functional working arrangements in a plant/ industry.

Fig. 12.2 shows an illustrative example of a functional organization. This is again based on Alloy Steels Plant (SAIL), Durgapur, where the author worked. It functionally shows the various departments, sections and shops etc. Functional organizations do not, necessarily, show the reporting system and is not considered for active use.

Some experts/ writers on Maintenance Engineering may consider 'Functional Maintenance Organization' as a sort of 'Line Organization', but the author considers 'Functional organization' more appropriate as mentioned above.

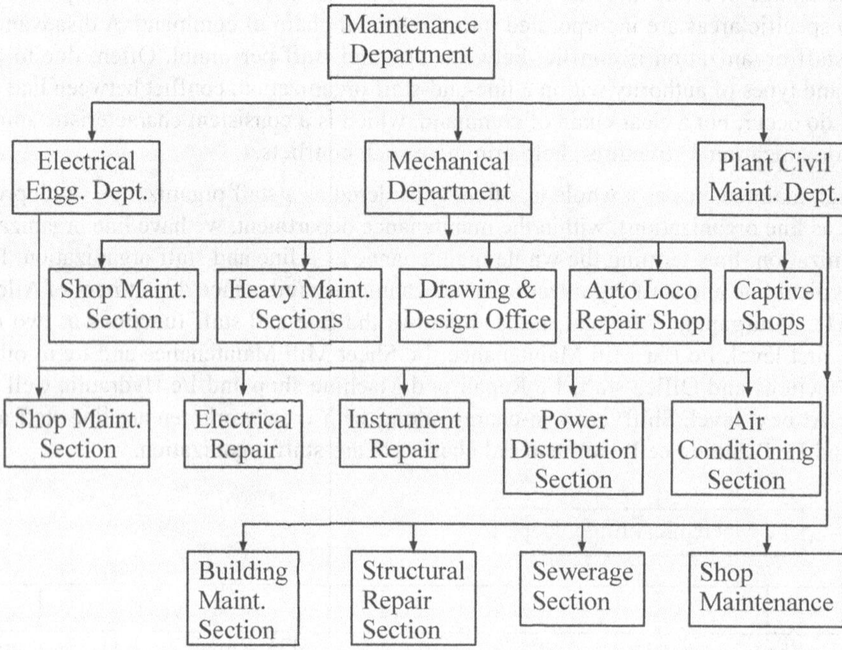

Fig. 12.2: Functional Organisation.

12.3.3 Centralized/ Decentralized Maintenance Organization

In this classification, there are three different types of organizations

- Centralized
- Decentralized
- Partially Decentralized (Mixed)

Centralized maintenance organization refers to that organizational structure and arrangement, in which all maintenance personnel and facilities, working anywhere in the plant industry, ultimately report to Chief of Maintenance (may be G.M. or D.G.M. or A.G.M. Maintenance). The maintenance facilities may be located in the same shop or in different production shops or different maintenance workshops or captive shops and the maintenance personnel may be working in shifts round the clock or in single shift, but at no stage they directly to report to or directly responsible to production in-charge.

The centralized maintenance concept seems to promote harmony between the maintenance personnel because they are closer together. When area concept is used with centralized maintenance, a small area near the process or production is set aside for maintenance personnel. If needed, they can go to central shops or call additional crews from there to do the heavier or specialized work that can only be done in the larger and better equipped place or by more workmen.

Centralized maintenance organization may be of two types—pure and amalgamated. In '**pure centralized maintenance organization**', the employees of individuals disciplines ultimately report to chief of that discipline (e.g. all mechanical workers and executives ultimately report to chief mechanical engineer, electrical employees should ultimately report to chief electrical engineer and so on) and these individual chiefs report to head of plant (e.g. GM-Works or Director-Works etc). In '**amalgamated type**', the lower levels (crew in-charges or sub-area in-charges) of individual disciplines (mechanical, electrical, instrumentation etc) may report to area in-charge of maintenance (senior manager or other such high level), who may be of any of those disciplines and who, in-turn, will report to chief of plant maintenance (GM/ DGM- Maintenance and so on).

The decentralized maintenance organization may base on two concepts——the **discipline-wise decentralization** and **area-wise decentralization**. Discipline-wise means mechanical crafts/ trades may have separate shop/ department, electrical have separate shop/ department etc. Area-wise means separate composite group of maintenance personnel for maintaining separate system or physical area.

Normally, the decentralized maintenance organization refers to that organizational structure and arrangement in which all the maintenance jobs of a shop are also the responsibility of in-charge of production of that shop and the maintenance personnel of that shop ultimately report to the production in-charge of that shop. In this type of organization, chief of maintenance or chief mechanical engineer is responsible only for central facilities or services, e.g. repair and machine shop, central drawing/ design office, utility (steam, air etc) generation and distribution and other captive shops.

As either centralized or decentralized organization is not suited for all types of organizations/ industries, the concept of 'Partially Decentralized" or "Mixed" organization is also used. The partially decentralized (mixed) organization, which is the modified type of centralized organization, is most suitable for bigger plants or plants having units at far away places. In this type of organization, maintenance personnel, attached to production unit, carry out day-to-day maintenance, routine maintenance and most emergency jobs. However, important maintenance functions, e.g. overhauling, planned (preventive/ predictive etc) maintenance, major jobs, procurement of spares etc, are kept under the charge of chief maintenance engineer (I/c of the centralized maintenance section), who will look after all the central and captive shops, maintenance planning, drawing and documentation and utility generation and distribution.

The above three types of organizations are, however, not always followed rigidly and some adjustments can be made to suit the working or available environment and needs. The advantages and disadvantages of all the three types are given in the following table:

Type	Advantages	Disadvantages
Centralized organization	1. Skill and technology easily disseminated, 2. Problem easily investigated, 3. Easy deployment from a more diversified and specialized craft group. 4. Lesser duplicating of workers and facilities, thus lesser manpower and facilities needed, 5. Excel in planning, scheduling 6. Costly and specialized equipments can be procured and used more effectively, 7. More specialized supervision, when needed	1. Collaboration with operation departments difficult, 2. Incomplete collection of operating data, 3. Higher transportation cost, both for men and material, 4. More time getting to and from work area,

	8. Better standardization and variety reduction of spares and materials for industry, 9. Generally more cost-effective,	
Decentralized organization	1. Good communication with operating dept 2. Speedy maintenance response and reduced travel time. 3. Better familiarity with machines and Processes; more production mindedness. 4. Interchangeability of operation-maintenance workforce, even at managerial level, is possible 5. Conflict amongst workmen is reduced due to better coordination.	1. Difficult to share technology and skills, 2. Requires more people, 3. Non-availability or under-utilization of specialized tools and facilities, 4. Generally more costly, Because of duplication of tools and facilities,
Partially Decentralized organization	1. Good communication with operating Dept, 2. Skills and technology dissemination and problem investigation easier 3. Many of the other advantages of the above two types.	1. Management somewhat difficult 2. Job rotation requires ingenuity

Fig. 12.3 shows the example of Centralized Maintenance Organization and Fig. 12.4 as Decentralized Maintenance. These are again based on steel plants under Steel Authority of India Ltd. (SAIL).

12.4 ROLES AND RESPONSIBILITIES OF CENTRALIZED MAINTENANCE GROUPS

Be it any type of maintenance organization, there are few centralized groups (not limited or not permanently earmarked to any particular operation shop/ department), and shop maintenance groups (permanently earmarked for maintenance of a particular operating shop/department. Few of these are shown in Fig. 12.3 and Fig.12.4. The roles/ responsibilities of such centralized groups are as given below:

1 Planning and scheduling of all major repairs, shutdowns and overhauls etc, in consultation with shop maintenance and other agencies,

2. Planning of spares, tools, tackles, specific consumables and other necessary materials, in consultation with shop maintenance etc,

3. Preparation of annual maintenance budget and periodical reviews,

4. Manpower planning and their training and development, in consultation with personnel department and training department,

5. Execution and supervision of all heavy repairs and major overhauls,

4. Assistance and technical guidance on addition, modification and replacement (AMR) schemes pf the plant,

5. Advise on maintainability and reliability of equipments, before purchase and on improvement/ changes etc,

6. Assist shop maintenance personnel in repetitive failure analysis and analysis of other major defects and suggest suitable improvements/ changes,

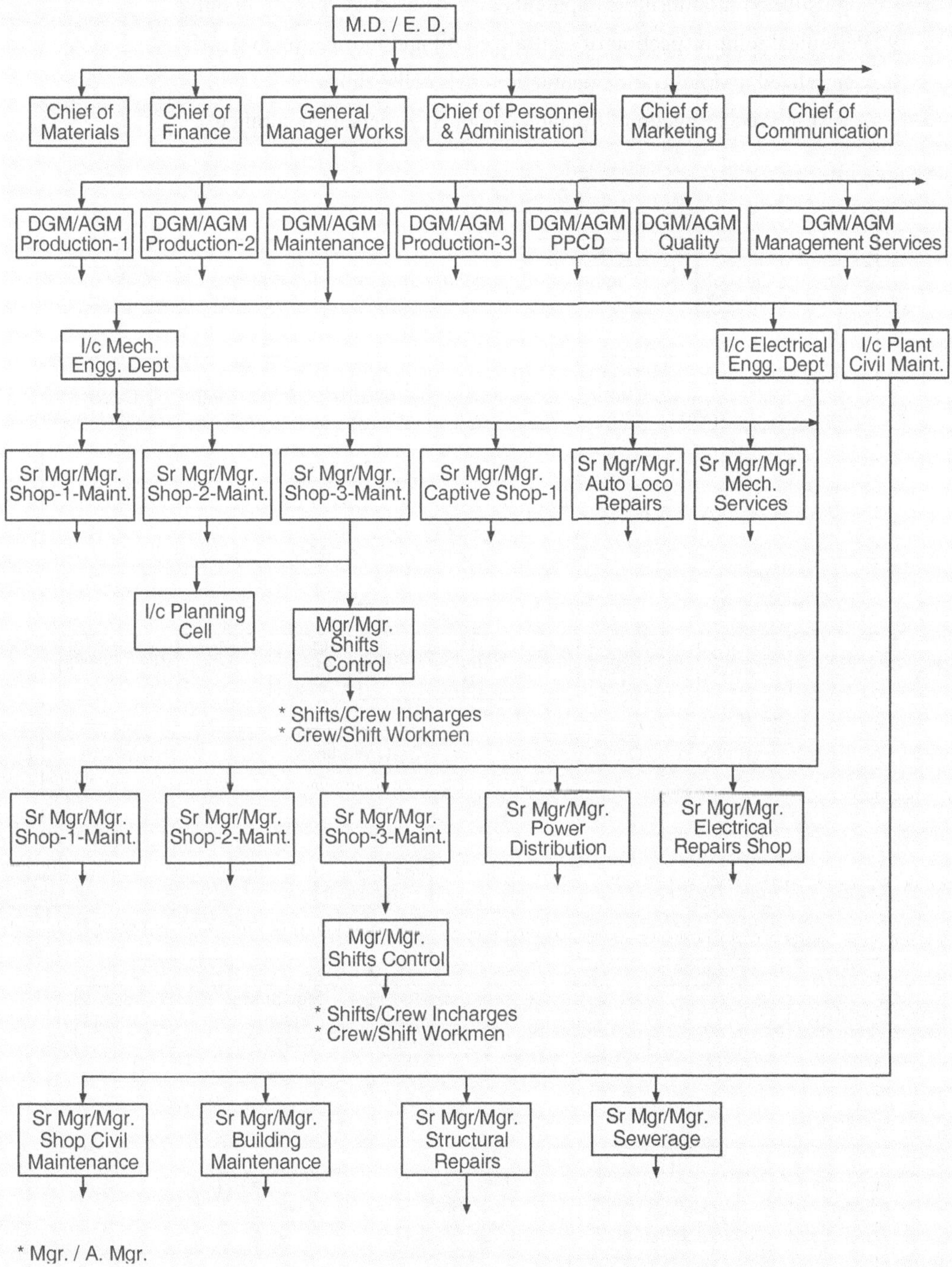

Fig. 12.3: Centralised Maintenance Organisation.

7. Improve maintenance productivity through better maintenance repair technologies (as discussed in chapter-3) or through systematic use of tools or better tools like torque or impact wrenches or hydraulic torque multipliers etc,

8. Assist in implementation of various condition monitoring systems; few hi-tech or sophisticated monitoring equipments may be used by this group only,
9. Periodical joint inspection of centralized oil and lubrication systems
10. Centralized hydraulic and pneumatic repair cells/ shops,
11. Assist in joint inspection of plant's columns, girders, structural etc,

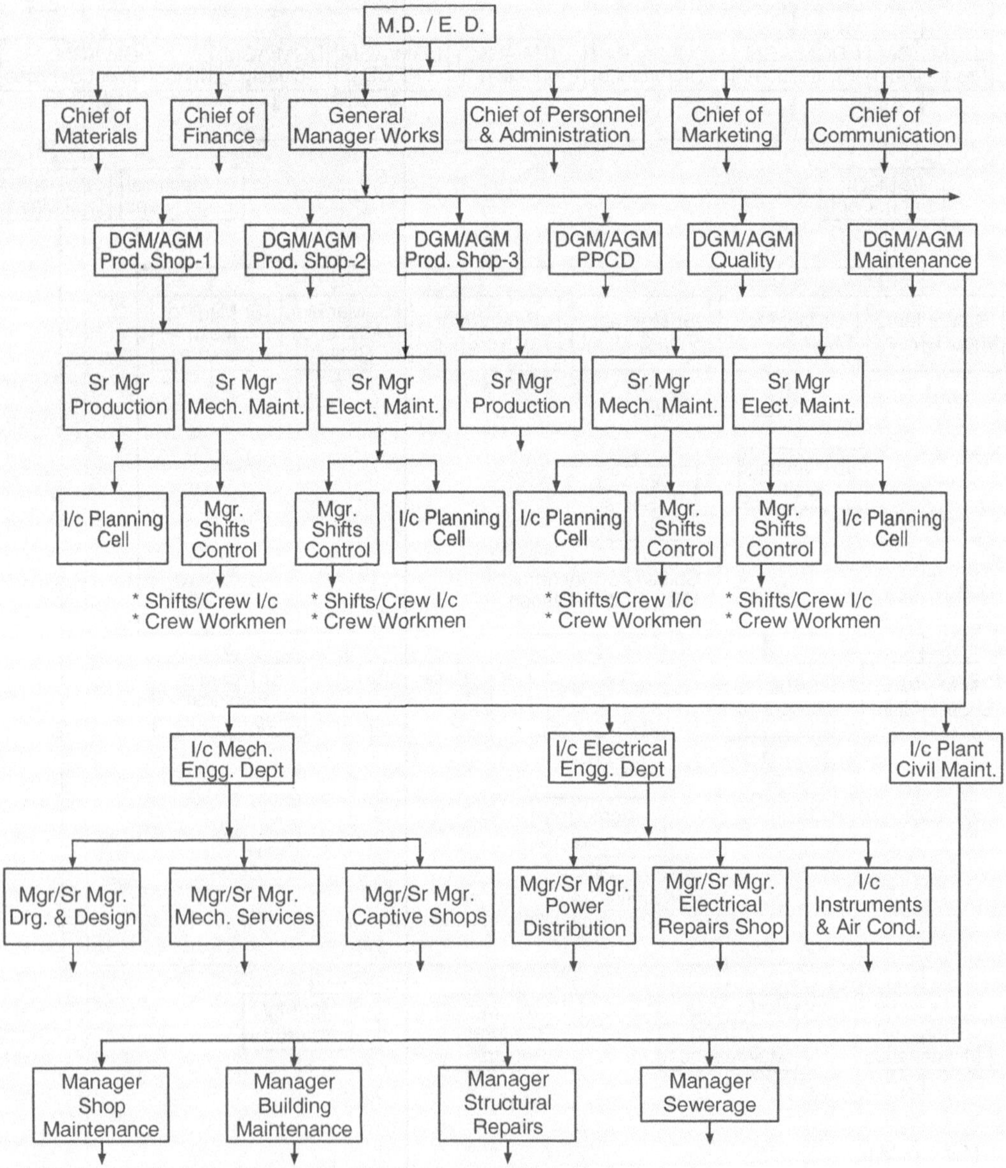

Fig. 12.4: Decentralised Maintenance.

12. Operation and maintenance of centralized services, which may include:
   - Captive shops, such as Central Workshop, Machine Shop, Structural and Fabrication shop, Electrical Repair Shop, Instrumentation Repair shop etc,

- Drinking and Industrial Water Supply System, including water treatment shops etc
- Boiler houses
- Central Compressor House, small Argon, Oxygen, Nitrogen gases manifolds,
- Auto and Loco Repair Shop,
- Centralized Hydraulic and Pneumatic Repair Cell/ Shop
- Wagon Repair Shop and Rail Track Maintenance Group (often this may be under separate Traffic Department),
- Storage and distribution of fuel oil and fuel gasses (occasionally this may come under separate Energy Department)
- Electrical power supply, distribution and billing etc,
- Plant Civil Maintenance, including sewerages and sewerage disposal units etc,
- Central Drawing Office, including archives and minor design works,

13. Assist Materials Department for inventory control, variety reduction, vendor analysis and vendor development and proper material procurement and inspection systems etc,
14. Participate in formulation of inter-plant and intra-plant standards of maintenance spares and stores etc; participate in inter-plant maintenance meetings and other such meetings to share the improvements for maintenance practices, systems, techniques and materials etc,
15. Coordinate waste recovery systems/plans of different shops/departments; Etc.

12.4.1 Roles/Responsibilities of Shop Maintenance Groups

As indicated earlier, shop maintenance groups are formed separately for each production shop and they may work in shifts and general shift, in line with the running of production equipments. Some of the responsibilities of these groups may be as followings:

1. Lubrication of equipments as per schedule and as per need,
2. Routine inspection of equipments, as per check lists and otherwise,
3. Attending reactive jobs (breakdown and emergency jobs), running repairs, adjustments and minor preventive maintenance jobs,
4. Starting and stopping of centralized grease systems, lube systems, hydraulic systems and motor rooms etc and giving electrical power to equipments, as per need,
5. Forecasting requirements of spare parts and other materials and intimating central groups or taking action otherwise as detailed,
6. Energy conservation and waste reduction measures pertaining to the shop; also initiate Suitable waste recovery plans in consultation with centralized maintenance groups,
7. Advise and help central maintenance groups on major repairs and overhauls etc,
8. Keep watch on on-line condition monitoring equipments/systems and carry out periodical Monitoring as per schedules,
9. Regular cleaning of valves, instruments and other critical items and their opening/closing as per need,
10. Ensuring necessary safety provisions (covers, guards, railings etc) and providing/ maintaining environmental control measures,
11. Delay/ breakdown/failure analysis and initiating recommendations such as modification or improvisation etc, for improving equipment's efficiency,

12. Coordinate with captive shops for repairs and reconditioning of broken-down components. Also coordinate with other central agencies (e.g. steam, compressed air, water etc) for need of their shops,
13. Controlling outsourcing (contractors and other outside agencies) expenditures for their shop and also helping budgetary control of their shops,
14. Liaison with operators and other personnel of operating/ production shop for upkeep of equipments and their standard operating procedures. Also arranging shutdowns for central maintenance groups; Etc.

12.5 EXTERNAL MAINTENANCE SERVICES

Such services are available in two forms

- Contract Maintenance Service
- Manufacturer's After Sale Service,

12.5.1 Contract Maintenance Service

Though this concept is used more in developed countries, few industries in India have also started using this concept and offloading total maintenance jobs of the industry to outside parties, a sort of outsourcing (discussed at Sec.12.8). However, in more cases, the total job is not off-loaded to contractors, but help of contractors and contractor's men are taken by most industries, to varying extents. Few of such cases are given herewith—

1. Hiring contract labours for cleaning and helping during major repairs, shutdowns, overhauls etc,
2. Yearly or periodical contract for cleaning of equipments and strategic places and for transportation of maintenance materials and structural etc,
3. Individual job contracts during major repairs and shutdowns, to supplement plant's own workforce,
4. Yearly or periodical contracts for maintenance and upkeep of high technology areas or for some other areas for which plant's own maintenance personnel are not trained or not capable or not available,
5. Building, owning, operating, maintaining and supplying few small but special services (e.g. argon, nitrogen and oxygen etc), within plant premises and captive to the plant,
6. Hiring few specialized maintenance consultants for advising and supervising specific areas; Etc.

12.5.2 Manufacturer's After Sales Services

Some of the equipment manufacturers/ suppliers offer their 'after sale service' for equipments supplied by them, which may include periodical or round the clock checking, maintaining and supply for spares for those. This is normally free for one or two years, immediately after initial commissioning of the equipment and may be extended for further period on chargeable basis. This is advantageous for purchasing industry as plant maintenance personnel may get acquainted with the equipment during that period and may become able to forecast jobs and spares for those.

Some industries do not want to increase their manpower for newly procured special equipments and, as such, enter into regular maintenance contract with the supplier for periodical or round the clock operation and maintenance of those equipments. Weigh bridges, weighing machines, weighing load cells, pneumatic conveyor systems, few special condition monitoring units, many electronic equipments are some of the examples in this category.

12.5.3 Advantages/Disadvantages of External Maintenance Services

The earlier concept of having more and more maintenance facilities within the plant, under direct control of maintenance department, was, probably, based on following apprehensions

- No adequate/ established external maintenance system was available,
- No adequate evaluation of performance and incentive were established for external Services; also it was presumed that they would be less reactive and proactive to emergencies,
- Lack of proper and quick communication,
- Lack of proper accountability and punitive measures for external services; Etc.

However, now, with the use of IT and quick communication and in today's competitive environment, much of the above mentioned apprehensions can be taken care-off. Such limited external maintenance services have many advantages over the conventional departmental maintenance, a few of those are mentioned below

1. It is more economical,
2. Skill preservation can be maintained better,
3. It is technically better as technicians, having enough experience on same or similar equipment, are only employed for the job,
4. Competition from other similar agencies leads to better service and upkeep from the existing external services; Etc.

However, there are few disadvantages in external maintenance services, such as—

(*i*) Lesser Liaison and rapport with operators and operating personnel,

(*ii*) Lesser familiarity of external service personnel with local practices and environment,

(*iii*) Creating an environment with offloading more and more to outside agency, without corresponding reduction of regular maintenance force,

(*iv*) With many outside agencies involved, effective coordination becomes difficult;

(*v*) Very often objections from worker's union; Etc.

12.6 CAPTIVE SHOP FACILITIES

As earlier mentioned in this book, some captive shops/ facilities are essentially needed in big industries for urgent repair/ reconditioning of the breakdown components and also making some new parts/ spares. Of-course, with more development of ancillary units/industries around a big industry, the need for such captive shops would be less. In centralized maintenance structure or mixed (partially decentralized)

Structure, relatively bigger captive shops are maintained and operated by centralized maintenance groups, but in decentralized maintenance structure, slightly smaller captive shops are attached with each or few production shops. Some of the captive shops, which are operated and maintained by maintenance department, are given herewith

12.6.1 Machine Shop

Caters to the need of different machining jobs, (e.g. turning, boring, drilling, shaping etc) for repairing, reconditioning and replacement of spares parts, and, thus, reducing bought-out items.

12.6.2 Machine-tool Reconditioning Section

Caters to the need of major repairs, overhauling, trueness checking of the machine-tools and equipments of captive machine shop and small machine tools used by production shops (e.g. guide conditioning shops), inspection department and Roll Turning Shops etc. Occasionally this section (in cooperation with machine shop) may also take up in-situ (during major repair in different shops) machining for removing severely jammed and/or broken big pins and shafts etc.

12.6.3 Lifting Tackles and Material Handling Section

Caters to the need of reconditioning/ repair of all lifting tackles used by maintenance and other sections in the plant, such as chain blocks, hooks, pulley-blocks, pull lifts, hydraulic and mechanical jacks, winches etc and also for splicing/making of wire-rope slings, chain slings etc. Often wire-rope slings, chain-slings and in-situ wire-rope testing and certification also becomes the responsibility of this section.

12.6.4 Hydraulic and Pneumatic Cell

Few plants have this section/ cell to cater the need of repair/ reconditioning and testing of different hydraulic and pneumatic components (e.g. pumps, motors and valves etc) installed all over the plant/industry. This cell is equipped with suitable test rigs for load-testing of different components. Often, the cell may take making of new hydraulic hoses for urgent requirements of any section of the plant/industry. Thus, the cell helps in reduced purchases of hydraulic and pneumatic components and hoses. As per need, the cell may help the different maintenance groups by advising or troubleshooting or taking up some jobs in hydraulic/ pneumatic systems during any major problem.

12.6.5 Fabrication and Structural Shop

Caters to the need of all heavy and critical fabrication jobs, associated with repair/ reconditioning of major components/ assemblies or for making new ones. The shop also makes various structural and fabrication items (e.g. tanks, reservoirs, columns, buck-stays, trusses etc) for whole plant. The shop also takes up brazing and specialized welding of few high-alloyed components. The shop is equipped with necessary testing facilities (e.g. pressure testing, ultrasonic and radiographic testing, die-penetrant test etc), either of shop's own or on contract basis.

12.6.6 Smithy and Foundry Section

Smaller plants do not have any foundry section for maintenance organization and their smithy section is attached either to machine shop or lifting tackle shop for recondition and remetalling of white metal/ babbitt metal bearings, making links for chain slings and also making/ repairing of forging tools (chisels, tongs, pockers, dies, punches, spoons etc) for maintenance and operation of whole plant. However, in bigger plants/industries, small foundry and smithy facilities are clubbed together in a separate section to do all such jobs and also to make smaller casting and forging spares for the plant. The foundry section is, naturally, attached with a small carpentry facility.

12.6.7 Electrical Repair Shop

Caters to the need of repair of various types of electrical equipments, machines and appliances, which requires special attention and facilities that can't be done in-situ, including rewinding of electrical motors, welding transformers etc.

12.6.8 Electronic Cell/Section

Looks after the repair and upkeep of various electronic equipments/ gadgets of the plant. Also occasionally advises on trouble-shooting and solving of the shop electronic equipments.

12.6.9 Instrument Repair Shop

Responsible for repair, testing and calibration of various industrial instruments of the plant, such as pyrometers, recorders, controllers, gauges (pressure, vacuum etc), and monitoring equipments etc. Occasionally the scope of this shop and electronic cell may appear somewhat similar, as such, in some industries these two are clubbed together.

12.6.10 Wagon Repair Shop

In bigger industries, like integrated steel plants, a separate wagon repair shop in provided, normally under maintenance in-charge, to repair and recondition different rail wagons, mould cars, match cars and other rolling stocks. The shop may also take up reconditioning of loco wheels and overhead cranes etc. In smaller plants, such facilities, in much smaller scale, are merged in machine shop.

12.7 WORKING ARRANGEMENTS

During and after finalization of maintenance organization, some healthy working arrangements and practices are considered and introduced so that the maintenance personnel of different categories and at different levels work with reasonable independence but supporting each other towards the common goal of improving maintenance effectiveness and equipment availability etc. Such arrangements are simple work practices, followed by most of the big industries. Followings are a few concepts, which should be considered

1. **Engineer-Scheduler-Executor Concept.** There should be separate agency to plan and schedule the jobs, separate to execute and supervise the job and yet separate agency to carry out monitoring and analysis of deviations and omissions, weak point analysis and other engineering functions to initiate suitable corrective measures. Same person doing any of the two or all three functions will create chances of omissions, deviations and improper adjustments. Depending on the workload, the persons doing each of the three functions may be exclusive for those functions or they may do other jobs also, but this should not increase manpower. Also occasionally the persons may be rotated,
2. **Delegation.** Delegation is a common topic for all types of jobs and establishments but is emphasized here in maintenance organization because of various categories and levels of maintenance personnel and their inter-dependencies. Proper delegation with suitable control should be introduced so that each individual and crew can work with reasonable freedom and at the same time support each other towards a common goal. Delegation in maintenance is not allocating different jobs to different tradesmen, but is allowing the individuals to take required decision about the job that has been allotted to him/her. The decision, of-course, has to be within the broad parameters of different job schedules, job manuals, SMPs etc necessary facilitating the other groups. Too much interference (poor or no delegation) or unlimited delegation, both would create lot of problem in carrying out maintenance jobs.
3. **Position Description/ Job Description.** Because of many varieties of disciplines, trades and levels in maintenance organization, the position description or rough job description of various levels and trades should be clearly defined and understood so that there is no

confusion, ambiguity or contradiction and not much overlapping. A limited amount of overlapping is, occasionally, introduced by design to take care of some specific situations.

4. **Multi-trade Concept.** Though this is somewhat contradictory to skill preservation and skill development, multi-trade concept is often introduced in most industries, except few critical areas and specialized sections. As one example, the jobs of fitter, welder and rigger may be merged and may be designated as 'millright' or any such designated for shop maintenance groups and few centralized maintenance groups, but the same individual trades may be maintained in few specialized and heavy maintenance groups and few captive shops. This concept leads to better utilization of manpower, without unduly sacrificing the quality. The concept is further discussed in detail in next section (Sec.12.7) as 'Multi-skilling'.

12.8 MULTI SKILLING

As briefly discussed in previous section (multi-trade concept), the concept of multi-skilling or multi-skill (multi-trade) workmen has, recently, assumed great significance and, as such, needs more deliberation. With increase in size, type, complexity and sophistication of equipments, maintenance personnel also got separated into different and specialized trades to be able to maintain those. This led to existence workmen of large number of trades today (fitters, riggers, welders, gas-cutters, millwrights, pipe fitters, machinists, electricians, instrument mechanics, electronic technicians etc), all having slightly differing knowledge and skill levels.

The confusion has now arisen that a complex equipment needs jobs of different trades/ skills but all those tradesmen can not work on that particular equipment at the same time. If many tradesmen work on one particular equipment, it results in many problems, such as:

- Need of more manpower,
- Need of more downtime,
- Possibility of improper job if one tradesperson does job of another trade without acquiring multi-skill,
- Possibility of some job being leftover as one may think the other might have done that,
- Nobody being clearly accountable for subsequent failure, etc.

As such, in today's age, craft lines do not need to remain so clearly drawn. The average industrial worker of today has qualification upto High School and developing them in multi-skill trade can solve may of the abovementioned problems.

Multi-skilling is the process of training maintenance employees in specific skills that cross the traditional trade or craft lines and, then, ensuring that the work is performed. The advantage of multi-skilling is that particular jobs, which historically require more than one craft, not necessarily more than one individual, can be performed successfully just by one person. In multi-skilling, individuals receive additional training, beyond the normal skills required by their trade. Multi-skilling give advantage to the company in the form of reduced manpower cost and ease in job scheduling which leads to time-saving and repair cost. The advantage to workmen come inform of some incentive or pay rise.

12.8.1 Multi-skilling Pitfalls

Many multi-skilling efforts have not worked as desired. They normally fail because of one or more of the following reasons

1. *Unrealistic Expectation*- Management expectation that most workmen can attain and retain many skills and that they can do anything and everything and also workmen expectation of substantial pay increase, both may be unrealistic.
2. *Vague Goals or Commitments*- Many multi-skilling efforts begin without clear goal or understanding of the company that what they expect or aim to gain. That results either finishing the training midway or workmen continue doing their job as before.
3. *Inadequate Definition of Multi-skilling*- Inadequate definition of the training efforts itself can leave the goal unclear in everyone's mind as to when and how the efforts will be accomplished.
4. *Failure to Implement*- This may be because of inadequate training facilities or inadequate skill development time or inadequate communication about work change etc.

12.8.2 Identifying Friction Areas and Multi-skill Opportunities

In industries where strong craft line distinction exists, it is imperative to identify the skills to be included in the programme. The most beneficial areas are those which involve jobs where two or more crafts are required to complete a job but only one or two workmen are required to do that job.

Such jobs are described as **Friction Areas**—jobs that are causing friction in productive deployment of workmen. Friction areas can be identified from completed work history cards, structured group interviews and multi-skill surveys etc. Although such areas may vary greatly from place to place and industry to industry, following are the few areas where multi-skilling can be considered as productivity improvement areas

1. Jobs combining electrical and mechanical skills (motor or mono-block pump changing),
2. Job requiring electrical/mechanical and simple welding skills (installing conduit /pipe supports and running conduits/ pipes and pipefitting jobs etc), also minor insulation removal and replacement on pipes/ conduits,
3. Oxyacetylene operations (heating, trimming, cutting, gas-welding, brazing, soldering),
4. Millwrights (fitting, rigging and general welding skills),
5. Forklift (or truck) operator cum material chaser; operating vehicle, procuring and handling smaller materials and minor loading/ unloading/ stacking etc.
6. Machine lubrication can be done by fitters or millwrights doing job there,
7. Machinists should do loading/unloading of their machines and do lubrication of m/c.
8. Millwrights or fitters should do minor and rough turning/boring on small machine tools; etc.

12.8.3 Identifying Potential Gains

After the possible training areas have been identified, the company can determine the potential productivity and financial savings to be achieved from the multi-skilling efforts. The maintenance workforce will become more versatile and adaptable by gaining values with new technical capabilities and multi-craft skills. Possible salary increase or other incentives should be negotiated with representatives of workmen also.

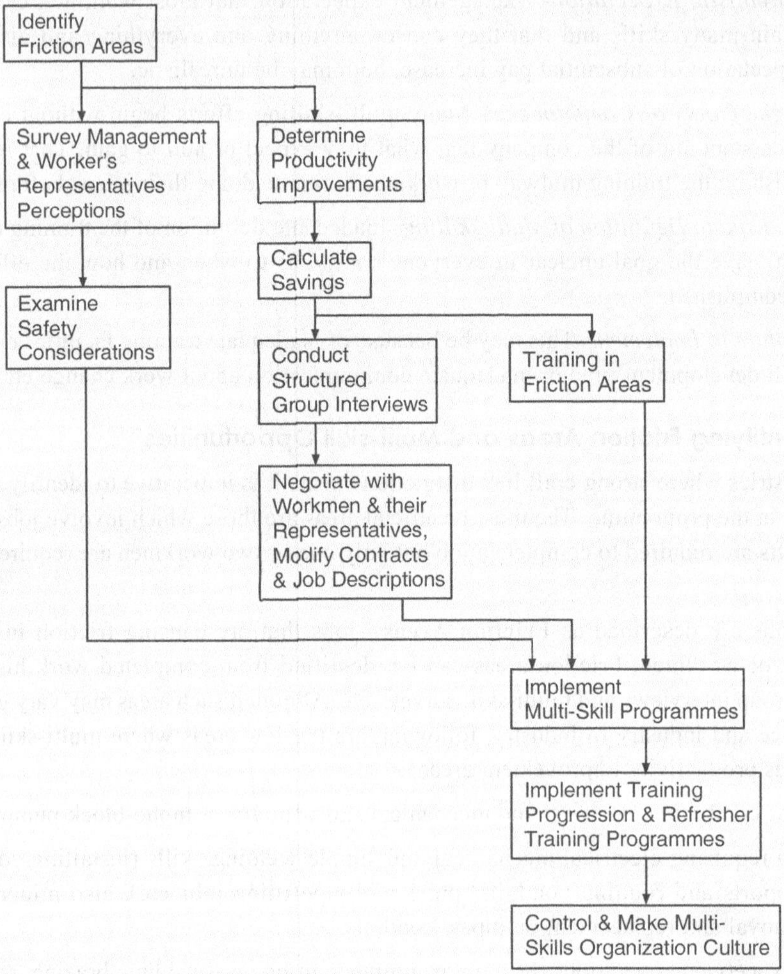

Fig. 12.5: Multi-Skill Implementation

12.8.4 Defining and Arranging Training

Defining and arranging multi-skill training is the most important step. It must equip the workmen with the specific skills they will need to safely and effectively perform duties formerly done by another craft or crafts. In addition to multi-skill training, a *training progression programme* (refresher courses etc) should be followed so that the workmen retain and refine their acquired skills.

12.8.5 Negotiating the Multi-skill Programme

Some time, during the training period, the management and concerned workmen (possibly along with workmen's representative) should sit together and discuss/ negotiate the benefits for both sides.

12.8.6 Implementing the Skills

Soon after completion of training, at an appropriate time, the multi-skills should be incorporated into the actual scheduling and work performance. The company's management and workmen both feel that they both are in "Win-Win" situation; as it offers real productivity and financial gains to the company and also offers job-enrichment and some other incentives to workmen.

Fig.12.5 shows the multi-skilling implementation in block diagram form.

12.9 OUTSOURCING IN MAINTENANCE

Some sort of outsourcing has already been discussed in Sec.12.4 (External Maintenance Services).There may be little repetition of the concept, but has been done to re-emphasize it. The term "Outsourcing" has recently assumed great significance due to emergence of BPO (Business Process Outsourcing). However outsourcing in small form is regularly used in form of awarding small maintenance contracts to outside contractors to augment maintenance workforce during shutdowns and major repairs.

Today, often it is preferred to increase the level of maintenance productivity by outsourcing for maintenance partners. Associating with professional maintenance people, give a company the advantage to address their issues with those experts who face and meet these challenges on a daily basis. In addition, outsourcing, also, enable a company to gain greater control over maintenance results. However, it is important to note that the following key elements have to be highlighted to achieve the optimum benefit from these outsourcing activities

1. Performance measurements are developed and clearly communicated,
2. Planned maintenance is established as the focus of the operation,
3. Maintenance becomes a competitive advantage,
4. Both manufacturer and maintenance contractor desire a mutually beneficial relationship,
5. Maintenance best practice are established and tracked.

In the outsourcing or maintenance partnership scenario, performance guarantees and continuous improvement goals provide greater control over maintenance results and assured production goals are being achieved. In addition to saving cost, expertise of the external agency is also achieved.

12.9.1 Types of Outsourcing

Today different types of outsourcing are followed in different organizations, depending on their specific needs. Few of those are given herewith

1 *Awarding small job contracts*: to augment maintenance force during shutdowns or other additional needs,
2. *Outsource Regular Diagnostic CM activities*: Suppliers or their agents carry and install permanently, maintain and operate diagnostic and monitoring equipments/ instruments and inform the monitoring results in different forms at regular intervals to plant personnel. Calibration of those equipments are also arranged by them, along with plant representatives, from external national or private laboratories or test houses.
3. *Outsourcing of Regular Maintenance*: All day to day and planned maintenance is awarded to specialized outside agencies and they keep their men suitably and submit checklists and other reports to plant regularly. Plant keeps only very few personnel to occasionally oversee the activities and monitor reports submitted by that agency or agencies. Spares etc are arranged by plant.
4. *Build, Operate and Maintain (BOM)*: This is mainly applicable for facilities, such as Oxygen supply, Air and other gas supply, Fuel (solid, liquid or gas) supply and Electrical

power supply etc. The contractor builds, operates and maintains the facility and recovers the cost from selling their supply to the plant. The facility may be built within the plant premises or nearby out side.

12.10 SELF-DIRECTED WORK TEAMS

Lean production is, now, displacing conventional mass production throughout the world. Human resource development practices play a critical role in any company's programme to develop and institutionalize lean method on the shop floor. One approach, that has been successful in many industries, involves organizing production workers into a self directed or self managed teams to do all connected jobs.

Teams, consisting of workmen, ranging between 5 to 15, depending size of process/ job, take responsibility of an integrated, customer driven, production processes/ jobs. The team members cross-train in many of the disciplines/ tasks within the defined process/ job and gradually expand their capabilities to include support and some administrative roles. Within the overall organization, the teams may attain some sort of autonomous nature.

Lean production and self directed work teams were developed historically, at different times and at different parts of world. Though the two concepts share somewhat common grounds, many lean production teams extended the boundaries of their works to encompass not only 'KAIZEN' but also some administrative and managerial tasks and, as a result of that, they created a truly self directed work teams, under the banner of lean production teams.

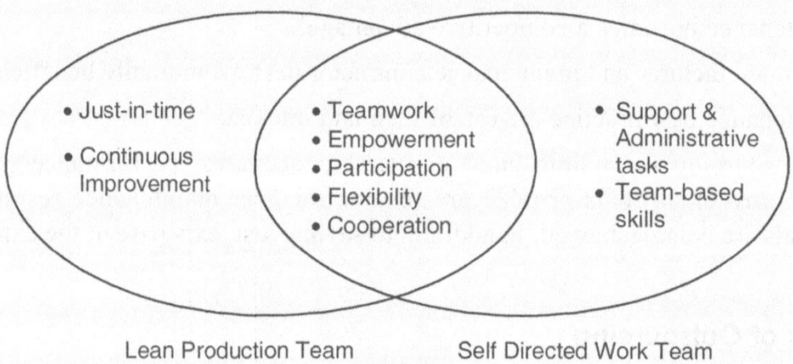

Fig. 12.6: Differences Between Lean Production Team and Self Directed Team

Developing self directed work teams consists of four major steps—(a)- Cross-training and multi-skilling, (b)- enhancing teamwork skill, (c)- participating in proactive improvement efforts and (d)- developing administrative and support function skills. The team members should be motivated employees, should overcome the resistance to change, should aim for continuous improvement and should have suitable incentive system. Fig. 12.6 shows briefly the difference between the lean production teams and self directed teams.

However, the concept of "self directed teams" appear to have some overlapping with "Total Productive Maintenance Teams" and involves lot of 'Multiskilling' and affects the maintenance organization accordingly. Still there are some differences between the two. Both direct towards economic excellence.

QUESTIONS

1. Discuss the role/responsibilities of centralized maintenance groups and shop maintenance groups
2. Describe relative merit and demerits of centralized and decentralized maintenance systems.

3. What is multi-skilling? What are advantages and disadvantages of multiskillind?
4. What process and cares to be considered for introducing multiskilling?
5. What is outsourcing in maintenance and what types of outsourcing are practices are practices in general industries?
6. What are the objectives of maintenance organization and what different types of organizations are in use in Indian industries?
7. Comment briefly on line and staff organization and functional organization.
8. What are the different captive shop facilities in most industries in India?
9. What are the factors that are to be considered for deciding maintenance organization in any industry?

CHAPTER

13 MAINTENANCE EFFECTIVENESS, PERFORMANCE EVALUATION/ INDICES AND AUDIT

As maintenance is basically a service type operation/ job, the measurement of maintenance effectiveness and evaluation of maintenance performance are not easy tasks. Cost alone is not an effective performance measuring tool as the cost depends mainly on type of service and criticality needed. In India, maintenance cost may be as low as 3% of cost of product in some industry and may be as high as 30% of cost of product in some other industry. Again equipment availability alone can not be a good measure of maintenance effectiveness and so on. As such, maintenance effectiveness and performance evaluation calls for more examination.

Maintenance organizational efficiency/ effectiveness depend upon many interdependent variables. Again, maintenance organizations function at three major levels: organizational level (functional and structural relationships), process level (work activities) and job performer level (individual worker). The presence (or absence) of these variables impact all levels.

Maintenance Effectiveness or "Effect of Maintenance" can be visualized from Table 13.1 (Effective Maintenance Vis-à-vis Non-effective Maintenance). Effective maintenance helps to improve ROA (Return of Asset Value) by reducing the need for expensive capital upgrades to increase output.

Table 13.1 Effective Maintenance Vis-à-vis Non-effective Maintenance

Effective Maintenance	Non-effective Maintenance
• Restore system productivity,	• Increased maintenance cost,
• Improves system/ equipment reliability,	• Reduced system/ equipment life,
• Avoid any unnecessary shutdown,	• Heat Loss by reduced efficiency of heat transfer
• Increase of equipment efficiency,	• Induced/ created unwanted waste,
• Prolong the system life,	• Additional cost to clean the equipment/ system,
• Improve the overall plant productivity,	• Can effect product quality,
• Necessary to maintain product quality,	• Reduced plant productivity,
• Increase of plant profit; Etc.	• Downgraded system effectiveness,
	• Decreased plant profit; Etc.

13.1 OVERALL EQUIPMENT EFFECTIVENESS (OEE)

It is an effective method to measure production floor performance. Capturing and recording accurate production floor information is critical for preparing reliable OEE reports.

Manual data collection, there is usually a second step of manually compiling the data. This is most commonly accomplished by entering in the information into spreadsheets. This manual compilation step also leaves room for both inconsistencies and inaccuracies. Another disadvantage to manually collecting data for your OEE reports, like with many maintenance and production improvement plans, is sustainability. With an automated data collection system, sustainability is no longer an issue. Once in place the automated OEE reporting stays in place and you need only act on the result.

The overall performance of a single piece of equipment or even an entire factory, will always be governed by the cumulative impact of the three OEE factors: Availability, Performance Rate and Quality Rate.

OEE = Availability x Performance Rate x Quality Rate

In above formula, the performance rate is the percent of parts produced per time frame (normally maximum or best known rate is used) and Quality Rate - Percent of good sellable parts out of total parts produced per time frame. Availability will be discussed in next paragraph.

OEE can be used to save companies from making inappropriate purchases, and help them focus on improving the performance of machinery and plant equipment they already own and to find the greatest areas of improvement etc. It can be used as an analysis and benchmarking tool for either reliability, equipment utilization, or both. While monitoring OEE per equipment brings focus on the equipment itself, it may not provide true cause of major costs, unless the cause is obvious. As such, you should also be calculating a production line OEE, and within a corporation, a facility OEE.

Also refer Fig.10.3— OEE Model for TPM

13.2 AVAILABILITY (COMMONLY KNOWN AS EQUIPMENT AVAILABILITY)

As also indicated in chapter-2, Availability (equipment availability, plant availability or system availability) is very important aspect of maintenance and plant/ industry. It is a measure of OEE and maintenance effectiveness. Availability is one of the most important reliability measures of repairable systems because it takes into account both failure and repair information. As a general definition, **Availability** is the probability that an item, under the combined influence of its reliability, maintainability and maintenance support, will be able to fulfill its required function over a stated period of time, or at a given point in time. The operating context of a piece of equipment will determine its performance requirements.

There are three basic approaches to define availability, which are the percent of "scheduled production", "calendar hours", and "equipment or line status". The schedule production approach defines availability by the production schedule for a piece of equipment, a cell or an entire production line. The calendar hours defines availability as the total time available to produce product which is usually week (7days), month (30 days) or year (365 days). The equipment or line status approach defines availability in terms that production line or piece of equipment is in a state to produce product. All three approaches are valid approaches. The percent of scheduled production and calendar hours provide a broad view of availability while the line status approach is a much more defined view.

Though the causes of un-availability may be many, such as equipment/ plant/ system not available for operation/ production, no power available, operating personnel not available, operating material and consumables not available or other infrastructure supports not available etc, generally the first cause i.e. "equipment/ plant/ system not available for operation/ production" is taken for availability calculations. Availability must be defined for each area or piece of equipment that will have OEE reports. In some of the instances there will be more than one element that needs to be monitored to determine if that piece of equipment or area is considered available.

Quantitatively, the **availability** for a single machine is as below

Availability = MTBF/(MTBF+MTTR)

As per this formula, high **availability** can be obtained either by increasing the Mean Time Between Failure (MTBF), and hence the reliability, or improving the maintainability by decreasing the Mean Time To Repair (MTTR). The MTTR would include the repair time and the logistic delay (obtaining labour and spares).

The availability of the system depends on its initial condition and both the failure and repair distributions specified for modeling its failure and repair processes. It is a common practice to assume that the system is initially in good condition. As shown in above formula, this availability value is a function of time, as such it is often called *time-specific availability* or *time-dependent availability* or *transient availability*. On occasion, it is also called *instantaneous availability* or *point availability* because it is evaluated at a particular instance or point in time.

Mean Availability, Steady-state Availability and Operational Availability etc are few other connected terms used for availability. Mean availability is the expected percentage of uptime within a time interval. It is also called *average availability* or *interval availability*. The cost of system downtime is directly related to its mean availability. Steady-state availability is calculated when time tends to infinity where the change in $A(t)$ with respect to time becomes negligible (converges to zero). Steady-state availability is the mean uptime (MUT) divided by the sum of the mean uptime (MUT) and mean downtime (MDT). An advantage of steady-state results is that they can be computed quickly and accurately. Under steady-state conditions, mean availability is equivalent to availability. Therefore, steady-state availability can also be used for *steady-state mean availability*.

Steady-state Availability, $A(\infty)$ = MUT / (MUT + MDT)

Maintenance or corrective repair of a component or system may not start immediately due to logistics and administrative delays. Such delays are generally grouped and classified as logistic delay times. In such cases, the mean downtime is the sum of the mean corrective/repair time (MCT) and the mean logistic delay time (MLDT). The availability value that considers the effects of logistic delay times is known as *operational availability*.

Operational Availability, A_o = MUT / (MUT + MCT + MLDT)

However, often the logistic delay can be controlled or eliminated and in that case A_o is almost similar to A(“). Again in general industrial situations there is not much difference in different availability mentioned above and in our further discussion we would mean availability as amalgamation of all.

13.3 MAINTENANCE EFFECTIVENESS ASSESSMENT/ SURVEY

The purpose of a Maintenance Effectiveness Assessment/ Survey is to determine where the maintenance organization’s strengths are to leverage improvements and identify areas of opportunity to correct. It provides a view of the structure, relationships, processes and people relative to good maintenance practice. We can survey by preparing questionnaires in following five categories

1. Questionnaire on Resource Management; e.g. overstaffed, inadequate skill, inadequate safety and housekeeping etc,
2. Questionnaire on Information Management, e.g. Information flow, CMMS and other reports, computerization of stores etc, Tracking of machine breakdown and expenditure,
3. Questionnaire on PM, PdM and Equipment Technology; Tracking of PM and PdM jobs, reports and costs etc,

4. Questionnaire on Planning and Scheduling; Priorities for maintenance jobs, Planning in time and advance,
5. Questionnaire on Maintenance Support; Timely availability of spares, special tools and facilities, reward system for good job etc

Suitable rating and weightage can be adopted for below average, average and above average and individual questions to be valued accordingly. The average of all five categories can be added to evaluate and measure maintenance effectiveness. With this calculated overall score and its analysis, we can grade the effectiveness as below average, average, above average or excellent.

13.4 KEY PERFORMANCE INDICATORS (KPI)

KPI or Key Success Indicators (KSI), help an organization define and measure progress toward organizational goals. It is yet another way to measure effectiveness. KPIs are quantifiable measurements, agreed to beforehand, that reflect the critical success factors of an organization. They will differ depending on the organization. A business may have as one of its Key Performance Indicators the percentage of its income that comes from return customers. Maintenance Service Department may have as one of its Key Performance Indicators, in line with overall company KPIs, percentage of equipment availability or reliability etc.

Whatever KPIs are selected, they must reflect the organization's goals, they must be key to its success, and they must be quantifiable (measurable). KPIs usually are long-term considerations. The goals for a particular KPI may change as the organizations goals change, or as it get closer to achieving a goal. It is important to define the KPIs and stay with the same definition from year to year. For a KPI of "Increase Equipment Reliability", you need to address considerations like whether to measure taken will actually lead to required availability and whether money spent will be covered by improved profitability and safety, etc. Similarly the Manufacturing Department may have a KPI of "Number of Units Rejected by Quality Inspection", while the Sales Department has a KPI of "Minutes a Customer is on Hold Before a Sales Representative Answers". Other departments may have their own KPIs. After finalizing KPIs, targets are to be set for each KPI. These performance Indicators will help the company meet its overall KPI.

We use Key Performance Indicators as a performance management tool.. KPIs give everyone in the organization a clear picture of what is important, of what they need to make happen. We use that to manage performance. We make sure that everything the people in our organization do is focused on meeting or exceeding those Key Performance Indicators.

Few typical **Maintenance Key Performance Indicators** are mentioned below

- Overtime worked against plan
- PM schedule compliance
- Time taken to answer maintenance calls
- Budget compliance
- PM Backlog man-hours
- Critical Equipment availability (Don't worry about non-critical equipment)
- Number of breakdowns (Unplanned maintenance)
- Production equipment performance (By output volume levels)
- Equipment performance (with respect to quality)
- Availability and Reliability
- MTBF; etc

13.5 MAINTENANCE PERFORMANCE ANALYSIS/ EVALUATION

The availability, maintenance effectiveness assessment/ surveys and KPIs, mentioned earlier in this chapter work as analysis for maintenance performance. Analysis of maintenance performance for shop level personnel varies slightly than senior management level personnel. At shop level, performance evaluation may be effective utilization of maintenance crews working there and total time taken for attending repairs and maintenance—machine-wise or category-wise. For such analysis criteria/ indices can be developed for tolerable backlog, waiting time, preparation time and actual working time (wrench time). Following techniques/ methods are helpful for such analysis

- Work Sampling Techniques
- Job Car review/ analysis
- Job Auditing; etc.

These techniques and other such analysis are quite common and do not need elaboration. Fig. 13.1 shows a sample job audit record-sheet.

Job Audit Record Sheet	
Manager/ Shift-in-charge —— ———— ——Date ———— Shift—— Crew and Designation ———— ——— Machine and Code——— —— Job/ Work ———	
Time (End of Activity) Hrs. Minutes	Activity
10 – 45	Job Card returned after completion of Job No. 1
12 – 05	No Job
13 – 00	Break for Lunch
13 – 20	Take job card for Job No.2 and wait for briefing
14 – 30	Waiting for material delivery
16 – 00	Working
16 – 30	Waiting for Shift I/c consultation and return of job card for Job No.-2
etc	etc

Fig. 13.1: Sample Job Audit- Record Sheet

Analysis at senior management level may concentrate more on availability, time spent on PM and PdM vis-à-vis reactive (breakdown etc) maintenance time, cost parameters, actual expenditure vis-à-vis budget, and maintenance productivity. Middle management level, in addition to many of above mentioned parameters, may also consider % overtime, chance failure ratio (chance failure per equipment operating hour), chance failure intensity ratio (chance failure shutdown hours / equipment operating hours) and spares/ consumables management etc.

True that the importance of maintenance is stressed mainly to keep the reliability and availability of machines/ equipments/ systems within acceptable, but, for measuring maintenance productivity, it is very important to analyze that at what cost and with how much resources and with what effectiveness such reliability and availability is obtained. The min reasons for developing such analysis and evaluation are

1. To monitor maintenance performance,

2. To compare performance of maintenance department and men with that of other departments of the plant and also with that of maintenance department of other similar plants,
3. To assess performance against established norms/ standards,
4. To assess impact of changes in methods, materials or personnel for maintenance; etc.

In order to carry out such analysis/ assessment, normally two models are used—Manufacturing Based Model and Service Based Model.

Manufacturing Based Model: In this method, the productivity of maintenance function of a particular class depends solely on two factors—Plant Output and Maintenance Inputs (man-hours, materials, facilities etc) associated with that class like mechanical, electrical or instrumentation etc. This method gives a set of well defined productivity measures on the basis of input costs occurred in carrying out various kinds of maintenance operations/ jobs in successive time periods. This is also called cost based model.

However, the main limitation/ disadvantage of this model is that it does not inform or help in determining the effectiveness of maintenance operations/ jobs in enhancing production outputs or reducing downtimes of the plants and equipments. Further, value of most of the measures are determined generally at the total plant level and, as such, productivity measures of individual group levels becomes little difficult.

Service Based Model: This model does not have the limitation of manufacturing based model and can be for plant/ organization level as well as for individual group or section level. The model is essentially based on monitoring and controlling separately for each of the two inter-related classes of activities i.e. activities as performed by maintenance department/ group and activities as needed by the production and other departments from the maintenance department/ group. These are, again, of two types— Quality based and Operation Based.

1. Quality Based Performance Parameters: These includes parameters/ attributes of quality characteristics which are related to production/ operation departments; e.g. availability, reliability, waiting time etc. These develop "Quality Based Indices".
2. Operation Based Performance Parameters: These include parameters/ attributes related to working of maintenance department or group; e.g. overtime, repair mean time etc. These develop "Performance based indices".

13.6 MAINTENANCE PERFORMANCE MEASURING INDICES

As indicated in the previous section and based on other sections of this chapter, the maintenance performance measuring indices/ parameters can be grouped in three basic categories—Quality Based Indices, Performance Based Indices and Cost Based Indices. Few commonly used indices of each category are given herewith

13.6.1 Quality Based Indices

Such as:

1. Availability (Equipment Availability, System Availability, Plant Availability): Already discussed in Sec13.2.
2. Equipment Reliability: These include MTBS (Mean Time Between Shutdowns), MTBF (Mean Time Between Failure), MTTF (Mean Time To Failure), MTTR (Mean Time To Repair), Failure Rate, MTBPL(Mean Time Between Production Loss), MTBM (Mean Time Between Maintenance) etc. There can be huge mathematical and statistical treatments for these but we will limit only to basics of these.

$$\text{MTBS} = \frac{\text{Total Operating Hours}}{\text{Number of Runs}}$$

More MTBS means better reliability and lower MTBS means lower reliability, may be due to operating stresses being higher or more corrosive conditions or inadequate PM and PdM etc. However, MTBS is not much used in repairable system in industries. Failure rate is normally defined as failures per unit time. MTBF is the mean (or average) time expected between failure of a given device and is normally measured in hours. Though MTBF is more used for repairable system, it is also used for non-repairable system. MTBF is reciprocal of Failure Rate.

$$\text{Failure Rate (Fr)} = \frac{\text{Number of Failures}}{\text{Total Operating Hours}}$$

$$\text{MTBF} = \frac{1}{\text{Fr}} = \frac{1}{\text{Fr1} + \text{Fr2} + \text{Fr3} \text{ -----}} \quad \left[\begin{array}{l}\text{Considering more}\\ \text{numbers of items}\\ \text{in circuit/ system}\end{array}\right.$$

MTTF is the mean time to system failure measured from a particular time and is normally used for non-repairable system. MTTR is the most common measure of maintainability. It is the average time required to perform corrective maintenance on all of the removable items in a product or system. It is also the average amount of time needed to repair a failed unit. This kind of maintainability prediction analyzes how long repairs and maintenance tasks will take in the event of a system failure.

$$\text{MTTF} = \frac{\text{Total number of operating hours for an equipment system/ complex}}{\text{Total number of system failures from a particular reference time}}$$

3. Equipment Maintainability: It is defined as he probability that a failed equipment can be repaired with in a given period of time. MTTR, discussed earlier, also comes in this category. Another indices is Mean Time Down (MTD), which can be calculated as below:

$$\text{MTD} = \frac{\text{Total hours Inoperable (Total Down Time)}}{\text{Number of Maintenance Actions}}$$

4. Equipment Utilization: It is different from equipment availability to the extent that an equipment may be available but may not be utilized for non-availability of operating personnel or operating tools, dies, facilities, consumables or non-availability of power or order etc.

$$\text{Equipment Utilization} = \frac{\text{Hours Equipment ran to capacity + off days + PM \& major overhauls}}{\text{Total Calendar hours during evaluation period}}$$

5. Manpower Efficiency: This may be calculated as below:

$$\text{Manpower Efficiency} = \frac{\text{Total Man-hours actually worked for the jobs}}{\text{Total Man-powers scheduled for those jobs}}$$

6. Reactive (Breakdown) Repair Ratio (%):

$$\text{Breakdown Man-hour \%} = \frac{\text{Total hours worked on breakdown \& emergency jobs}}{\text{Total hours worked in all maintenance jobs in that period}}$$

13.6.2 Performance Based Indices

There may be many such indices, but four such indices are as followings—

1. $$\text{PM and PdM Job Ratio} = \frac{\text{Number of equipments/items covered under PM \& PdM}}{\text{Total number of equipments/ items in that area}}$$

2. $$\text{Work-order Turnover \%} = \frac{\text{Total number of work-orders (jobs) completed}}{\text{total number of work-orders (jobs) received/handled}}$$

3. $$\text{Maintenance Overtime \%} = \frac{\text{Total maintenance overtime during a period}}{\text{Total plant/ department overtime during that period}}$$

4. $$\text{Indirect Man-hour \%} = \frac{\text{Total Indirect Man-hours during a period}}{\text{Total man-hours worked for those jobs during period}}$$

13.6.3 Cost Based Indices

There may be many indices/ parameters in this category and a few are mentioned below:

1. Maintenance cost per unit production, i.e. Cost/unit,

$$\text{Cost/ Unit} = \frac{\text{Total Maintenance cost per time period}}{\text{Total finished units produced during that period}}$$

2. $$\text{Maintenance Cost/ hour} = \frac{\text{Total Maintenance cost during a period}}{\text{Total direct man-hours applied during that period}}$$

3. Total Plant Maintenance Productivity (TPMP):

$$\text{TPMP} = \frac{\text{Plant output during a given period}}{\text{Total maintenance cost during that period}}$$

4. % cost scheduled maintenance

$$= \frac{\text{Cost of scheduled maintenance jobs/services}}{\text{Total production cost during that period}}$$

5. % Maintenance cost to Investment

$$= \frac{\text{Total maintenance expenditure}}{\text{Total Investment value}}$$

13.7 MEASURING MAINTENANCE EFFECTIVENESS

Today's environment requires measurements that can predict outcomes, not just measuring outcomes themselves. We need to be able to affect the final outcomes for the month by measuring interim indicators. In addition to the earlier mentioned techniques/ indices of maintenance effectiveness evaluation, few other measures can be considered for maintenance effectiveness; e.g.

1. Estimated Backlog in Crew Weeks,

2. Percentage of preventive, predictive, corrective and reactive (breakdown) jobs—This should be better than what is shown in Sec. 5.13 of this book.

3. (%) to schedule Compliance (SC): Ideally this should be at least 90%

4. Schedule Loading Factor (SLF): This measures the percent of available man-hours that are scheduled on a weekly basis. Having high schedule compliance, but few people scheduled makes little sense. A 90% load assumes that you are having about 10% emergent work on a continuing basis.

5. Wrench Time (WT): (Refer to Sec.7.3.2 of this book)- Wrench time is the measure of the percent of a workday that the craftsmen spend performing actual work, and is in reality a lagging indicator, but it is part of the overall measure of work or maintenance management effectiveness. It should ideally be around 65% but very often it is found to be around 30%.

From points (3), (4) and (5), a **Maintenance Work Capacity Index (MWCI)** can be worked out as below:

$$\text{MWCI} = (\%\ \text{SC}) \times (\%\ \text{SLF}) \times (\%\ \text{WT})$$

Considering SC as 90%, SLF as 90% and WT as 65%, MWCI should ideally be 0.53 (0.9 x 0.9 × 0.65), but very often the MCWI is 0.11, taking SC as 65%, SLF as 50% and WT as 35%.

13.8 MAINTAINABILITY INDEX (MI)

MI was originally proposed by Oman and Hagemeister and is comprised of weighted Halstead metrics (effort or volume), McCabe's Cyclomatic Complexity, lines of code (LOC), and number of other indices. It is a composite metric that incorporates a number of traditional source code metrics into a single number that indicates relative maintainability. It *has evolved into numerous variants and has been successfully applied to a number of industrial strength software systems and continues to be used. It is a q*uantitative measurement of an operational system's *maintainability* and an instantaneous measure and as a predictor of maintainability over time. It indicates when it becomes cheaper and/or less risky to rewrite the code than to change it. The US Air Force and few big industries used Maintainability Index (MI) as a measure of their success in their reengineering efforts.

However, MI can also be used for equipment and systems in industries to periodically check for maintainability and to compare or evaluate equipments/ systems. Comparing the MIs of known quality equipment/ system and a third-party system can provide key information in a make-or-buy decision.

13.9 MAINTENANCE AUDIT

Apparently, this section may appear somewhat similar or overlapping to Sec. 13. 7, but is discussed here to emphasize the topic further.

The Maintenance Audit is not necessarily a financial audit but is mainly a performance audit. It is yet another way to find out the performance and effectiveness of maintenance. However, it differs from the measuring parameters/ indices mentioned earlier to the extent that it does not give any percentage or ratio or rating of maintenance performance but it lays more emphasis on finding out shortcomings, losses and wastages etc in maintenance system so that corrective actions can be taken. Maintenance audit provides a formal, insightful evaluation of maintenance personnel compliance practice. An audit of a maintenance department will determine the efficiency and effectiveness of the existing operations, and will highlight the strengths and weaknesses. An audit is an invaluable planning/improvement tool, that allow us to most cost effectively and efficiently maintain and operate our industrial facility.

The Audit is divided into two segments. First is the Maintenance Operation, which measures maintenance personnel's compliance to the maintenance operation procedures and the maintenance process flow and the second part is Maintenance Plan Audit, which measures maintenance personnel's compliance to preventive and predictive maintenance programs and other repairs.

Maintenance Operation Audit: It mainly includes the followings

1. Management compliance to maintenance operation procedures,
2. Maintenance organization and communication,
3. Planning and scheduling compliance to planning process flow,
4. Plant Shutdown/ Turnaround procedure compliance,
5. Maintenance Personnel compliance to procedures and process flow,
6. Maintenance Personnel compliance with their roles and responsibilities,
7. Inventory Management compliance to process flow and procedure,
8. Cost/ Finances in compliance to the approved budget,
9. Contractor performance and procedures compliance,
10. Regulatory and statutory authority's compliance to inspection frequency and statutory/ regulated requirements,
11. Reliability Team in compliance to the maintenance operation procedures; Etc.

Maintenance Plan Audit: It includes mainly the followings:

1. Work Request /Orders,
2. Equipment failure history accuracy
3. Equipment life cycle profile tracking up-date,
4. Failure and Root Cause Analysis practice and compliance
5. Equipment Maintenance Plans/ schedules compliance,
6. Equipment Spare Parts Consumption and Rationalization compliance,
7. Maintenance improvement in compliance with the Reliability Improvement Plan,
8. Asset performance evaluation monitoring is in compliance with specifications; Etc.

Apparently, these two audits may look similar and are overlapping to some extent. All audits provide detailed and comprehensive documented reports that will supply you with a "snap shot" and facts about compliance. This enhances your decision-making. Each report may provide suggestions also, that will assist your solution process.

Conducting Audit: We can ask simple questions for such maintenance audit and assign values for assessing. Few of the questions are given below, as example:

1. Does the department maintain a complete machinery history of the facility and equipment? (All—10 points, Major—5 points, none or intermittent- 0 point)
2. Does the department use work standards, standard maintenance procedures and review past work to determine and implement time saving work methods? (91% ~ 100%— 10 points, 70% ~ 90% —5 points, Less than 70%— 0 point)
3. Is there a departmental safety program of safety training courses, including tagout / danger tags, lift-truck driving, lifting, electrical hazards, electrical explosive hazards, electrical arc flash hazards, welding, compressed gases, and other state mandated safety courses.? (All or 100%— 10 points, less than 100%- 0 point); Etc.

The logical summation of all the points will indicate the health and effectiveness of the department. Companies can either conduct their own maintenance audits or outsource the job to one of the many maintenance audit specialists available. Those conducting the audit should observe demonstrations of all systems applications, and attend any functions or meetings pertinent to plant maintenance. The auditor should also assess how processes, storage, and documentation are managed.

13.9.1 Technical Audit

Technical audit in maintenance mainly relates to audit of centralized facilities, services and captive shops of the maintenance organization. The generation and/or storage, as well as distribution of such facilities/ services are, normally, the responsibility of maintenance organization. Such audits are very useful for shop level and middle management level.

Followings are some of the areas where such technical audits are used:

1. Thermal Audit (Steam Supply etc),
2. Electrical Audit (for electrical energy),
3. Other Energy Audits; e.g. Fuels (Solid, Liquid and Gasses), Oxygen Supply, etc
4. Industrial Water Usage Audit,
5. Compressed Air Audit etc,
6. Plant Hydraulic System Audit,
7. Plant Centralized Lubrication System Audit; etc.

The main steps for such audit are finding out all input sources, all output sources, estimating evaporation losses, condensation losses and other such known losses and balancing all these data in the form of a equation to find out the unaccounted losses. Considering a system of two input sources and two output sources, the balancing equation can be as follows—

[Input No. 1] + [Input No. 2] = [Output No. 1] + [Output No. 2] + [Evaporation/ Condensation Losses] + [**Unaccounted Losses**]

The next step would be to find out the sources of unaccounted losses and take suitable remedial measures. Efforts are also made to reduce evaporation and condensation losses through innovation, minor redesign or other possible measures.

13.9.2 Benefits of Maintenance Audit

Few benefits of Maintenance Audit are given below:

- Develop a corrective plan for improving personnel compliance,
- Enhance your decision-making; helps in prioritizing or justifying decisions,
- Gives tools to negotiate at higher management levels,
- Provides input for future budgeting,
- Motivate maintenance personnel to achieve the level of professionals,
- Shows chances of increase in resource utilization;
- The audit report allows you to target areas of opportunity without re-engineering; Etc.

However, audit normally means financial audit. As such, 'Maintenance Financial Audit' will be discussed in next chapter.

13.10 MAINTENANCE BENCHMARKING

Another modern concept for evaluating maintenance effectiveness is maintenance benchmarking. A benchmark is a point of reference for measurement. It is a very versatile tool that can be applied in a variety of ways to meet the range of requirements for improvements. It starts with having a current benchmark as to where you are with your maintenance processes and then successively evaluating as you go on improving and adopting CMMS, EAM and other advanced techniques. Maintenance benchmarking can be of following types:

1. Strategic Benchmarking: This improves an organization's overall performance by examining the long term strategies and general approaches that have enabled the high performers to succeed. It involves high level aspects such as core competencies, developing new services/ products, improving capabilities by changing departmental work culture and environment etc. This type of benchmarking is difficult and will take more time materialize.

2. Performance benchmarking or competitive benchmarking: It normally considers the physical asset management. The total organization/department considers its position in relation to performance characteristics of key product and services as compared to benchmarking competitors, such as contract maintenance providers or other similar organizations. Such benchmarking is not done by the in-house maintenance organization, but is done by third party (trade association or others) to protect confidentiality.

3. Process benchmarking: This focuses on improving specific critical processes, procedures and operations. Benchmarking partners (competitors) are sought from best practice organizations that perform similar jobs or deliver similar services. This type of bench marking can result in both, short-term and long-term benefits. This benchmarking should go beyond the analysis parameters but should go upto successful implementation and measured results. Predictive maintenance technology, CMMS, TPM, Lean Maintenance etc can, again, be used in this.

4. Functional benchmarking or generic benchmarking: This benchmarking is done amongst the partners (competitors), drawn from different organization (public/private/ vendors), which may have similar and also different maintenance environment. Comparing parameters from different organizations (e.g. process industries, pure facility maintenance, Fleet/ transport industries, healthcare facility maintenance etc), having different maintenance environment, may lead to innovation and dramatic improvement.

5. Internal Benchmarking: It involves seeking the partners (competitors) from the same big organization, with units located at different places. The main advantages of such benchmarking are twofold; (a)- access to all required data and information and standardized data are readily available and (b)- generally less time and resources are needed. However, in this case, real innovation may be lacking and best in category performance may be found through external benchmarking outside the organization.

6. External benchmarking: Organizations, that are known to be best in the category, give opportunities for learning to those who may not be that good. External benchmarking keeps in mind that not all best practices and solutions can be transferred to others. However, such benchmarking may take more time and resources to ensure the comparability of data and information, credibility of findings and development of sound recommendations.

7. International benchmarking: This is similar as external benchmarking, except that the competing partners are selected from the known best practitioners all around the world, with only few organizations from the same country, to produce valid results. However, working with this type of benchmarking is very difficult.

QUESTIONS

1. What is equipment availability and what are the three basic approaches to define and quantify availability?
2. How do you quantify availability in terms of MTBF and MTTR? What are Steady-state Availability and Operational Availability?
3. What is Overall Equipment Effectiveness (OEE)? What are other ways to measure maintenance effectiveness?
4. What do you mean by Key Performance Indicators (KPI) and mention few KPIs for maintenance?
6. What do you mean by Maintenance Performance Measuring Indices? Mention few such indices.
7. Explain very briefly MTBS (Mean Time Between Shutdowns), MTBF

   (Mean Time Between Failure), MTTF (Mean Time To Failure), MTTR (Mean Time To Repair) and Failure Rate.
8. What is maintenance audit and how does it differ from financial audit? Explain briefly 'maintenance operation audit' and 'maintenance plan audit'.
9. What is technical audit in maintenance? Give one sample example.
10. What do you understand by 'maintenance benchmarking'?

CHAPTER

14 MAINTENANCE BUDGETING, COSTING AND COST CONTROL

One universal measurement of maintenance performance, and perhaps the measure that matters most in the end, is the cost of maintenance. Maintenance Costs is generally the total cost in doing and getting done the maintenance jobs and keeping a maintenance organization. In addition to maintenance and upkeep plant's equipments, facilities and utilities, often, maintenance department is also charged with the responsibility of construction, modification, rearrangement and buildings, equipments and other assets. The treatment of costs, associated with such activities, depend the nature of jobs undertaken.

Costs of inadequate maintenance or no-maintenance are in the nature of "**Opportunity Lost Cost"**, which are, normally, not recorded in any system of cost accounting as part of regular reporting system. "Opportunity Lost costs" is the differences between what that company could have produced and earned throughout the lifecycle or designated shorter period and what is actually earned as a result of no or improper maintenance action during that period. It may also include cost/ loss due to lower rate of output, poor quality of products, wastages, defectives, damage to equipments and reduction in useful life of equipments, arising from inadequate maintenance. These are, often, likely to be overlooked in exercising As Such, a well conceived maintenance policy must minimize the total maintenance cost and also cost of inadequate maintenance.

14.1 SOURCES OF MAINTENANCE COSTS

Sources of maintenance costs are all the actions taken in connection with the maintenance and upkeep of various equipments, systems and plants in any industry/ organization, many of which already mentioned earlier I this book. It also includes the results of no-maintenance or opportunity-lost costs as mentioned earlier. Concerned plant's representatives should be vigilant enough to identify those sources.

As a guideline, some of such sources of maintenance costs are shown in Fig. 14.1.

14.2 CLASSIFICATIONS/ TYPES OF MAINTENANCE COSTS

Different maintenance costs are of complex nature and, so, generally following three classifications are used to identify and account those costs for different purposes

14.2.1 Fixed and Variable Costs

For planning and control purpose, maintenance costs are generally divided into two types/ categories—"Fixed" and "Variables"; in the same way as they are used in other departments like production and sales etc. Of-course, many maintenance costs fall in grey or overlapping area of "Semi-variable".

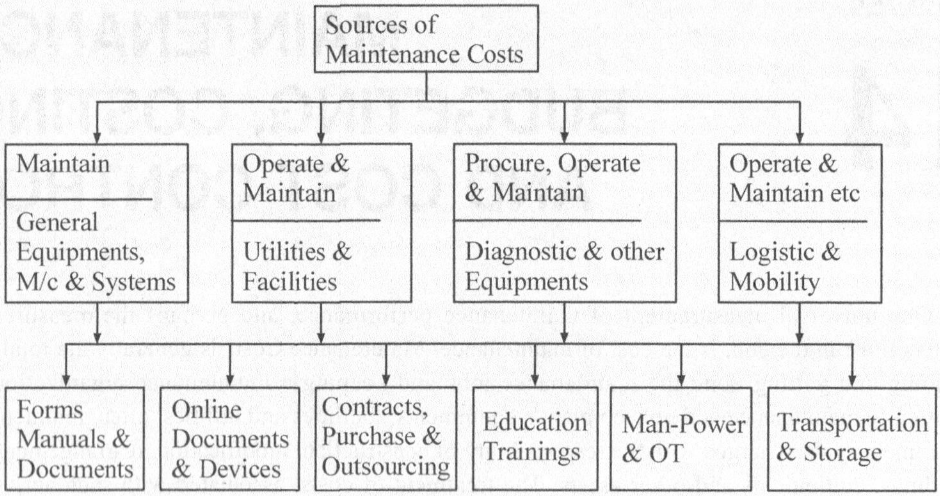

Fig. 14.1: Sources of Maintenance Costs

A plant/ industry, generally, has to have a maintenance department/ section/ service, with necessary tools, facilities/ equipments with necessary workmen/ technicians and supervisors in readiness to attend breakdowns, urgent repairs/ adjustments and to render other services, demands of which may arise in course of operation of equipments or as demanded by the operating personnel the equipment. Such "ready to serve" men and facilities are "Fixed costs", which have little relationship with the actual amount of work done during a given period of time. However, other maintenance costs, such as cost of major repairs and overhauls and costs of maintenance spares, materials and facilities etc, tend to vary directly with the level/ amount of maintenance activities. These costs are termed as "Variable costs".

Some maintenance actions like small PM, PdM, and condition monitoring etc fall in "Semi-variable cost" type/ category, as they vary only little and again not in proportionate to the level of production.

This classification can be further divided depending upon two types of assets; one is "*Maintenance cost for Fixed Assets*" which covers buildings, shop structures, drinking water, electrical power for buildings and shops etc, which has little relationship with scale of operation/ production or output. The other one is "*Maintenance cost for Production Assets and Facilities*", which covers maintenance costs of all other equipments and facilities, including raw water or production water, electrical power for equipments and fuel costs etc and they tend to vary with the level of production. This cost need to be analyzed more seriously and controlled.

Often, for shorter period, the relationship of maintenance cost to production cost gets complicated because of management's deliberate policy to defer maintenance activities in dire times owing to the lack of funds or with a view to avoid higher maintenance cost in a period of low productivity. As such, the consideration period may be large.

14.2.2 Direct/ Indirect (Apparent/ Hidden) Costs

In this classification all costs are divided into "Direct" or "Apparent" costs and "Indirect or Hidden" costs. Direct cost can be defined as budgeted cost, affecting the cash flow of an organization. " Direct Maintenance Costs" are those which are essentially required to keep the equipments and systems operable. The results from periodic inspection, condition monitoring, PM, PdM, corrective maintenance, major repairs and overhauls etc come in this category. These are easily visible and apparent. In direct maintenance cost may be defined as loss related costs as well as opportunity costs, affecting the profit of an organization. Indirect cost may be labour and material overheads that contribute to overall maintenance costs through activities such as administration, controlling, monitoring, planning, testing, human resource development and tools preparation etc. "Indirect maintenance costs" are again of two types; "Extra costs" and "Penalty costs" and are generally not clearly visible. Extra include extra costs for standby equipments to put on line when primary equipment is inoperable or is under maintenance, standby equipments for regular rotation and many other such costs. "Penalty costs" include production loss when primary equipment is out of service and there is no standby, excess inventory, waste generation and many others.

Indirect costs are also called "**Maintenance Burden**".

The interactions between these three costs are complex. Again, the direct cost and extra cost are controllable but the penalty cost is not so controllable.

Direct cost of maintenance is, probably, a tip in the iceberg, as compared to indirect cost. Fig. 14.2 shows the maintenance cost iceberg.

Direct Costs	Materials Maintenance Labour Overtime etc Contract Services Overheads etc	Mostly Apparent
Indirect Costs	Start up losses (Set-up, Start-up) Missed Schedules Ripple effect on production Cost of idling or running at reduced speed Excess Inventory Crisis management Emergency purchases Waste generation Extra cleanup Customer complaints Regulatory/ statutory Fines Lost Business Any extra environment control Any product liability; Etc.	Mostly Hidden

Fig. 14.2: Maintenance Cost Iceberg

14.2.3 Revenue/ Capital Costs

Another way of classifying the total maintenance cost as either "Revenue Expenditure/ Cost" or "Capital Expenditure/ Cost". Revenue expenditure includes all costs which are covered by revenue budget allocated for various normal maintenance activities for a group of equipments / sections or shops. In revenue budget, some minor adjustments of cost /expenditure is possible from one equipment to another or from one type of job to another. Revenue budget is normally allocated for one year and almost all normal maintenance jobs are to be done within that. Any leftover revenue budget is, normally, not carried over to next year.

Capital expenditure /cost are covered by capital budget, which are specifically sanctioned for specific job on specific equipment, generally in connection with "Addition, Modification and Replacement (AMR)" schemes, other new jobs or few jobs of heavy nature, which are generally not taken in regular course of maintenance. No adjustment of capital budget is permissible from one equipment to another or from one job to another.

14.3 MAINTENANCE COST COMPONENTS

Basing on classification and types of costs, mentioned in Sec. 14.2 and also shown in Fig.14.1 and Fig. 14.2, maintenance cost includes huge number of activities, few of which are again enumerated herewith

1. Cost of "ready to serve men, tools and facilities etc,
2. Cost of men, equipments and facilities for CM and other diagnostic monitoring,
3. Cost of PM, PdM, and CBM etc, intended to minimize breakdowns,
4. Cost of corrective maintenance and other such repairs,
5. Cost of major repairs, overhauls, revamping and other such activities,
6. Cost of materials (spares, consumables, tools, tackles); cost of storing and transportation,
7. Cost of diagnostic and monitoring equipments/ tools,
7. Cost of Overtime for different reasons,
8. Cost of captive shops and other central services/ facilities, attached to maintenance organization/ department,
9. Cost of external agencies and services for monitoring, testing, calibration and guiding,
10. Cost of other direct and indirect men, including clerical staff,
11. Cost of contractors and other maintenance outsourcings,
12. Cost of, books, literatures, standards, guides purchased, cost of stationary and printing various formats; Also cost of PC, lap tops, palm tops and other such connected devices.
13. Cost of training and development of maintenance personnel (inside or outside plant),
14. Regulatory and Statutory expenditures and fines and cost for safety and environmental Protection;
15. Insurance liability for automobiles and other equipments, health and life of workmen and workers compensation;
16. Non-working" maintenance labor (e.g. maintenance safety meetings, waiting time, etc)
17. Maintenance work done by Production employees, e.g. cleaning, inspections, replacement of "production" components (e.g. filter media, etc) and perhaps some lubrication. These are normally considered "Operation Cost" but are considered as maintenance cost when making comparison,
18. Non-capital plant improvements- This includes both process and reliability improvements, etc.

14.4 MAINTENANCE COST ANALYSIS

Maintenance cost analysis is a type of failure analysis. The ultimate measure of the reliability of equipment is the cost to maintain it. Maintenance cost typically includes the cost of labor and materials to perform repairs and, sometime, cost of downtime, production loss and idle employees etc. It also identifies the failure modes that contribute most to maintenance costs. As one example examining how maintenance man-hours were distributed on their shovel fleet (over a period of two years), it was found that the maximum (44%) of the maintenance hours were allotted to the shovels were spent maintaining the hydraulic system, others being low volt electrical, air system, track, track frame and wear components etc. After further, in-depth analysis, it was found that the maximum cost (71%) was spent on maintaining the pumps, valves and cylinders. Having identified the largest cost center for the hydraulic shovel fleet, the reliability team's next move was to determine the most efficient way to reduce expenditures on pump, cylinder and valve maintenance. To do this, the root cause of the accelerated pump, valve and cylinder wear had to be determined. Database containing three years of oil sample data for each shovel was also dugout. It was discovered that the shovel fleet was averaging an ISO fluid cleanliness level of 21/20/16. Since the system operates at pressures in excess of 275 bar, Ii immediately became obvious that steps had to be taken to reduce the contamination levels, because the typically recommended fluid cleanliness levels for a high-pressure system with spool valves and piston pumps are around 16/14/11 to 17/15/12. This solved the problem.

The advantages of maintenance cost analysis may be summed up as below

1. Understand the costs to maintain equipment,
2. Identify failure modes that contribute most to maintenance costs,
3. Identify opportunities to reduce cost,
4. Develop a realistic view of the value versus cost of spares inventories,
5. Develop reliability model directly from maintenance records; Etc.

14.5 MAINTENANCE COST-BENEFIT ANALYSIS (CBA)

It is somewhat similar to cost analysis. Cost-Benefit Analysis is a relatively simple and widely used technique for deciding whether to make a change. As its name suggests, to use the technique simply add up the value of the benefits of a course of action, and subtract the costs associated with it. In its simple form, cost-benefit analysis is carried out using only financial costs and financial benefits. Maintenance actions can be programmed as a control problem in which the expected net present value of an infrastructure during the project or equipment life cycle should be maximized. However in maintenance scenario, the intangible benefits and other such benefits can be converted back to some sort of financial benefit to use this technique. Even the loss of human life or environmental hazard etc can be converted and assigned financial value. The purpose of a CBA is to promote efficient resource allocation through well-informed decision-making system by higher level management. It estimates and totals up the equivalent money value of the benefits and costs to the community of jobs/ projects to establish whether they are worthwhile.

Cost-Effectiveness Analysis is, again, another name for same or similar technique.

14.6 COST CONTROL IN MAINTENANCE

Cost control is not only cost reduction. The **purpose of cost control** in maintenance may be as followings

(*i*) Cost reduction, wherever possible,

(*ii*) Equitable distribution of repair costs over the department serviced,

(*iii*) A source of information necessary for sound administration of the maintenance department. Also works as one of the indicators for measuring maintenance effectiveness.

(*iv*) Compliance with legal requirements for taxes and earnings, (*v*) A source of information for plant accounting group in its function of recording and reporting financial position of the plant.

One of the most effective cost control system is the conventional job cost method which accumulates expense items for labour, supplies, services etc on a specific job number which, in turn, is the liability of a specific department.

Companies today are placing greater emphasis on controlling costs in all aspects of its activities. While focusing on human capital cost control opportunities in the areas of benefits, compensation, utilization and other key human capital areas, it also focuses on other areas like materials, services and administration etc. It also helps in implementing appropriate strategies to maximize those areas and helps companies improve efficiencies, eliminate waste and enhance the bottom line.

Maintenance plays a great part in any industry and has enough, rather continuous scope for cost control. It basically needs strict discipline (technical and otherwise) in carrying out the jobs. Evolving good standard operating practices (SOP), standard maintenance practices (SMP), work instructions and job manuals are prelude to that. Few specific principles/ areas of maintenance cost control are given herewith

1. Keep breakdown cost separate from PM, PdM and CBM costs and ensure that ratio is not exceeded. Breakdown maintenance cost tries to maintain the status-quo and do not add any value to the equipment/ process/ product. If the distinction and the ratio between the two types of cost are not controlled, tendency will be to take more break-down or ad hock jobs and defer other jobs to reduce total cost and that would be very harmful in the long run.
2. Extra cost-consciousness on part of maintenance engineers does not often bring in good result. Extra cost-conscious persons tend to use inferior (but cheaper) materials and replace the parts only it has completely worn-out and beyond reclamation, both being bad in the long run. They should be more quality-conscious, which automatically leads to cost-consciousness.
3. Cost prevention must precede cost control in maintenance. The first task for maintenance personnel is to prevent breakdown cost by improving technical systems and practices and taking preventive action in time. Again all preventive and predictive maintenance cost to be analyzed for their essentiality. Thus, breakdown cost to be prevented and preventive/ predictive maintenance costs to be controlled.
4. Planning and installing new cost reduction measures, which are also known as cost control measures on the basis of performance analysis, failure analysis and RCA etc. Such cost reduction measures call for improvements over existing equipments/ systems, where earlier mentioned cost prevention has limited scope.
5. Waste Reduction and waste recovery: Lot of waste materials are generated in course of maintenance, e.g. used engine oil, lubrication oil, welding electrode ends/ butts, used tool-bits, small cut plates and industrial water etc. Various measures can be taken to reduce such wastes; such as using high performance and long drain interval oils to change oils less frequently, using more automatic and semiautomatic welding with wire feeder in place of manual welding, using proper planning and templates before cutting plates, necessary inline cleaning and recycling of industrial water etc.

   Even after taking waste reduction measures, some wastes will be generated. Waste Recovery measures include collection of such wastes and recycling/ recovering those so that those can be used subsequently either in place of original application or for other applications. Recycling (refining and regenerating used industrial oils (lubricating, engine or hydraulic oils) and reclamation of tool-bits etc are few examples.

6. Salvaging of spare sub-assemblies: Though repair of components and subassemblies is a regular maintenance job, salvaging involves bigger and intricate assemblies or subassemblies which, normally, can not be done within the plant. Today many costly subassemblies can be salvaged by techniques mentioned in chapter3 or otherwise. Salvaging brings huge cost saving. However, salvaging jobs are only taken if the salvaging cost is not more that 45% of cost of new part or the part is "difficult to procure" or has become obsolete; etc.

14.6.1 Cost Control Presumptions

However, before attempting any cost control measures, certain presumptions must be kept in mind, such as:

1. There is certain minimum amount of money, that must have to be spent on maintenance process to keep the plant going,
2. The key issue is not 'how much money we spend' but 'what we spend it on' and 'why'?
3. Not all variables in maintenance environment can be clearly quantified. In this category, knowledge and skill can, somehow, be quantified but attitude can not be quantified expect assessing as positive or negative.
4. Random elimination of 'wastes' or 'buffers' or 'spares' or 'peoples' will tend to reduce the maintenance process 'insurance variables' but will also increase the 'risk'.
5. Cost represents only a very partial (financial) viewpoint of the whole issue, i.e. 'performance' or 'performance improvement'. This may include items like 'better results with same men and inputs or eliminating waste (process model', 'speed gain or quality gain (stratification model)' etc.
6. A number of 'key variables' in the maintenance environment are, often, completely outside of effective control of maintenance management. This implies limited control over their cost. This may include items like, 'labour practices', 'interdepartmental relations', 'accidents or environmental hazards' etc.
7. Cost is not a stand-alone variable. It forms an inseparable part of the 'trilogy'—Cost-Risk- Performance. Change in any of the three will result in change of the others also.

14.7 MAINTENANCE BUDGET

Budget is a forecast of expenditures and revenues (i.e. itemized estimate of income and spending) for a specific period of time It is the list of proposed expenditures and expected receipts of any person, enterprise, industry or government for a specified period, usually one year. Budget estimates are based on the expenditures and receipts of a similar previous period, modified by any expected changes.

Budgeting of maintenance costs is very essential and also very beneficial to control maintenance costs as maintenance is a service organization and, probably, no better way is available to control its cost. Other benefits of maintenance budgeting are

1. It improves the system effectiveness and efficiency of maintenance organization. Increase of expenditure over budget is. Generally, because of inadequate/ wrong/ untimely planning.
2. As maintenance personnel know their budget in advance, they plan their expenditures (spares, materials, contracts, jobs etc) judiciously and timely so that no job is held up for shortage of funds,

3. It is very effective technique for projecting future and additional requirements of fund.

However, disadvantages and problems are faced if budgeting has not been done properly and with foresight and if the budget is not flexible. In that case, maintenance in-charge has to run from pillar to post to carry some urgent job and also faces lot of delay.

14.7.1 Types of Maintenance Budget

As already mentioned in Sec.14.2.3, maintenance budget are of two types; Revenue and Capital. As a further classification, budget can be of following three types

1. Appropriation budgets, which sets a lump-sum as the maximum amount that can be spent for a given item,
2. Fixed budgets, which specify the allowable amount of cost for a period of time,
3. Flexible budget, which relate the allowable cost to some measure of activity.

Budgeting for maintenance cost would require a combination of all the three types mentioned above. Some maintenance jobs (e.g. capital repairs, major overhauls, rearrangement of shops/ plant's equipments etc) are best planned as "projects" and appropriation budgets should be used for such jobs. This is somewhat similar to "capital budget", mentioned earlier. Cost of breakdown jobs can not be forecasted exactly. Bulk of such costs take place in form of maintaining labour and facilities as "ready to serve" costs and these are best budgeted as "fixed budget". Costs for routine maintenance, corrective maintenance and to some extent, PM, PdM and CBM etc, which can be expected to vary with respect to level of production/ operation, should be budgeted as "flexible budget". The combination of fixed budget and flexible budget may be treated as "revenue budget" mentioned earlier.

14.7.2 Few Tips for Maintenance Budget

Based on the findings of some private survey of Maintenance costing and budgeting practices, following tips are recommended

1. Budgets are prepared based on previous year's budget with adjustments to meet Management cost targets, changes in production throughput and adjustments for price/ cost inflation etc. It can start zero-based also, if no previous history available;
2. In general, maintenance budgets appear to be prepared with a moderate level of detail,
3. Maintenance Managers or maintenance in-charges are most involved in the budgeting process, with maintenance planners and supervisors being likely to help the maintenance manager/ in-charge in the process. As such, maintenance manager/ in-charge own the budget.
4. Majority of budgets are prepared or revised annually, but many organizations update their budgets more often than this.
5. Majority of cost reports are issued monthly, with weekly being the next most popular reporting time..
6. Maintenance managers/ in-charges are most likely to receive regular maintenance cost reports. Shop floor maintenance personnel are more likely to be informed by maintenance in-charge.
7. It forecasts items by revenue, personnel, operating expenses, assets; again it forecasts by cash or units.
8. It creates rolling forecasts using multiple scenarios,
9. It considers user definable revenue recognitions schedules,

14.7.3 Preparation of Maintenance Budget

Generally following three basic approaches are used for preparation of maintenance budget

- Allocate revenue and expenses using top-down/bottom up,
- Production/ maintenance Schedule Approach,
- Labour Allocation Approach
- Integrated Approach

The three approaches are self explanatory. Individually, the production/ maintenance approach and labour allocation approach may not be very precise as, working alone, both approaches may face some uncertainties, such as hidden waiting time from one process to another, variation in skill requirements and skill available etc. As such, a combination or integrated approach is, often, used for preparation and presentation of the maintenance budget. In addition to various connected cost components, it also includes some cash budget for contingent expenditures which may come up unexpectedly.

Maintenance budget may be in the form of a sheet or booklet, presenting in tabular form. Table 14.1 shows one such budget for as revenue budget for Mechanical Engineering Dept. of one small steel plant.

14.7.4 Misfit items in maintenance budget

Every budget contains some items that could be called 'misfit' or 'out of place'. However, whether an item can be called 'misfit' or not, depends on how the budget is planned, how specific areas are categorized and how certain items are defined. For example, the area of 'outside repairs' could easily be placed under the heading 'miscellaneous purchases' and, thereby, not clearly defining the service to be rendered. Thus, an item, which should normally not be misfit or out of place, becomes so by virtue of the definition given to it. As such, while preparing or reviewing budget, we should guard against such misfits by proper definition or categorization of different expenses.

14.8 BUDGETARY CONTROL

It is a **system of financial control, i.e.** a system for controlling the finances of a business by comparing actual income and expenditure with forecasts in order to monitor whether budgets are being maintained.

Management is often faced with a situation in which funds are limited. In maintenance, this requires that some work be deferred. In many cases, due to the time framework for the decision, not much thought is given to the procedure for addressing which projects should be deferred.

14.8.1 Few Tips on budgetary Control

Few such tips are given herewith

1. Your budget must be a vibrant document, one that is well considered and that is used on an ongoing basis. Use it as a historical reporting system. Check the important elements (direct costs as one example) far more frequently.
2. Use a line item method. For the purpose of budget or cost control, you may want to break down general cost items into subcategories; e.g. utilities in gas, electricity, water and sewerage etc and office expenses into stationeries, equipment leases, postage and temporary help etc.
3. Breakdown yearly budget in small measurable interval (generally monthly)
4. Monitor on regular basis; the period may vary in case to case, monthly or weekly.

Table 14.1 Revenue Budget for MED, SAIL, for 1994~95

Sl.	Item-wise description	Budget (in Rs.)
1.	Contractor's expenditure for general fabrication, repair and maintenance	—
2.	Contractor's expenditure for major fabrication and repair (ladles and charging buckets etc)	—
3.	Contractor's expenditure for regular cleaning and repair jobs (running contracts) e.g. periodical cleaning of M/cs, surroundings and pits, cleaning of vacuum ducts, diffuser tubes and nozzles, condensers and heat exchangers etc	—
4.	Contractor's expenditures for statutory overhaul, repair and verification jobs; e.g. boilers, pressure vessels etc	—
5.	Regular services (Regular/ periodical inspections jobs — * Weigh M/cs, load cells, standard weights etc (m/s. Avery) — * Pneumatic carrying system (m/s. Jost) — — — * Specific engine overhauling jobs (camshafts, crankshaft grinding, cylinder sleeve fitting, fuel injection pump calibration etc) * Vibration monitoring and Lube monitoring etc— — —	 — —
6.	Consumption of grease, lube and hydraulic oils, special lubricants etc	—
7.	Consumption of acids, alkalis etc	—
8.	Consumption of mechanical spares (this has to be broken down at shop/ section wise and major equipment wise for detailed monitoring	—
9.	Consumption for electrical. electronic and other items for MRS etc	—
10.	Purchase of stationeries, computer peripherals, manuals, standards etc	—
11.	Consumption of fuels - * Coke Oven Gas — — — * Furnace Oil — — — — — * LPG and Petroleum coke — — — — * Oxygen etc — — — —	 — — —
12.	Overtime —	—
13.	Administrative Expenses — * Cycle, motor cycle, car advance, — * Leave travels (LTC and LLTC) and conveyance allowance etc — * Leave encashment and festival advance etc — — —	— — — —

Compare the actual expenditure with average and production pro-rata basis etc.

5. Delegate and identify the amount of money budgeted for each area or section manager and ask them to create a list of priorities,
6. Be Prepared to Sacrifice. Don't make the mistake of choosing short-term satisfaction at the risk of long-term stability; etc.
7. Periodical meetings with concerned agencies to correct deviations etc.

Table 14.2 shows the budgetary control of Mech. Engg. Dept of one steel plant.

Table 14.2 shows the budgetary control of Mech. Engg. Dept of ASP

Sl.	Expenditure/ Consumption	Average Proportion-ate Budget	Production Pro-rata Budget	Actual Consump-tion	Deviation	Remark
1.	Lubricating Oil	—	—	—	—	—
2.	Hydraulic Oil	—	—	—	—	—
3.	Furnace Oil	—	—	—	—	—
4.	Mechanical Spares	—	—	—	—	—
—	— etc —					

14.9 MAINTENANCE FINANCIAL AUDIT

An audit is a professional, independent examination of a company's or department's financial statements and accounting documents according to generally accepted accounting principles. Maintenance audit is already discussed in some details in Sec 13.9. Accordingly, maintenance financial audit should review and assess the maintenance budgets and budgetary controls measures etc.

Maintenance financial audit should lead to optimize maintenance cost rather than just minimizing it and should consider the expenditures on knowledge based systems (CMMS, TPM and others) as essential tools.

However, financial audit is the function of separate financial department and will not be a part of this book.

QUESTIONS

1. What is maintenance cost and what are its sources?
2. Explain briefly different types/ classes of maintenance cost.
3. Explain briefly what do you mean by 'Maintenance Cost Iceberg'.
4. What is cost control? Give few specific areas/ principles for cost control in maintenance.
5. What is maintenance budget? How maintenance budget is prepared?
6. What is budgetary control in maintenance and give some idea how is it done in industries?

CHAPTER

15 TRAINING (HRD) FOR MAINTENANCE PERSONNEL

Today many companies are finding it difficult to attract qualified technicians for manufacturing technology, maintenance, repair and service functions. The shortage can be attributed to many factors; one is that not enough young people pursuing careers in manufacturing/ maintenance technology-related fields as it appears little dirtier in nature and IT related jobs or other softer and glamorous jobs looks more attractive. As such, few companies often resort to stealing them from original equipment manufacturers (OEMs) or other companies with similar equipment, because they also do not want to take trouble of training their workforce. But, if more companies would begin training maintenance, repair and service technicians, the problem would eventually solve itself.

Management Consultant Peter Druker defined the knowledge worker as a person who has formal education but may require manual dexterity skills to perform a job. This description fits the maintenance worker of today or tomorrow, who will see the knowledge content of the work continue to increase and who will thrive only if he or she uses their wits and keeps adding to their skill base.

Training or human resource development (HRD) was not given due importance earlier as maintenance was considered as secondary job. But lately it is being appreciated that training is very essential for maintenance personnel because

- Maintenance personnel have to encounter more uncertainties and unpredictable situations,
- Maintenance jobs are, generally, not repetitive,
- Maintenance is a multidisciplinary function,
- Maintenance repair technology is changing very fast and new developments are taking place regularly in that direction,
- Often maintenance personnel are confronted with situations where many necessary materials/ inputs are not available and ingenious or makeshift ways have to be used.
- Often the mistakes/ omissions by maintenance personnel is not visible immediately, but affects more seriously later.

Training increases the effectiveness of maintenance personnel in both, the human behaviour aspect and technical and skill aspect. It gives a clear picture of what is expected of them. Training also creates a feeling in maintenance personnel that their skills are important, needed and highly valued. For maintenance manpower effectiveness and manpower potential, following formula can be used

$$\text{Manpower Effectiveness} = \text{Manpower Potential} \times \text{Application Factor}$$

The application factor is the product of moral, skill, motivation and other management factors and is always less than one. The job of training and human resource development is to increase this application factor to as near to one as possible, so that the effectiveness of maintenance manpower improves towards their potential and they are able to give more and better output. If our full manpower potential is not being used, we are wasting a part of our resources of skill and time and that may add up to considerable loss to our organization. The calculation of above formula is somewhat hypothetical, as it is difficult to quantify different factors, but it surely highlights the importance of training functions.

15.1 HUMAN FACTOR IN MAINTENANCE

Human error in maintenance can impact the safety and performance in a number of ways. A maintenance person, who is well trained, motivated, under no undue pressure, given correct information and working with equipment which has been designed to be maintenance friendly, will most likely complete all specified maintenance jobs to a high standard. As with most types of work, the scope for human error in maintenance operations/ functions is vast. These can range from becoming distracted and forgetting important checks to knowing deviating from a proper work procedure in order to save time or to get the job done in unexpected circumstances. There are a number of factors which influence the behaviour of maintenance crew and the likelihood of human errors need to be considered, both in terms of safety of people and damage to equipments/ systems or subsequent breakdown.

The human errors may be in form of Slips and lapses, Mistakes and Violations etc. There are many human factor issues which can impact the safety and maintenance performance, e.g. safety commitments and perceived impracticality of any safety rules, communication, knowledge and skill, plant and equipment ergonomics, working condition, job designs, standards and monitoring of supervision, complacency and management style etc. Most of these issues can be identified, reviewed and improved by suitable training and awareness.

Like in few other areas, maintenance workmen must have following physical attributes

1. Natural or corrected near-distance visual acuity,
2. Colour vision sufficient to discern system and component colour coding,
3. Hearing sufficient to respond properly to audible alarms and use communication system,
4. Sufficient strength, power, range of motion and dexterity to allow access to facility etc.

15.2 BROAD MAINTENANCE TRAINING LEVELS AND PROFILE OF PERSONS

Meeting the diverse training needs of the maintenance workforce is impossible to accomplish through one "canned solution". As illustrated in Fig 15.1, the training needs of the maintenance workforce are multi-tiered, specifically of following four levels

1. **Level-1 (Core Competencies), which** include knowledge and skills such as mathematic, reading, print reading, hand and power tools, and mechanical and electrical theory etc,
2. **Level-2 (Craft Fundamentals), which** include skills and knowledge in basic hydraulics, pneumatics, pumps, PLC introduction, motors, etc. Much of this training can be delivered via self-study material. This training should be supplemented with hands-on training and structured on-job-trainings (OJTs),
3. **Level-3 (Advanced Craft Training), which** can include topics on advanced hydraulics, PLC programming and troubleshooting, and advanced techniques associated with pneumatics and facility-specific equipment etc. Historically, the only cost-effective way to deliver this training was via an instructor with print-based materials; however, with the

technology currently available, this training may be delivered via computer-based training (CBT) or distance learning. For maximum effectiveness, this training must be supplemented with hands-on training and structured on-job-trainings (OJTs)

4. **Level-4 (Equipment-Specific Training) which** covers the particular operation and maintenance of equipment used at the facility. Training materials can be developed from vendor literature and prints and delivered by an instructor. Equipment walk-downs should supplement the classroom training coupled with detailed OJTs or supervisor-led hands-on training.

Irrespective of plant or industry, following are the broad profile of the maintenance personnel need to be trained. These may enter at bottom level or other suitable level (depending on their previous skill etc) of Fig. 15.1.

1. Fresh entrant in industry (unskilled workmen, artisans, junior operatives, etc)
2. Fresh entrant as senior operatives or fresh engineers etc,
3. Working workmen, supervisors or executives, having little or no experience in maintenance,

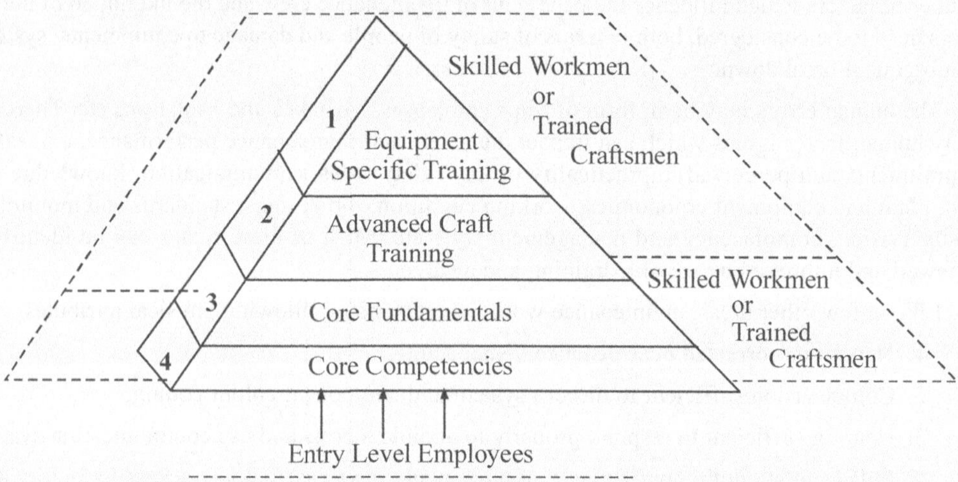

Fig. 15.1: Maintenance Training Levels

Irrespective of plant or industry, following are the broad profile of the maintenance personnel need to be trained. These may enter at bottom level or other suitable level (depending on their previous skill etc) of Fig. 15.1.

1. Fresh entrant in industry (unskilled workmen, artisans, junior operatives, etc)
2. Fresh entrant as senior operatives or fresh engineers etc,
3. Working workmen, supervisors or executives, having little or no experience in maintenance,
4. Executives/ supervisors on change of functions (e.g. operation or project department to maintenance); Need refresher course for shorter duration,
5. On promotion to higher grade, involving change in responsibility (e.g. unskilled to semiskilled workmen etc)
6. Skilled workmen for shorter duration refresher course to increase their effectiveness and remove deficiencies if any,

7. Multi-skilling training,
8. Supervisors/ engineers for short term advanced management courses or for specialized equipment maintenance etc, etc.

15.3 OBJECTIVES OF MAINTENANCE TRAINING

As already mentioned in this chapter, training and development of maintenance personnel is, primarily, designed to achieve following objectives

- *Knowledge Objective*: To learn about maintenance policies, systems/ strategies, condition monitoring and diagnosing, failure analysis and record keeping etc,
- *Attitudinal Objective*: Attitude is the approach towards job and job related person. Attitude influences behaviour, which affects the output and effectiveness of the maintenance person. As such training should also aim to modify the job attitude.
- *Job-behaviour objective*: Attention is focused opn the extent to which the knowledge, skilled and abilities acquired during the training can be generalized or transferred to maintenance problem solving efforts,
- *Skill Objective*: Skill generally refer to the ability and expertise of a person (generally physical) to perform his/ her job.
- *Advanced Technological Objective*: These objectives refer to short term trainings, aimed to acquaint the maintenance engineers/ supervisors about the modern developments taking place in the field of repair and maintenance.

The extent of these objectives vary from level to level. These objectives and aim should not remain on paper only. Management should also communicate to trainees that—

1. Results are expected from the maintenance training,
2. Support for implementation of the relevant course will be provided,
3. There will be a follow-up on the results or expertise achieved,
4. In case of multi-skill training, they would be scheduled for other trades in addition to their original trade; and so on.

15.4 TYPES OF MAINTENANCE TRAINING PROGRAMMES

As per one classification, maintenance trainings, it can be of three types— Classroom Training, On-job Training and a combination of the two. Most maintenance trainings are a combination programme, the portions of classroom and on-job trainings vary depending on the topic of the training and level of the workmen.

Again, the training programmes can be general or customized. Most maintenance training programmes are customized, i.e. specifically designed to meet some specific requirements. Few of such programmes are given herewith—

1. **Multi-Skilled Technician Programsmes:** These training programmes are designed for "cross-training" your craft and operations personnel. Many organizations are taking advantage of the ability of workers to perform multi-discipline tasks. This decreases the time to perform many jobs, resulting in significant cost savings. Refer Sec.12.7.
2. **Maintenance Management Courses:** Only recently the management of a maintenance organization has been given the attention it deserves. The reality is that maintenance organizations have changed from simple repair facilities to actual business units responsible for the pro-active care of equipment assets. Maintenance supervisors require

skills in people management, maintenance management, and project management to effectively run a maintenance service business. Courses are designed to provide the skills required of a maintenance supervisor and improve their potential for higher responsibilities. Topics include team building, leadership, project management, planning and scheduling, autonomous maintenance, predictive maintenance, and 5-S etc.

3. **Customized Maintenance Training Programs:** Customized maintenance training programme is a total maintenance and operation solution, tailored to the particular needs and circumstances of its users, and brings better optimization between various connected elements. Such trainings give one-stop opportunity for all maintenance and operation needs for that topic/ unit. Such trainings can be a regular course in plant's training calendar or can be specially made for few special occasions. These trainings are, generally, of shorter or medium duration. Huge numbers of such trainings are used in any plant/ industry. Few examples are given herewith
   - Vibration Monitoring, Tools, Techniques and Applications
   - Steam Turbines- Operation and Maintenance
   - Precision Bearing Maintenance
   - Maintenance Planning and Scheduling
   - Commercial Aircraft Carrier Maintenance
4. **Performance Test Trainings**- Such trainings are used for job performance evaluations in conjunction with knowledge evaluations as screening devices for
   - Hiring/ recruiting new employees, Also a hands-on evaluation of the skills required to perform the job (depending on the workforce diversity and culture) can be seen as less threatening than written evaluations/ tests,
   - As determinants of training that is needed (gap analysis),
   - For advancement and pay-for-skills,
   - For certification testing, etc.
5. **Accelerated Apprenticeship Training:** Many maintenance departments are staffed with an aging workforce. As these craftspeople retire, the maintenance departments lose a wealth of institutional knowledge. In many cases, substituting employees/ apprentices cannot be trained fast enough under the current programs to effectively replace retiring workers. In such cases, special trainings are arranged or existing trainings can be augmented to accelerate the training while increasing its effectiveness.
6. **Continuing Training (Refresher courses):** In this rapidly changing world, learning cannot stop at any stage. A semiskilled craftsperson learns to become a skilled craftsperson, and the skilled craftsperson has to constantly learn new technologies to stay at the top of his or her field /skill. As such refresher courses (continuing trainings) are arranged for existing craftsperson at some intervals to update their skill/ knowledge in their field.
7. **Induction and Orientation Training:** These trainings are arranged for new employees to acquaint them about equipments/ systems and facilities of the plan/ industry, organizational structure of the industry and the role of participants in the organization, general rules and regulations of the industry etc. Generally classroom trainings.
8. **Safety and Environmental protection Training:** This training is essential programme of most of the trainings. However, small independent trainings are also arranged to highlight the safety and environment protection measures and general statutory obligations etc. Few

such topics are— Lock and Tag procedure, Electrical safe work practices, Working in confined space, Personal protective equipments, How any job can be done safely etc.

9. **Maintenance Troubleshooting Training:** It teaches employees a fundamental sequence of steps to apply when attempting to locate and repair problems in electrical, mechanical equipment and other equipments. Refer to Chapter 16.

Huge numbers of "**Onsite Maintenance Courses**" or '**Composite** (both onsite and classroom) Courses" are available in industries or special training institutes to impart or enhance the skill and effectiveness of maintenance workmen. These are generally in three categories—mechanical, electrical and controls. Mechanical courses are like Pipes and Pipefitting, Pump Maintenance, Hydraulics and Pneumatics Fundamentals, Shaft Alignment, Valves and Valve Actuators, Rotating Mechanical Equipment, Air Conditioning Refrigerant Regulations etc. Electrical courses are Circuit Breaker Maintenance, Basic Protective Relaying, Generators and Voltage Regulators, Motors and Motor Controllers, Grounding and Shielding Electronic Systems, Power Distribution and Switchgear Maintenance, Power Transformer Maintenance and Testing etc. Controls courses are Digital Electronics Principles, Engine Control Fundamentals, Instrumentation and Controls Overview, Process Measurement Instrumentation and Variable-Speed Drives etc.

Again Industrial Safety trainings (e.g. working safely with chemicals, material handling safety etc), general courses (e.g. preventive maintenance techniques, use of measuring and monitoring gadgets etc) and composite courses like HVAC (heating, ventilation and air conditioning) etc are few important ones.

The duration of such trainings vary depending upon the content of course and type of person attending.

15.5 TRAINING SOURCES/ AGENCIES/ INSTITUTIONS IN INDIA

Training for maintenance personnel can be organized by/ from different sources/ agencies/ institutions, such as

1. Department's or plant's own training organizations/ institutes, taking faculty from training department, maintenance or other departments or from outside sources,
2. Organization's central training/ development facilities- Most big organizations/ industries, having their plants at different locations, often have their own training and development facilities. As one example, Steel Authority of India Ltd (SAIL) has its Management Training Institute at Ranchi.
3. Training Institutes/ facilities from various manufactures and equipment suppliers; e.g.
   - School of Hydraulics, Vickers System International, Mumbai
   - TELCO, at Jamshedpur and Pune, for maintenance of their vehicles and components,
   - Ashok Leyland, Chennai, for maintenance of their vehicles and components,
   - M/s Cummins, at different places, for maintenance of automobile engines,
   - M/s MICO, at different places for maintenance and calibration of fuel pumps etc; etc.

   Many of these firms and others have mobile training facilities and arrange different trainings at plant's premises.
4. Various government, semi government and autonomous, a few of which given herewith-
   - Industrial Training Institutes (ITIs);over 4000 spread it is, throughout India
   - National Power Training Institute, Sector – 33, Faridabad – 121 003; regional institutes at Regional Institutes of NPTI located at Nagpur (Maharashtra), Neyveli (TN), Badarpur (New Delhi) and Durgapur (WB).

- Dr. Ambedkar Institute of Productivity (AIP), (training wing of the National Productivity Council, NPC, of India), Chennai-600098,
- Central Training Institutes for training of Instructors under the Directorate General of Employment and Training, establishment at seven locations.
- Advanced Training Institute, Chennai,
- Indian Institute of Production Management, at Kansbahal (Orrisa) and Banglore,
- Central Staff Training and Research Institute, Calcutta,
- Advanced Training Institute in Electronics, Hyderabad,
- National Centre for Technical Development, Mumbai; etc.

5. Universities, engineering colleges and other academic institutions also arrange some vocational trainings for craftsmen, supervisors and executives of industries.

6. Small Seminars, organized by different government, semi-government, private and other autonomous institutions or maintenance consultants, on special topics.

15.6 TRAINING TOOL, MODES AND METHODS

As can be seen from any course of training, there are three distinct **learning styles**— "**Auditory**" (learn by hearing, "**Visual**" (learn by seeing) and "**Kinesthetic**" (learn by doing). A good training session combines all three of these learning styles so that everyone gains the maximum benefit from the session. A trainer gives the in-depth presentation of the material/ system/ programme so that you are hearing how to use or work with those. The trainer will also demonstrate how to work or use those so that you learn by seeing. Finally the trainer will allow you to hands-on exercises, so that you learn by doing. Experiences have shown that, in most cases, learning by doing is the most effective way to retain information because you programme both, your mind and your body, with the new knowledge.

Most maintenance trainings can be classified as "Formal" or "Less Formal". **Formal training** is given by universities, colleges and recognized government, semi-government or private institutions, as mentioned in Sec. 15.5, and institutions award degrees, diplomas and certificates, which is recognized by all.

Less-formal trainings are those which are acquired by the individuals from books, seminars, on-job trainings, trainings within their own organizations or through manufacturer's /supplier's representatives/ premises, interactions and visits at manufacturers/ suppliers place, exposures to different technical workshops and exhibitions and periodical meetings/ discussions with representatives of different industries/ plants and radio/ TV broadcasts etc. These trainings have limited recognitions.

Again both, formal and less-formal trainings can be obtained by through conventional tools, techniques and methods or through distance learning (mainly e-learning) modes/ methods. These are explained in Figure 15.2. In most industries, the conventional methods/ modes are used and include both, classroom training and on-job trainings. Different tools, techniques, modes and methods are briefly given in the Fig.15,2 and need no elaboration. Distance learning, mostly in form of e-learning is a recent phenomenon in India.

15.6.1 Computer Based Training (CBT)

Presently CBT or Computer Based Learning (CBL) is used extensively, both for formal or class room training and for distant learning. It is a type of technology supported education/ learning, where the medium of instruction is through computer technology, particularly involving digital technologies. CBT is especially effective for training people to use computer applications because the CBT program, along with internet, can be integrated with the applications so that students can practice using the application as they learn and also give online tests.

Maintenance seminars and various maintenance trainings like lubrication, gas and air compressors, electrical safe working practices, AC motor controls, Hazardous area instrumentation and Shutdown, turnaround and outage etc, are being done through CBT.

15.6.2 Web Based Training (WBT)

This is the modern version of e-learning. It is an innovative approach to distance learning in which computer-based training (CBT) is transformed by the technologies and methodologies of the World Wide Web, the Internet, and intranets. Web-based training presents ***live*** content, as fresh as the moment and modified at will, in a structure allowing self-directed, self-paced instruction in any topic. WBT is media-rich training fully capable of evaluation, adaptation, and remediation, all independent of computer platform.

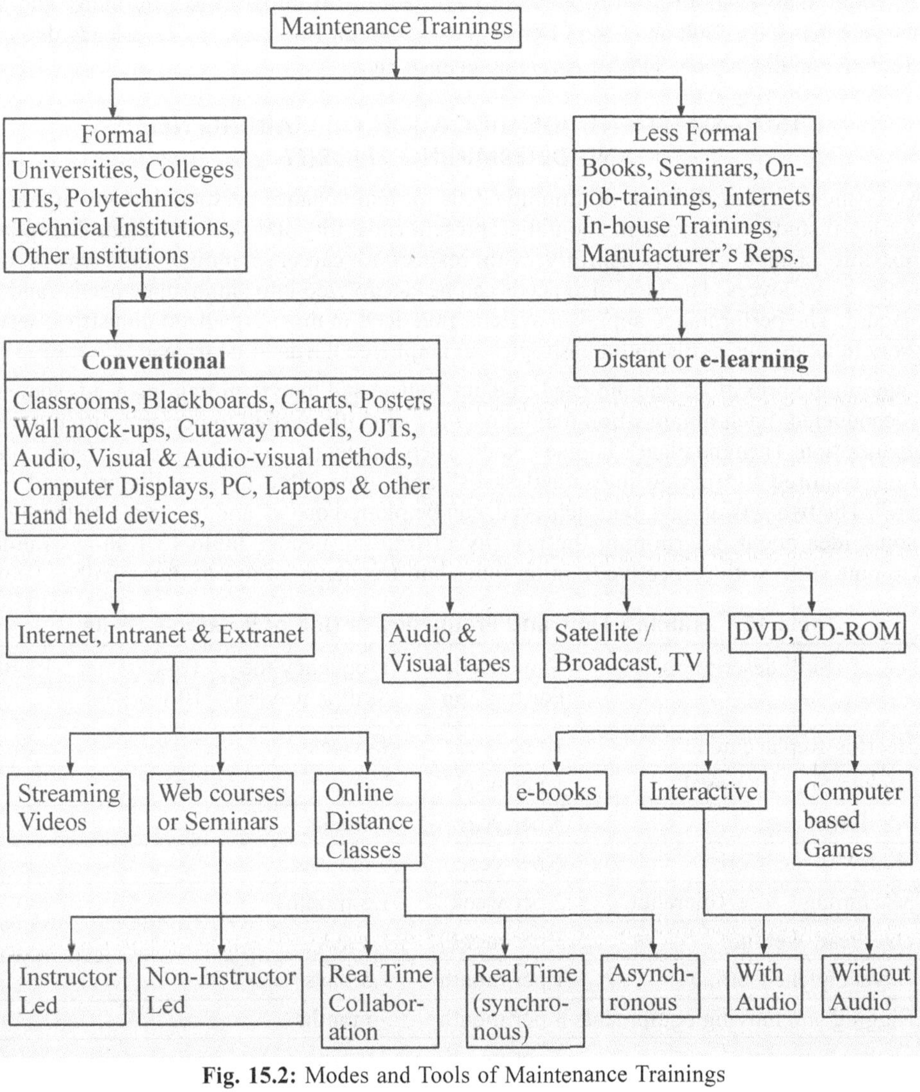

Fig. 15.2: Modes and Tools of Maintenance Trainings

WBT is an ideal vehicle for delivering training to individuals anywhere in the world at any time. Web browsers that support 3-D virtual reality, animation, interactions, chat and conferencing, and real-time audio and video will offer unparalleled training opportunities. With the tools at hand today, we can craft highly effective WBT to meet the training needs of a diverse population. Web-based performance support systems (WBPSS) further help today's busy workers perform their jobs by integrating WBT, information systems, and job aids into unified systems available ***on demand***. However, WBT designers are still struggling with issues of user interface design and programming for high levels of interaction.

WBT has many advantages over existing CBT, e.g. WBT has easy delivery of training to users, opportunities for group training (asynchronous and synchronous) as well as individual training, multi-platform capabilities (Windows, Mac, UNIX, PDA, phone, other wireless devices), easy updating of content, requires less technical support etc. However, it has few disadvantages or limitations, e.g. bandwidth/browser limitations may restrict instructional methodologies, leading to slower performance for sound, video, and intense graphics etc.

Today many vendors or providers are available who impart WBT on many and varies maintenance items, e.g. Belt and Chain Drives; Fuses, Circuit Breakers, and Overloads; Preventive and Predictive maintenance; TPM, Lean maintenance etc.

15.7 ASSESSMENT/ IDENTIFICATION OF TRAINING NEEDS AND DETERMINING PRIORITY

Assessment or identification of training needs for maintenance personnel in any organization calls for lot of study and judgment about their present capability and skill and about their potential and possible change of responsibility with respect to career planning and growth of the maintenance employees. The identified needs can be categorized into "immediate needs" and "long term needs". The performance appraisal system, practiced in most organized industries, serves in a big way to assess the skills and areas/topics an employee needs to be trained.

Determining priority of training needs, i.e. immediate and long term training needs some study can be conducted. Each organization maintains a list of skills needed to perform different jobs in that organization. Then the plant's human resource department (training department) has to find out the "time required to training individual skills" and "how often those skills are used in that industry". The two sets of data, thus achieved, can be plotted on "x" and "y" axis and the plot will give some idea about determining the priority of training needs. Table 15.1 shows randomly selected nine skills with respective training times and frequency of use of those skills.

Table 15.1 Training Time and Frequency of Use of 9 Sample Skills

Sl.	Skill description	Time needed For training	Frequency the Skill is needed	Trg.cost per 10 workers	Remarks
1.	Bearing Replacement	10 per year	1- week		
2.	Broken Stud Removal	1 per week	2-days		
3.	Fan Balancing	3 per year	3 weeks		
4.	High Voltage Work	4 per year	6 months		
5.	Machining Close Tolerance	2 per week	1.5 months		
6.	Overhead Welding	3 per month	1 week		
7.	Oxyacetylene Cutting	1 per month	1.5 days		
8.	Rigging and moving equipments	3 per month	3 months		
9.	Shaft Alignment	2 per month	2 weeks		

By analyzing the time and frequency columns and plotting those on x and y axis, one can decide which training gives best result and to be taken immediately. Those activities, giving less return, can be taken later as long term measures. Cost for training 10 person in one batch has also been shown in the table and that, in addition to priority, will help if the training should be given in-house or from outside. Adding a column of total number of workers in that skill in the industry will further help in such decisions.

15.8 EVALUATION OF TRAINING

Evaluation, including feedback and follow-up, of trainings (including learning, OJT, other training courses and programmes) is very important as it determine how far the training has been implemented at trainee's plant/ work and how effective the training has been. Post training evaluation includes different feedback forms, action plans, follow-up reports, discussions, interviews, observation on jobs and, occasionally, tests. Evaluation has to be done from following four angles/ sources

1. *Validation*: In many formal training programmes, evaluation starts with validation session, which is normally at the end of session. Effect of validation evaluation depends upon who is responsible for the validation/ evaluation processes and what resources of time, people and money are available for validation/evaluation purposes.
2. *By Trainer(s)*: Trainer(s) evaluate their trainees/ participants either at the end of programme or at the end of individual sessions, to monitor the learning as the programme progresses and to determine how effective the training has been. This may be done in the form of questionnaires about learning levels achieved etc, asking for action plan to reinforce, practice and implement the learning and in other forms, like making models etc. The trainer submits suitable reports to management.
3. *By Trainee/ participants*: Often trainees/ participants are involved to some extent, before or at the start of training, in planning and design of the programme. Although the principal role of the trainee in the programme is to learn, the learner must be involved in the evaluation process. This is essential, since without their comments much of the evaluation could not occur. Neither would the new knowledge and skills be implemented nor there be much improvements in subsequent training programmes. This again may be done in form of questionnaires or otherwise.
4. *By Management/ Organization*: This level involves two set of people— Maintenance Dept or concerned production executives and Human Resource (HR) Dept. executives. Often the two do not see the training evaluation from the same angle. HR Dept may not have sufficient resources - people and money - to do post training evaluation and maintenance and concerned production head may not feel much concerned. Good methodical evaluation produces a good reliable data. As such, the post training evaluation should be the responsibility of both. They both should find out
   - To what extent were the identified training needs /objectives achieved by the training programme,
   - To what extent were the learners' objectives achieved?
   - How successful were the trainees in implementing their action plans or skills?
   - To what extent were they supported in this by their line managers?
   - To what extent the trainee's action plants or skills helped in achieving Return on Investment (ROI) for the organization, either in terms of identified objectives satisfaction or, wherever possible, a monetary assessment?

These evaluations will give immediate benefits in the form of decision on effective utilization of acquired skill and knowledge, finding the strength and weaknesses of the programme, trainees and trainers and assessing cost-benefit analysis and ROI on training expenditure. It will also guide for future training needs.

QUESTIONS

1. Why training is essential for maintenance personnel? What are the objectives of maintenance training?
2. What are broad maintenance training levels and profile of persons?
3. Describe briefly the training tool, modes and methods for maintenance personnel.
4. What is on-job training or on-site training for maintenance personnel? Name few such trainings.
5. Why evaluation of training is essential in maintenance? What are the different evaluating agencies?
6. How do you identify and prioritize the maintenance training needs?
7. Explain briefly the CBT and WBT in training and the difference between the two.

CHAPTER

16 MAINTENANCE ENGINEER AND MAINTENANCE TROUBLESHOOTING

16.1 MAINTENANCE ENGINEER

Much have been mentioned about maintenance engineering objectives and functions etc in chapter 1 and have been explained in some details in subsequent chapters.

16.1.1 Definition

In nutshell, a maintenance engineer can be defined as a person who is responsible for the maintenance, repair and upkeep of the machinery and equipments in any industry and also for the supervision of its engineering personnel.

Maintenance engineers play a vital role in efficiency, development and progress of manufacturing/processing, transportation, mining and other industries. They work with other professionals, such as manufacturing systems engineers and production managers, quality engineers etc to improve production facilities, reduce the incidence of costly breakdowns, and develop strategies to improve overall reliability and safety of plant, personnel and production processes.

High-risk industries, such as nuclear, petrochemical and aviation, have well-developed reliability strategies and reliability engineers and maintenance engineers work in close cooperation with them. In traditional manufacturing, maintenance engineers promote the improvement of overall reliability, in addition to responding to breakdowns.

Maintenance engineers may also be called upon to develop plant asset care policies, which mean liaising closely with suppliers, customers, other departments and senior managers.

16.1.2 Responsibilities and Accountabilities

A maintenance engineer may be responsible for: directing and monitoring maintenance technicians; maintaining, troubleshooting, and repairing equipment, utilities, and facilities; sharing knowledge of electrical and mechanical concepts and principles with other personnel; designing, modifying and reviewing equipment.

Maintenance engineers plan the routine maintenance of plant, equipment and machinery. They also attend breakdowns, diagnose faults and oversee time-critical repairs. They are also involved in all stages of manufacturing. In equipment development, they work to incorporate efficient methods of maintaining new equipment or plant, and may be involved in the installation and commissioning process. Through the production phase, they work to improve the useful life of plant, equipment and machinery.

The Maintenance Engineer will also be responsible to provide technical diagnosis and analysis, implementation of technical improvements and management of preventative/ predictive and other maintenance activities for the plant mechanical equipment. The core responsibility of the Maintenance Engineer is to ensure maintainability and reliability of equipment, buildings and facilities and achieve a high level of asset preservation at a minimum total operating cost.

To fulfill these responsibilities, in modern complex plants, maintenance engineers employ sophisticated computerized systems (refer chapters 9, 10 and 11 etc) to schedule and monitor the work and also to oversee the work of team of maintenance personnel, such as fitters, millwrights, welders, electricians and other technicians.

Maintenance engineer will be accountable for equipment integrity and improvement of unit/ plant, equipment reliability and OEE improvements, maintaining a safe and environmentally sound working condition and budgetary control. However, the responsibilities and accountabilities of a maintenance engineer will overlap to a great extent.

16.1.3 Activities and Duties

The activities and duties of maintenance engineer may be vast and may cover most of the chapters mentioned earlier in this book. Again, the activities may vary to some extent from industry to industry. However, few common ones of those are mentioned here-below:

1. Designing maintenance strategies, procedures and methods,
2. Planning and scheduling planned and unplanned work,
3. Diagnosing breakdown problems; troubleshoots electric, hydraulic, pneumatic and other circulating oil systems and equipment as necessary.
4. Maintain/ monitor codification of drawings, spares and jobs and standard maintenance procedures etc; monitor CMMS and other computerized/ manual scheduling techniques,
5. Monitoring and improving various condition monitoring systems and gadgets and ensuring the use od their data,
6. Directing, instructing and supervising all maintenance technicians and workmen,
7. Liaising with client departments (operation, inspection, material, finance etc) and customers and contractors/ vendors,
8. Arranging specialist procurement of jigs, fixtures, fittings and other helping devices/ components,
9. Controlling maintenance tools, tackles, consumables, stores and spares/ subassemblies.
10. Monitoring and controlling maintenance costs,
11. Developing and improving the skills of maintenance personnel by hands-on, internal or external trainings,
12. Writing maintenance strategies to help with installation and commissioning guidelines.
13. Help in development of ancillary industries/ units for quicker and better availability of spares and other outside helps; Etc.

N.B.: This is not the job description of a maintenance engineer. The job description may be more quantified.

16.1.4 Qualification needed

As already indicated in Sec.1.7, maintenance engineering has not yet got its due recognition as an independent discipline, on par with other disciplines as Mechanical and Electrical etc. As such, a maintenance engineer can either a mechanical or electrical or instrumentation engineer (or even a process engineer for specific process plant), with some basic knowledge of mechanical, electrical and instrumentation engineering. In addition to basic qualification, they should have:

1. Some knowledge of hydraulic, pneumatic and circulating oil systems
2. Some knowledge of electric circuits, control, instruments and PLCs etc.
3. Reasonable computer and internet appreciation to handle CMMS, Power Point presentation and other computerized planning and maintenance systems,
4. General appreciation of equipment faults and condition monitoring.
5. Good troubleshooting and analytical skills,
6. Some ability to read and interpret drawings, catalogs, codes, data sheets, and vendor quotes etc,
7. Other normal requirements for a maintenance engineer, as discussed earlier.

16.2 MAINTENANCE TROUBLESHOOTING

The term 'Troubleshooting' means resolving diplomatic or political or industrial or other disputes, impasses, etc., as in business, national, or international affairs. It also means discovering and eliminating the cause of trouble in machines and equipment, power lines, etc. In addition to engineering and other industries, troubleshooting is effectively practiced in administrative system, medical diagnostic and many other fields.

Maintenance troubleshooting is application of troubleshooting for maintenance and upkeep of machineries and equipments. Hereafter, we would be mentioning 'trouble-shooting' as 'maintenance troubleshooting' only.

16.2.1 Definition

Troubleshooting is a form of problem solving, almost invariably, used for repair of failed products or processes. It is a logical, systematic search for the source of a problem so that it can be solved, and the product or process can be made operational again. It is the identification of, or diagnosis of "trouble" or "problem" in a system or component caused by a failure of some kind. The problem is initially described as symptoms of malfunction or defect, and troubleshooting is the process of determining the causes of these.

Usually troubleshooting is applied to something that has suddenly stopped working, since it's previously working state or condition forms the expectations about its continued behavior. So the initial focus is often on recent changes to the system or to the environment in which it exists. (For example an electric table fan, that "was working when it was plugged in over there"). However, there is a well known principle that correlation does not imply causality. (For example the failure of a table fan shortly after it has been plugged into a different outlet doesn't necessarily mean that the events were related. The failure could have been a matter of coincidence.) Therefore troubleshooting demands **critical thinking** (purposeful and reflective judgment about what to believe or what to do in response to observations, experience, verbal or written expressions, or arguments) rather than **magical thinking** (a nonscientific causal reasoning that often includes such ideas as the ability of the mind to affect others or simple guess-work).

Troubleshooting can also take the form of a systematic checklist, troubleshooting procedure, flowchart or table that is made before a problem occurs. Developing troubleshooting procedures in advance allows sufficient thought about the steps to take in troubleshooting and organizing the troubleshooting into the most efficient troubleshooting process. Troubleshooting tables can be computerized to make them more effective and efficient for the user.

16.2.2 Objective of Troubleshooting

The objective of troubleshooting is built-in in its definition. The primary objective of conducting machinery troubleshooting is to prevent the equipment and machinery from repeat incidents/ failures. The objective can be elaborated as below:

1. Identify the abnormality or symptom based on normal operation behavior,
2. Determine the faulty element or component based on symptoms,
3. Plan/schedule a course of action to repair the equipment component,
4. Safely perform repairs on the equipment/component,
5. Apply observation techniques to check and prevent reoccurrence once the problem is repaired/ solved.

It may be argued that troubleshooting does not include repair, but a good trouble-shooting must include repair and feedback, then only it may be known if the troubleshooting was correct.

16.2.3 Troubleshooting traps

Often we might not be as effective at accomplishing this objective as desired. One of the reasons may be that in our troubleshooting efforts, we don't always take the time to document our efforts. Too often, what may prevent us from fully realizing our troubleshooting potential are some common troubleshooting traps. Few of the traps are:

1. "Everyone's job is no one's job", where the Operations group decides it is too difficult for them to fix and the Maintenance group decides it is too minor for them to get involved. The result is often just replacing the item or addressing the symptoms only and not the causes for those symptoms.
2. The specialized maintenance man or small group, on whom a facility/ industry may rely, because he or they have been at that facility for a long time. This is not necessarily adverse, but it can be a problem if this individual(s) has not kept up with current technology, or become indisposed.
3. "Telephone troubleshooting" can also be a trap, where the troubleshooter attempts to solve the problem by interviewing people over the phone and not taking the opportunity to look at the failed equipment or machinery in person.
4. "Too much familiarity" can also be a trap, which can arise if a person becomes too familiar with a piece of equipment or machinery and begins to assume that the current problem is the same as the last time, whether or not in fact it is; or he thinks only he knows the best about that equipment.

16.3 TROUBLESHOOTING PROCESS

A basic principle in troubleshooting is to start from the simplest and most probable possible problems first. This is illustrated by the old saying "When you see hoof prints, look for horses, not zebras", or to use the **KISS (Keep it simple and short) principle**. Implicit in maintenance troubleshooting activity is a fundamental understanding of the systems functioning within equipment. With that understanding of the principles and the sequential process involved, maintenance professionals can develop and apply effective step-by-step maintenance troubleshooting procedures.

Efficient methodical troubleshooting starts with a clear understanding of the expected behavior of the system and the symptoms being observed. From there the troubleshooter forms hypotheses on potential causes, and devises (or perhaps references a standardized checklist of) tests to eliminate these prospective causes.

16.3.1 Few common strategies

Used by troubleshooters are:

1. To check for frequently encountered or easily tested conditions first (for example, checking to ensure that a printer's light is on and that its cable is firmly seated at both ends),
2. Use experience, i.e. if the trouble shooter has seen somewhat similar problem on this machine or nearly similar machine. This is like "on the job training skill".
3. Use some guesswork, which can also be called as "gut-feel" or "intuition". The brain parallel processes the information known about the problem on a subconscious level.
4. Divide and conquer- Use experience to divide/ bisect the system in two parts, examine the symptoms and check to see which section/ part is performing correctly and which is not.
5. Replace and conquer- In electronic systems or such complex interconnected systems, it is sometimes difficult to determine what box/ component is causing problem. One method to solve this kind of problem is to remove the boxes/ components, one at a time, until the problem goes away or located.
6. Walk through- Again in electronic, electrical or such complex systems, often it is not known where to start the testing. In such cases, start at one point in a signal path, testing that point, then going to another point downstream of first point and testing and so on.

Troubleshooting requires identification of the malfunction(s) or symptoms within a system. Then, experience is commonly used to generate possible causes of the symptoms. Determining which cause is most likely is often a process of elimination - eliminating potential causes of a problem. Finally, troubleshooting requires confirmation that the solution restores the product or process to its working condition.

Usually a system can be described in terms of its expected, desired or intended behavior. Events or inputs to the system are expected to generate specific results or outputs. (For example selecting the "print" option from various computer applications is intended to result in a hardcopy emerging from some specific device). Any unexpected or undesirable behavior is a symptom. Troubleshooting is the process of isolating the specific cause or causes of the symptom. Frequently the symptom is a failure of the product or process to produce any results.

16.3.2 Categories of symptoms

Such symptoms may, basically, be grouped in following four categories:

(*a*) **Clear Symptoms:** These are simplest form of symbols, which can easily be seen or felt and for which, the cause can also be easily ascertained. As an example, oil leakage (symptom) from hydraulic cylinder can easily be seen and cause can be said as damaged seal, which can be further investigated for primary cause as dirt in hydraulic oil or worn cylinder/ plunger surface etc.

(*b*) **Reproducing Symptoms:** These symptoms can, normally, be reproduced during investigation. One of the core principle of troubleshooting is that reproducible problem can be reliably isolated and resolved. Often considerable effort and emphasis in troubleshooting is placed on reproducibility, i.e. on finding a procedure to reliably induce the symptom to occur. Once this is done then systematic strategies can be employed to isolate the cause or causes of a problem; and the resolution generally involves repairing or replacing those components which are at fault. Hydraulic and pneumatic system malfunctioning are some of the many examples for such symptoms.

However, reproducing the symptoms may not be possible or may not be desirable in some equipments or industries.

(*c*) **Intermittent Symptoms:** Some of the difficult troubleshooting issues relates to the symptoms that are only intermittent. In electronics this often is the result of components that are thermally sensitive (since resistance of a circuit varies with the temperature of the conductors in it). Compressed air can be used to cool specific spots on a circuit board and a heat gun can be used to raise the temperatures; thus troubleshooting of electronics systems frequently entails applying these tools in order to reproduce a problem.

However, in cases where intermittent symptoms are very difficult to be reproduced, various techniques can be used to force the particular function or module to be called more rapidly than it would be in normal operation (analogous to "heating up" a component in a hardware circuit) while other techniques can be used to introduce greater delays in, or force synchronization among, other modules or interacting processes.

Loose connection in some electrical joint may also cause intermittent symptom because of occasional vibration etc.

The heat phenomenon can also be felt in few hydraulic or circulating oil systems where viscosity and flow-ability of oil changes because of heat or cold.

(*d*) **Multiple Problems and Multiple Symptoms:** Many problems only occur as a result of multiple failures or errors and often exhibit multiple symptoms. This is particularly true of fault tolerant systems, or those with built-in redundancy. Even in simple systems the troubleshooter must always consider the possibility that there is more than one fault. Replacing each component, using serial substitution, and then swapping each new component back out for the old one when the symptom is found to persist, can fail to resolve such cases. More importantly the replacement of any component with a defective one can actually increase the number of problems rather than eliminating them.

16.3.3 Systematic Troubleshooting Process

In order to conduct troubleshooting effectively and avoid the earlier mentioned traps, you should follow a logical process. Followings are the steps for logical process to determine "what" happened, "why" it happened, and then develop effective fixes for the "why" it happened.

1. Identifying Symptoms: We may discuss and look (and measure) for symptoms as mentioned below and look for abnormality, i.e. deviation from normal working condition. Also we have to consult the right person and connected persons about the symptoms, which may be the deviations in followings:

1. Observation—Appearance and feel,
2. Temperature,
3. Vibration,
4. Pressure,
5. Sound/ noise,
6. Smell/ odor; etc.

Other diagnostic measurement, to identify/ measure the symptoms may include stress, strain, displacement, torque, force, displacement, current, voltage, resistance etc. Major deviations in any of these have to be examined.

2. Determining Faulty Components: As one example, if we examine bearing failure/ damage, the reasonable list of possibilities may include:

- lubrication problems,
- loading problems,
- misalignment problems,
- clearance problems, and
- friction problem, etc.

For determining the exact cause, we may have to isolate the fault, and adopt Component failure analysis and other tools as mentioned subsequently in this section. We may refer to existing checklist or develop our own checklist to determine the "path of failure".

After the Failure Mode, we proceed to the Failure Agents. The Failure Agent is the catalyst that allowed the Failure Mode to occur. The Failure Agents consist of: Force, Reactive Environment, Time, and Temperature etc. One of these will be the primary, which often creates secondary and tertiary Failure Agents that are exhibited.

After the Failure Agent, we proceed to determine the Failure Classification. This is where we determine whether the issue is strictly an 'Equipment Difficulty' or whether there may be a 'Human Performance Difficulty' associated with it. Too often, many troubleshooters stop at this point. They are missing a valuable opportunity to take their troubleshooting to another level. This is where we now involve the "people talking back", as part of the investigation.

3. Planning a Course of Action: This step considers the followings also:

- Component access,
- Process interruption,
- Maintenance economics.

There may be three levels of responses/ solutions for any problem/ fault:

1. Emergency Response/ solution: Dedicated to plan jobs to prevent plant downtime,
2. Rapid Response/ solution: Dedicated to plan jobs for conditions that place the unit into a Limited Condition of Operation (LCO),
3. Long-Term Response/ solution: Dedicated to plan jobs for follow-up actions identified by the Emergency Response, Rapid Response or as per 'Root Cause Analysis' for equipment.

4. Repair and Follow-up Observation: Once the proper course of action is planned, rest is straight forward. It consists of basically following the courses:

1. Replacement preparation,
2. Component or sub-assembly replacement,
3. Operational observation and feedback/ follow-up.

Although follow-up is not an official step of the troubleshooting process, it nevertheless should be done once the equipment has been repaired and put back in service. You should try to determine the reason for the malfunction. Did the component fail due to age? Did the environment, the equipment operates in, cause excessive corrosion? Are there wear point that caused the electrical wiring to short out?

16.3.4 Troubleshooting Tools

As already explained elsewhere in this book, many inspection, monitoring and other tools and techniques are available for troubleshooting, a few of those are again given herewith——

1. **Bathtub Curve:** It describes a particular form of the hazard function or defect which comprises three parts: decreasing failure rate, known as early failures, a constant failure rate, known as chance failures and an increasing failure rate, known as wear-out failures,
1. **Cause and Effect Diagram:** (Ishikawa diagram) or Fish-bone diagram,
2. **Forensic engineering:** It is the investigation of materials, products, structures or components or even the process that fail or do not operate/function as intended, causing personal injury or damage to property. The consequences of failure are dealt with by the law of product liability.
3. **Root cause analysis:** RCA, as a maintenance troubleshooting method that would anticipate and control the systemic causes of maintenance problems, has occupied the thinking and planning of maintenance professionals for decades. FTA, FMEA and other analytical methods can be used as per need.
4. **5 Whys:** The 5 Whys is a question-asking method used to explore the cause/effect relationships underlying a particular problem. Ultimately, the goal of applying the 5 Whys method is to determine a root cause of a defect or problem. It is repeating why five times. The technique was originally developed by Sakichi Toyoda and was later used within Toyota Motor Corporation during the evolution of their manufacturing methodologies. **Three Ws** (What, When, Where) can be used for preliminary analysis.
5. **Scientific method:** It refers to bodies of techniques for investigating phenomena, acquiring new knowledge, or correcting and integrating previous knowledge. To be termed scientific, a method of inquiry must be based on gathering observable, empirical and measurable evidence subject to specific principles of reasoning. A scientific method consists of the collection of data through observation and experimentation, and the formulation and testing of hypotheses.
6. **Trial and error**, or **trial by error:** It is a general method of problem solving for obtaining knowledge, both propositional knowledge and know-how.
7. **Debugging:** It is a methodical process of finding and reducing the number of bugs, or defects, in a computer program or a piece of electronic hardware thus making it behave as expected. Debugging tends to be harder when various subsystems are tightly coupled, as changes in one may cause bugs to emerge in another.
8. **A flowchart:** It is a common type of chart, that represents an algorithm or process, showing the steps as boxes of various kinds, and their order by connecting these with arrows. Flowcharts are used in analyzing, designing, documenting or managing a process or program in various fields,
9. **RPR (Rapid Problem Resolution):** deals with failures, incorrect output and performance issues, and its particular strengths are in the diagnosis of ongoing recurring problems. The method comprises:

- Core Process (Discover, Investigate, Fix, etc)
- Supporting Techniques- it detail how the objectives of the Core Process steps are achieved, and cite examples using tools and techniques that are available in every business.

16.4 TYPICAL TROUBLESHOOTING TIPS/ PROCESSES

While applying the earlier mentioned sections of troubleshooting, following three different cases are presented here to illustrate the process:

16.4.1 Fluid-Power System/ Equipment Troubleshooting tips

The first prerequisite for troubleshooting for hydraulic or any other fluid power or circulatory oil systems is the technical knowledge, but equally important is the mental attitude. In addition to understanding machine operation and hydraulic components and circuitry, the individual must also have training and discipline necessary to troubleshoot a hydraulic system in a logical and orderly manner.

Quick and efficient hydraulic trouble shooting requires to follow logical steps to locate the malfunctioning in the shortest possible time. The challenge in resolving this problem consists of two steps:

(*i*) Think through and identify the elements that might contribute to a system malfunction,

(*ii*) Avoid jumping to incorrect and illogical conclusions which might result in improper and inappropriate actions.

Initial trouble shooting procedures can be in following steps:

1. Any fluid power actuator is operating improperly if you notice no movement, movement in wrong direction, incorrect speed, or creep etc.
2. Using the circuit diagram, identify each component in the system and determine its functions,
3. Compile a list of components that could possibly cause the problem,
4. Arrange the list in rough order of priority, based on your past experience and the ease in which they could be tackled,
5. Conduct a preliminary check of each component in the list. Also determine if any component shows abnormal symptoms; e.g. excessive temperature, noise, vibration etc,
6. Use **5 Why** method or at least **three Ws** (What, When, Where); e.g. **What** is or is not occurring in the system's operation? **When** did the problem begin? Was it a sudden failure or a gradual failure? **Where** in the machine cycle does the problem occur? Was it at startup or after the system has been operating for a while? Etc.
7. If preliminary check reveals no obviously faulty component, conduct a more exhaustive repeat test of each, using additional instruments, without removing any components from the system
8. The instrument check should identify the faulty component. Someone then must decide whether to repair or replace,
9. Take corrective action
10. Before restarting the corrected machine, stop and consider the cause and consequence of the failure. For example, if the failure was caused by contaminated or overheated fluid, you should anticipate additional component failure until you take the remedial action against that cause.

Similar trouble shooting procedures can be followed for pneumatic equipments and other such systems. Thinking safely should be build-up in each step.

Troubleshooting Tips for Lubricants and Lubrication Systems: This also calls for nearly the similar steps as mentioned above. However, as the system is normally very long/ vast, special attention is given to following aspects:

(*i*) Lubricant incompatibility,

(*ii*) Lubricant discolouration,

(*iii*) Lubricant contamination at various places from different sources,

(*iv*) Filter inspection and inspection from windows at different sub-reservoirs (big gearboxes etc)

(*v*) Taking representative samples from different locations/ sub-reservoirs,

16.4.2 Troubleshooting tips for automotives and other fuel engines

Noises, smells, leaks and operating or performance changes are all signs that something could be wrong with your vehicle or engine. The basic troubleshooting tips will be similar as mentioned earlier. Here we would see two different sets of tips; first general tips and then specific tips.

(A) General tips: This may include the followings:

(*i*) **Performance Problems:** Knowing something about how your vehicle/engine performs and makes communications with service advisers and technicians, fast and easy; e.g. cuts-out (Temporary loss of power), Detonation, Dieseling, Hesitation, Missing, Rough idle, Stall, surge etc.

(*ii*) **Noises:** What's your car or engine trying to tell you when it clicks, clunks and grinds? Translating these noise signals is key information for automotive technicians who are trained to identify problems via sound. The important thing to remember with odd sounds is to be specific, narrow that 'rattles' down, and then detail when it happens; e.g. grinding brake sound means worn pads and shoes, growling from air-conditioner means worn compressor etc.

(*iii*) **Smells:** Learning to pinpoint strange vehicle/ engine odors can help your automotive technician sniff out the real cause of the problem. Here are a few common odiferous car/ engine complaints. E.g. Acrid odor, like burned toast means electrical short circuit etc

(*iv*) **Leaks:** If your vehicle is leaving strange stains or other visual signs behind on the driveway, you need to pinpoint the signals like Black stains with a heavy, thick consistency means Axle leak or Clear liquid with a smell means brake fluid leak etc.

(B) Specific and detailed Tips: This may include as followings:

1. Engine will not crank: Check starting system, Check belt-driven engine accessories for seizure, Check air compressor for seizure, Check engine for fluid-locked pistons. Etc
2. Engine cranks but will not start: Check fuel level in fuel tanks, Check air cleaner indicator, Check for white exhaust smoke during cranking, Check for restricted fuel filter, etc.
3. Starter cranks engine slowly: Check belt-driven engine accessories, In cold weather, ensure proper engine oil is being used and/or replace oil etc,
4. Engine stops during normal operation: Check air cleaner indicator for restrictions, Check fuel system for contamination and restrictions, Check for restrictions in exhaust system. Ensure exhaust system is not bent, restricted, or damaged, Check engine idle speed (Adjust if idle speed is below specification). etc
5. Engine stops when accelerator is returned to idle position: Checks as mentioned earlier
6. Engine fails to stop: Check for defective fuel shutoff solenoid, or leaking or excessive fuel at injectors,
7. Engine misfires during operation: Check air cleaner indicator for restrictions, Check for air or water in fuel system etc,

8. Poor acceleration and/or lack of power: Check air cleaner system for restrictions, Check fuel system for contamination and restrictions, Check exhaust system for restriction, Check vehicle for dragging brakes, Check turbocharger hose and intake tube for leak, etc
9. Engine surges: Check fuel system for contamination and restrictions, Inspect accelerator cable for proper operation and adjustment etc,
10. Excessive engine oil loss or consumption during normal operation: Check oil for overfilling, Check for external oil leaks at oil pan, drain-plug, oil filter, oil filter etc
11. Engine oil pressure is too low or too high at normal operating condition: Check that engine oil grade is correct for vehicle use and climate conditions, Check all external oil lines for restrictions or leakage, Check for fuel in oil etc.
12. Excessive vibration or clunking: Check engine mounting brackets and pads for looseness or damage, Check vibration damper for looseness etc,
13. Excessive fuel consumption: Inspect fuel lines, hoses, and connections for leaks and damage, Check maximum engine idle speed is set within specifications etc
14. Excessive exhaust noise: Inspect exhaust pipes, stack, and muffler for secure connections, cracks, breaks, and excessive rust, Inspect turbocharger for secure mounting and exhaust leaks. Etc
15. Engine does not reach normal operating temperature: Remove surge tank filler cap and visually check coolant for proper circulation, if the If coolant is circulating below desired temperature (say 38°C), replace thermostat, etc.
16. Coolant loss during normal operation: Pressurize cooling system and check for leaks, Check for coolant in oil. etc.
17. Transmission shifting irregularly: Check transmission fluid level, Check for clogged or dirty transmission oil filter etc.

16.4.3 Electrical equipment/ system troubleshooting tips

To expertly troubleshoot electrical equipment, problems must be solved by replacing only defective equipment or components in the least amount of time. One of the most important factors in doing this is the approach used. Expert troubleshooters use a system or approach that allows them to logically and systematically analyze a circuit and determine exactly what is wrong.

This approach differs slightly from other troubleshooting procedures in that it does not tell you step by step how to troubleshoot a particular kind of circuit. It is more of a thinking process that is used to analyze a circuit's behavior and determine what component or components are responsible for the faulty operation or no-operation.

However, general troubleshooting steps in such cases are also somewhat similar, i.e.

1. Gather information: This is by observation, discussing with concerned personnel, access to circuits and other necessary diagrams, by review of the equipment's current signature analysis (CSA) and other monitoring devices and any previous history of that,

2. Understand the malfunction: It means that you understand how or what the process/ circuit is and what portion of the process/circuit is not operating or operating incorrectly. Answer these questions:

- How is it or the system supposed to work?
- What is not functioning as it should?
- What would cause these results or malfunction?

Examine all possible causes. The installed condition monitoring might have trended a possible anomaly. As an example, in case of a electric motor, rotor defects, bearing misalignment, magnetic offset, or abnormal load fluctuations were determined to be possible causes of the pole pass frequency trending upward.

3. Identify which parameters need to be evaluated: It requires a clear understanding of the discrepancy and which signals affect the suspected component. Which input signals control the component? What is the expected output from the suspect circuit? Is there a timing delay, sequence, or set point that can be verified?

Identify the parameters that need to be recorded which could either confirm or refute your suspicions regarding the problem. Identify the following:

- What parameters can you measure?
- What are the expected values for any measurements that are to be taken?
- What test equipment is needed?
- Is there access for the required readings?
- Is there an alternative method to gather the required readings?
- Could other components have been affected by this fault?

Taking the example of the same motor, gaining access to the high voltage cables supplying the boiler feed pump motors may prove to be difficult. However, testing from the current and potential transformers (CTs and PTs) offers an easy alternative method to gather the required voltage and current signals to assist in troubleshooting.

4. Identify the source of the problem: It requires the technician to isolate components and evaluate circuit parameters, to isolate the circuit by group when dealing with a complicated circuit (half-step approach), and to identify the malfunctioning component using the recorded data.

One test recommended for confirming a possible anomaly and establishing a severity is a current profile comparison between two like machines. This is sometimes referred to as a process analysis test. Additional testing can be done vibration, motor circuit analysis, and current signature analysis etc.

Many times after the first round of troubleshooting, the first three steps may need to be repeated to find out exactly where the defect is and what is to be done.

5. Correct/repair/replace the component: Correct or repair the component identified as damaged or not working, based on the recorded data. Perform the required repairs to the circuit. Jobs at this step can range from simple adjustments to a complete component replacement.

6. Perform root cause analysis: This began from the first step of the troubleshooting process. You should use the knowledge gained throughout the troubleshooting process in determining what could have possibly caused the component to fail.

Did the component fail prematurely? Why are the motor windings failing after only three years of service? These are just a couple of the questions that may come to light when evaluating the whole repair process. While working through the troubleshooting process, ask yourself, "Is this the root cause or just a symptom of the problem?" Without identifying the real/root cause that led to the failure, the repair will always be only temporary. In one such example, the troubleshooter concluded that the motor was undersized for the varying load it was driving.

7. Verify the repair: Check and verify the repair after completion. Ensure the equipment is operating as designed. Perform another round of testing to verify the equipment is in fact running correctly and that no other discrepancies exist. Retest to ensure the installation will, normally, not result in the same failure mechanism in the future.

Another simple example: Symptom is 'No power due to Fuse blow (open)

Possible causes:

1. Excess current, caused by overload on the load,
2. Excess current, caused by short circuit between the wires,
3. Excess current, caused by grounded wire,
4. Excess current, caused by short circuit in the load,
5. Excess current, caused by ground in the load,
6. Excess current, caused by voltage spike,
7. Excess current, caused by voltage droop; etc.

If the maintenance person does not troubleshoot the circuit prior to replacing the fuse and restoring power, negative consequences may arise.

Based on the cases/ examples given in this section, troubleshooting chart/ process/ tips can be made for each and every machines, equipments, systems and appliances.

QUESTIONS

1. What are the responsibilities and accountabilities of a maintenance engineer?
2. What is maintenance troubleshooting and why is it so important for each equipment/machine?
3. Mention the different process/ steps followed in general for troubleshooting of machines.
4. What do you understand by the symptoms of failures and what may be the types of symptoms in machines?
5. What are "three why" and "five why" approach in relation to troubleshooting?
6. Give troubleshooting tips, in some detail, of any particular type of equipment.

BIBLIOGRAPHY

Sl.	Name of Book or Internet Website	Writer	Publisher	Year
1.	Maintenance Planning, Control and Documentation	E. N. White	Gower Press U.K.	
2.	Integrated Maintenance Management—Concept to Computerization	B. N. Saha	SBA Publication, N.Delhi	
3.	Maintainability	B.S.Blanchard, D. Verma, E L Peterson	John Wiley and Sons, N.York	1955
4.	Physical Condition Monitoring of Rotating Machines	O.T.Sangerud	Norwegian Maritime Research, Vol.4	1975
5.	Review of Vibration Monitoring of Rolling Element Bearings by High Frequency Resonance Technique	P. D. McFadden and J. D. Smith	Tribology International	1994
6.	Detection of rolling element bearing damage by statistical vibration analysis	D. Dyer and R. M. Stewart	Journal of Mech. Design	April 1978
7.	Wear Debris Monitoring	George W Smith	—	—
8.	National Convention of Maintenance Management for Higher Productivity	Indian Institute of Plant Engineers	At Chennai and Jamshedpur	1987 1988
9.	Proceedings; Maintenance Management	N. P. C., New Delhi	Held in Dec.	1975
10.	www.maintenanceresources.com/referencery/	—	Internet	2004
11.	:www.ewi.org/technologies/microjoining/	—	Internet	2004
12.	www.key-to-steel.com/Articles/Art129.htm	—	Internet	2004
13.	www.efunda.com/processes/surface	—	Internet	2004
14.	www.fmeca.com/	—	Internet	2004
15.	www.isixsigma.com/	—	Internet	2004